Lecture Notes on
Knot Invariants

Lecture Notes on
Knot Invariants

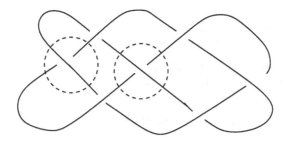

Weiping Li

Southwest Jiaotong University, P. R. China
Oklahoma State University, USA

W **World Scientific**

NEW JERSEY · LONDON · SINGAPORE · BEIJING · SHANGHAI · HONG KONG · TAIPEI · CHENNAI · TOKYO

Published by

World Scientific Publishing Co. Pte. Ltd.

5 Toh Tuck Link, Singapore 596224

USA office: 27 Warren Street, Suite 401-402, Hackensack, NJ 07601

UK office: 57 Shelton Street, Covent Garden, London WC2H 9HE

Library of Congress Cataloging-in-Publication Data
Li, Weiping, 1963–
Lecture notes on knot invariants / by Weiping Li (Southwest Jiaotong University, China &
Oklahoma State University, USA).
    pages cm
  Includes bibliographical references and index.
  ISBN 978-9814675956 (hardcover : alk. paper) -- ISBN 978-9814675963
  (softcover : alk. paper)
  1. Knot theory. 2. Braid theory. 3. Low-dimensional topology. 4. Alexander ideals.
  5. Knot polynomials. I. Title. II. Title: Knot invariants.
  QA612.2.L5 2015
  514'.2242--dc23

                          2015021070

British Library Cataloguing-in-Publication Data
A catalogue record for this book is available from the British Library.

Printed in Singapore

To my family

Diana, Oliver, and Xiaoli

Preface

Knot theory represents a rich mixture of many branches of mathematics, including topology, algebra, and geometry. It is also rich in its interactions with chemistry, physics and, more generally, technology. One of the fundamental problems in knot theory is the question of how to tell when two knots are really different. The main idea is to assign some sort of invariants to knots, so that if the invariants for two knots are different, then so are the knots. It would be best to have a complete set of easily computable invariants, so that we could go the other way, saying when the knots are equal - a much harder question.

There are many books that deal with knot invariants from various points of view. We do not try to cover everything here, but rather emphasize basic calculation skills and particular invariants. The reader will be introduced to this beautiful subject by means of some classical knot invariants, which have geometrical or topological origins aspects. We will explain the Jones polynomial from the original approach using braids and representations of Hecke algebras. We will understand the Casson-Lin invariants via representations of braids.

By proceeding in this way, we are forced, unfortunately, to skip many other interesting topics in knot theory. On the other hand, we feel strongly that it is better for students and readers to learn how to get their hands on the topics, rather than to know some fancy words. All the materials presented in this book can be checked by fairly direct methods, or at least can be followed by some basic steps toward the understandings. Of course, some basic knowledge of algebra and topology is required.

We start from the basic knot presentations, their equivalence classes, and the well-known Reidemeister moves. In Chapter 2, we introduce braids and their relationship to knots and links. Some classical invariants of knots and links are given and discussed in Chapter 3. By no means is this a complete list. There are many textbooks dealing with knots and their invariants. We choose to concentrate on the original definition of the Jones polynomial and the proof of Tait's conjectures. The Casson-type invariants of knots are constructed from braids and their representations. It is one of the typical and important problems in knot theory to find a geometrical or topological interpretation of the Jones polynomial which is constructed from algebra or to find a combinatorial interpretation of the Casson-type invariants which is constructed from geometry and topology.

This book is based on lecture notes that were originally prepared for the Chinese Graduate Summer School at Sichuan University in July–August, 2000. Some parts were also taught at Oklahoma State University. I would like to thank Xiao-Song Lin for many helpful conversations and communications, which helped me to understand the subject better, and to thank my collaborators Weiping Zhang and Qingxue Wang for many stimulating discussions. The students at the Chinese Graduate Summer School and Oklahoma State University also provided useful comments.

I would like to thank World Scientific Publishing Company for publishing these lecture notes, and also Rebecca Fu for her patience and help. Thanks also go to my students Zhili Chen, Xiaowei Yang and Bin Xie for various assistance in preparing this book.

Last but not least I want to thank my family for being there all the time.

Weiping Li

Contents

List of figures

Chapter 1

Basic knots, links and their equivalences

1.1 Definitions and equivalences

Definition 1.1.1. Let S^1 be a unit circle. L is a *link* in a closed 3-manifold Y if there is an embedding map $L : S^1 \bigsqcup \cdots \bigsqcup S^1 \to Y$ from disjoint unit circles to Y. Let $\mu(L)$ be the number of S^1-components in $S^1 \bigsqcup \cdots \bigsqcup S^1$. If $\mu(L) = 1$, then L is called a *knot* in Y.

Remark 1.1.2. (1) We do not distinguish knots and links as images of embeddings (Im (L)) or the embeddings themselves (L). There are higher dimensional knots (embedding $S^k \to M^{K+2}$ for $k \geq 2$) in [Rolfsen, 1976] which we are not taking into consideration in this book. We restrict ourselves only to knots and links in 3-manifolds in this book.

(2) One can consider knots and links in different topological categories. If the embedding L is (C^∞) smooth, then L is a smooth link or smooth knot. If L is piecewise linear, then L is a *PL* link (*PL* knot) (sometimes called polygonal link or polygonal knot).

Definition 1.1.3. Two knots or links L and $L^{'}$ are equivalent if there is a homeomorphism $h : Y \to Y$ such that the following diagram is commutative:

$$
\begin{array}{ccc}
S^1 \bigsqcup \cdots \bigsqcup S^1 & \xrightarrow{\ L\ } & Y \\
\Big\downarrow {\scriptstyle =} & & \Big\downarrow {\scriptstyle h} \\
S^1 \bigsqcup \cdots \bigsqcup S^1 & \xrightarrow{\ L^{'}\ } & Y.
\end{array} \qquad (1.1)
$$

If Y is an oriented closed 3-manifold and h is an orientation-preserving homeomorphism, then L and $L^{'}$ are oriented equivalent.

Remark 1.1.4. (1) Two equivalent knots and links have the property that there is a homeomorphism identifying their images in the 3-manifold Y.

(2) The map h is dependent upon the category one wants to work with. If two knots or links are C^∞, then the equivalence is given by a diffeomorphism $h : Y \to Y$ in Definition 1.1.3. Similarly for PL equivalence, h must be a piecewise linear homeomorphism.

Exercise 1.1.5. Prove that the relation defined in Definition 1.1.3 is an equivalent relation.

Let $\mathcal{L}(Y)$ $(\mathcal{L}^+(Y))$ be the space of (orientation-preserving) equivalence classes of links and knots in Y (oriented) under the above equivalent relation.

Definition 1.1.6. Two links (or knots) L and L' are *ambient isotopy* if there exists a map $H : [0,1] \times Y \to Y$ such that (i) $H(0,\cdot) = Id_Y$, (ii) $H(t,\cdot) : Y \to Y$ is a homeomorphism for each $t \in [0,1]$, and (iii) $H_1 \circ L = L'$.

Any two ambient isotopy links are certainly equivalent. But it is not true in general that any pair of equivalent links or knots is ambient isotopic. It depends on the mapping class group of Y. Let $\mathcal{L}_{ai}(Y)$ $(\mathcal{L}_{ai}^+(Y))$ be the space of equivalence classes of (orientation-preserving) ambient isotopy knots or links in Y (oriented). Thus we have $\mathcal{L}_{ai}(Y) \subset \mathcal{L}(Y)$ and $\mathcal{L}_{ai}^+(Y) \subset \mathcal{L}^+(Y)$.

Proposition 1.1.7. *If the mapping class group of Y has only one path-connected component, then $\mathcal{L}_{ai}(Y) = \mathcal{L}(Y)$; If the orientation-preserving mapping class group of oriented Y has only one path-connected component, then $\mathcal{L}_{ai}^+(Y) = \mathcal{L}^+(Y)$.*

Proof: There is always a path-connected component of the identity map. By the hypothesis, any equivalent (orientation-preserving) homeomorphism h can be connected by a (orientation-preserving) homeomorphism path from Id_Y. Thus we obtain $\mathcal{L}(Y) \subset \mathcal{L}_{ai}(Y)$ and $\mathcal{L}^+(Y) \subset \mathcal{L}_{ai}^+(Y)$. Therefore results follow. □

Example 1.1.8. If $Y = \mathbb{R}^3$ (noncompact oriented) is the Euclidean space of real 3-tuples, then the orthogonal group $O(3)$ (3×3 matrices with determinant ± 1 form a compact submanifold of $\mathbb{R}^9$ of dimension 3) is a deformation retract of $Diff(\mathbb{R}^3)$ (the diffeomorphic mapping class group of $\mathbb{R}^3$). Note that $O(3)$ has two components, the component of the identity is the subgroup $SO(3)$ of orthogonal matrices of determinant 1. Therefore $Diff_+(\mathbb{R}^3)$ (orientation-preserving diffeomorphisms of $\mathbb{R}^3$) has only one path-connected component. By Proposition 1.1.7,

$$\mathcal{L}_{ai}^+(\mathbb{R}^3) = \mathcal{L}^+(\mathbb{R}^3).$$

The orientation-preserving diffeomorphism equivalent knots or links in $\mathbb{R}^3$ are also C^∞-ambient isotopy to each other.

Example 1.1.9. If $Y = S^3$ is the (compact oriented) 3-sphere, then $Diff_+(S^3)$ has only one path-connected component. Let $f : B^{n+1} \to B^{n+1}$ be an orientation-preserving diffeomorphism, where B^{n+1} is the $(n + 1)$-dimensional ball with a restriction map $r : B^{n+1} \to \partial B^{n+1} = S^n$. Hence the restriction $f|_{\partial B^{n+1}} : S^n \to S^n$ is an orientation-preserving diffeomorphism. Such a restriction map r defines a group homomorphism from $Diff_+(B^{n+1})$ to $Diff_+(S^n)$. Let G_n be the image of this group homomorphism. If $g \in Diff_+(S^n)$ is isotopic to Id_{S^n}, then g can be extended to an orientation-preserving diffeomorphism of B^{n+1}, i.e., $g \in G_n$. So G_n is the path-connected component of the identity of S^n. Let $\Gamma_n = Diff_+(S^n)/G_n$ be the quotient group (always Abelian). There is a short exact sequence

$$0 \to Diff_+(B^{n+1}) \xrightarrow{r} Diff_+(S^n) \to \Gamma_n \to 0.$$

In fact every Γ_n is important in classifying differential structures. The set of diffeomorphism classes of oriented differential structures on S^n forms a group H_n under connected sum. $H_n \cong \Gamma_n$ except perhaps for $n = 4$. It is still an open question whether there is an exotic smooth structure on S^4 (*smooth Poincaré conjecture for 4-dimensional sphere S^4*). It is known that the Γ_n's are finite for all n except the case $n = 4$, by the computation of Kervaire and Milnor. The first nontrivial group is $\Gamma_7 \cong \mathbb{Z}_{28}$ by Milnor [1956]. $\Gamma_2 = 0$ by Smale [1959] and Munkres [1960]. $\Gamma_3 = 0$ is an interesting and difficult Morse-theoretic proof by Cerf [1974]. Therefore we have

$$\mathcal{L}_{ai}^+(S^3) = \mathcal{L}^+(S^3).$$

Example 1.1.10. For $Y = \mathbb{R}P^3$, the set of 1-dimensional subspaces of $\mathbb{R}^4$ through the origin can be identified with the real projective space $\mathbb{R}P^3$. Hence S^3 is a 2-sheeted covering space of $\mathbb{R}P^3$. By the result of Cerf [1974], $Diff(S^3)$ has two components that are path-connected to Id_{S^3} and $-Id_{S^3}$ respectively. Under the identification in $\mathbb{R}P^3 = S^3/\pm Id$, the mapping class group of $\mathbb{R}P^3$ has only one path-connected component containing $Id_{\mathbb{R}P^3}$. Hence we obtain $\mathcal{L}(\mathbb{R}P^3) = \mathcal{L}_{ai}(\mathbb{R}P^3)$ by Proposition 1.1.7.

1.2 Polygonal (PL), smooth (C^∞)-links and knots in $\mathbb{R}^3$

It is convenient to present knots and links in $\mathbb{R}^3$ through their perpendicular projections on a plane. Call a link segment of a polygonal link L in $\mathbb{R}^3$ an *edge* of L, and an end point of an edge a *vertex* of L. Let $p : \mathbb{R}^3 \to \mathbb{R}^2$ be an orthogonal projection , and $p(L)$ be the image of L in the projected plane.

Definition 1.2.1. A point $x \in p(L)$ is a multiple point if the cardinality of the set $p^{-1}(x) \cap L$ is greater than or equal to 2.

If the cardinality $|p^{-1}(x) \cap L| = n$, then x is called an n-multiple point, or a point of order n. Any 2-multiple point x is also called a double point.

The projection should be generic in the sense that (i) $n \leq 2$ for any n-multiple point, (ii) the number of double points in $p(L)$ is finite, and (iii) every vertex of the link L has order 1.

Definition 1.2.2. A projection $p : \mathbb{R}^3 \to \mathbb{R}^2$ is a regular projection for L in $\mathbb{R}^3$ if for every $x \in p(L)$, then $o(x)$ (the order of x) has the following property:

(1) For any $x \in p(L)$, $o(x) \leq 2$;
(2) The set $C_p(L) = \{x \in p(L) : o(x) = 2\}$ is finite;
(3) For any $x \in C_p(L)$, $p^{-1}(x) \cap L$ does not contain any vertex of L.

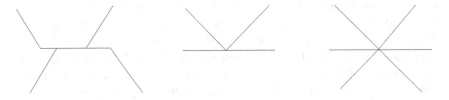

Fig. 1.1   Irregular multiple points

The first condition for regular projections is to rule out any multiple point other than double points, the second is to present the link with only finite crossings, the third is to avoid non-transversal double points. Hence any multiple point of any regular projection has a local image which looks like a letter X. The multiple points in Figure 1.1 are not contained in any regular projection of Definition 1.2.2.

Proposition 1.2.3. *Any polygonal link L has a regular projection.*

Proof: Any projection $p : \mathbb{R}^3 \to \mathbb{R}^2$ has the same image for two parallel planes, and it is completely determined by a fixed point and perpendicular direction to the projection plane. If we fix the point to be the origin, then the space of projections is the space of straight lines through the origin.

This is the two-dimensional projective plane $\mathbb{R}P^2$. There is one-to-one correspondence between projections $\mathbb{R}^3 \to \mathbb{R}^2$ with $p(0,0,0) = (0,0)$ and straight lines in $\mathbb{R}^3$ considered as elements in $\mathbb{R}P^2$.

Let S be the set of non-regular projections. Then it corresponds to a set of straight lines through the origin which do not satisfy Definition 1.2.2. Let S_1 be the subset of non-regular projections that have order 2 non-transverse points (Definition 1.2.2 (3) invalid). Let S_2 be the subset of non-regular projections which have order ≥ 3 multiple points (Definition 1.2.2 (1) and (2) invalid). We have $S_1 \cup S_2 \subset S$. Any non-regular projection must have either that the vertex is a double point (projection in S_1) or that a multiple point has order ≥ 3 (projection in S_2). Therefore $S = S_1 \cup S_2$. The set S_1 consists of finite line segments in $\mathbb{R}P^2$, and the set S_2 consists of finite many curve segments of second order. Hence S is a one-dimensional subset of $\mathbb{R}P^2$ (see [Crowell and Fox, 1977] for more details). The result follows. $\square$

Fig. 1.2   An elementary move

Definition 1.2.4. (1) An elementary move of a polygonal link L is a re-placement of L by a new link L_1 in the following way: there is a triangle ABC formed by three vertices which do not intersect any other point of L, two edges AC and BC are replaced by AB in a link L_1 (a disk remove) (see Figure 1.2).

(2) Two links L and L' are equivalent if they can be joined by a fi-nite sequence of links $L, L_1, \cdots, L_n = L'$ in which each subsequent link is obtained from the previous one by an elementary move or its inverse.

Exercise 1.2.5. Show that the relation between two links defined in Defini-tion 1.2.4 is indeed an equivalence relation.

Exercise 1.2.6. Let $P\mathcal{L}(\mathbb{R}^3)$ be the set of equivalence classes of polygonal links in $\mathbb{R}^3$. Prove that the equivalent relation defined by elementary moves is the same as the relation defined in Definition 1.1.3. Hence $P\mathcal{L}(\mathbb{R}^3) = \mathcal{L}(\mathbb{R}^3)$.

Let $f : S^1 \to \mathbb{R}^3$ be a smooth embedding, i.e., $f(t) = (f_1(t), f_2(t), f_3(t))$ has the following properties: (a) each $f_i : S^1 \to \mathbb{R}$ is a smooth function, and (b) $df : T_{t_0}S^1 \to T_{f(t_0)}\mathbb{R}^3$ is injective for every $t_0 \in S^1$. The property (b) is the same as saying that the linear transformation $df(t_0)$ from $\mathbb{R} = T_{t_0}S^1$ to $\mathbb{R}^3 = T_{f(t_0)}\mathbb{R}^3$ has zero kernel, where

$$df(t_0) \cdot v = \begin{pmatrix} f_1'(t_0) \\ f_2'(t_0) \\ f_3'(t_0) \end{pmatrix} \cdot v, \quad v \in T_{t_0}S^1.$$

Exercise 1.2.7. Verify $f : [0, 2\pi] \to \mathbb{R}^3$ is a smooth knot, where

$$f(\tau) = ((2 + \cos 3\tau)\cos 2\tau, (2 + \cos 3\tau)\sin 2\tau, \sin 3\tau).$$

In fact this is a smooth parametrization of a trefoil knot.

Theorem 1.2.8. *[Burde and Zieschange, 1985, Proposition 1.10] There is a bijective map from the equivalence classes of polygonal links in $\mathbb{R}^3$ to the equivalence classes of smooth links in $\mathbb{R}^3$.*

Remark 1.2.9. (1) For links in general 3-manifolds, we have to add an extra Riemannian metric, and replace each edge by a geodesic. Theorem 1.2.8 is also true for the 3-manifold S^3.

For a closed compact oriented 3-manifold $Y(\neq R^3)$, we can use the Riemannian metric g_Y and geodesics to define a *PL*-link in Y, where an edge of the link is a geodesic with respect to the metric g_Y. Similarly we have elementary moves and its corresponding relations. Hence the set of equivalence classes is $P\mathcal{L}_{g_Y}(Y)$, the *PL*-links in the 3-manifold (Y, g_Y). So $P\mathcal{L}_{g_Y}(Y) \subset \mathcal{L}_{g_Y}(Y)$.

Does there exist a metric g_Y such that $\mathcal{L}_{g_Y}(Y) \subset P\mathcal{L}_{g_Y}(Y)$? This is equivalent to asking whether there exists a metric g_Y such that the embedding $L : S^1 \bigsqcup S^1 \bigsqcup \cdots \bigsqcup S^1 \to Y$ can be represented by finitely many geodesics. Conjecturally $\mathcal{L}_{g_Y}(Y) \subset P\mathcal{L}_{g_Y}(Y)$ if there is a uniform lower bound for the injective radius.

(2) For usual knots and links in $\mathbb{R}^3$, the *PL* theory and C^∞ theory provide the same result. This is why people often use these two approaches interchangeably.

1.3 Continuous links in $\mathbb{R}^3$

We have seen that both smooth and polygonal links and knots can be presented by diagrams (regular presentations) with a finite number of crossings by Proposition 1.2.3 and Theorem 1.2.8. Artin and Fox [1948] showed that there exist topological embeddings of $S^1 \bigsqcup \cdots \bigsqcup S^1$ in $\mathbb{R}^3$ whose projections always have infinitely many crossings and are not isotopic to any smooth (polygonal) links or knots.

Definition 1.3.1. A C^0-embedding of $S^1 \bigsqcup \cdots \bigsqcup S^1$ in Y is said to be tame if it is isotopic to a smooth link in Y; otherwise it is called wild.

Note that $C^\infty - \mathcal{L}(Y)$ is always a proper subset of $C^0 - \mathcal{L}(Y)$, and its complement consists of wild links and knots. We only deal with tame links, or C^∞, PL-links and knots in this book.

1.4 Reidemeister moves and equivalences

Definition 1.4.1. The Reidemeister move of type I is given by twisting a crossing in a half way as in Figure 1.3.

Fig. 1.3 Reidemeister move of type I

The Reidemeister move of type II is given by pulling away of two unlinked arcs as in Figure 1.4.

The Reidemeister move of type III is given by shifting an arc as in Figure 1.5.

Definition 1.4.2. Two link diagrams are regularly isotopic if they can be transformed into each other by a finite sequence of the Reidemeister moves.

The following result characterizes the equivalent relation between two links in the standard 3-sphere.

Theorem 1.4.3. *For links L and $L^{'}$ in S^3, the following statements are equivalent :*

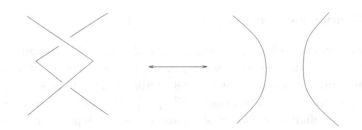

Fig. 1.4   Reidemeister move of type II

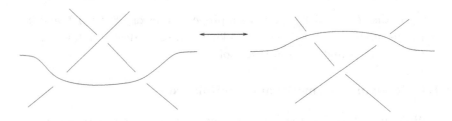

Fig. 1.5   Reidemeister move of type III

(1) L and L' are equivalent in the sense of Definition 1.1.3.

(2) L and L' are equivalent in the sense that there exists a PL-homeomorphism $h : S^3 \to S^3$ such that $h(L) = L'$, and $h : S^3 \to S^3, h|_L : L \to L'$ are orientation-preserving.

(3) L and L' are equivalent in the sense of Definition 1.2.4.

(4) L and L' are ambient isotopic as in Definition 1.1.6.

Proof: (1) $\Rightarrow$ (2) This essentially follows from the PL-approximation theory of a topological homeomorphism. To that end, let $f : (S^3, L) \to (S^3, L')$ be a C^0-homeomorphism. By a PL-approximation, $f|_P : P \to S^3$ is a PL-embedding for any compact PL-subspace of $S^3 \setminus L$. Let $N(L)$ be a regular neighborhood of L in S^3, and let (m, l) be a meridian-longitude system of L in the fundamental group of $\partial N(L)$. Since $f|_{\partial N(L)}$ is a PL-embedding, $f(N(L))$ is a PL-submanifold of S^3. Take a regular neighborhood $N(L') \subset f(N(L))$ of L'. Then each component of $f(N(L)) - int\{N(L')\}$ is PL-homeomorphic to $S^1 \times S^1 \times [0, 1]$ since it has the fundamental group $\mathbb{Z} \oplus \mathbb{Z}$. Hence $f(N(L))$ is a regular neighborhood of L' in S^3. Using $(f(m), f(l))$ as a meridian-longitude system of L' in $\partial f(N(L))$, we extend the PL-homeomorphism $f|_{S^3 \setminus int\{N(L)\}} : S^3 \setminus int\{N(L)\} \to S^3 \setminus int\{f(N(L))\}$ to a PL-homeomorphism $(S^3, L) \to (S^3, L')$. The orientation is preserved through the extension. So (2) follows.

(2) $\Rightarrow$ (3) For any 3-ball $B' \subset S^3$ with $B' \cap L = \emptyset$, we choose 3-balls B_0 and B such that

$$B_0 \subset int\{B'\} \subset B' \cup L \subset int\{B\} \subset S^3.$$

Let $S = \partial B$. There is a homeomorphism $f : B \setminus int\{B_0\} \to S \times [\varepsilon, 3\varepsilon]$ for $\varepsilon > 0$ such that $f(B' \setminus int\{B_0\}) = S \times [\varepsilon, 2\varepsilon]$. Let p_i be the ith-factor projection of $S \times [\varepsilon, 3\varepsilon]$ $(i = 1, 2)$. Denote $f_1 = p_1 \circ f : B \setminus int\{B_0\} \to S$ and $f_2 = p_2 \circ f : B \setminus int\{B_0\} \to [\varepsilon, 3\varepsilon]$. The map $f_1|_L : L \to S$ is a local homeomorphism and $f_2|_L : L \to (2\varepsilon, 3\varepsilon]$ since $f_2(B' \setminus int\{B_0\}) = S \times [\varepsilon, 2\varepsilon]$. Let us define $F : L \times [0, \varepsilon] \to S \times [\varepsilon, 3\varepsilon]$ by $F(x, t) = (f_1(x), f_2(x) - t)$ for $x \in L$ and $t \in [0, \varepsilon]$. For any $t \in [0, \varepsilon]$, there exists a closed interval $N(t)$ of t in $[0, \varepsilon]$ such that $F|_{L \times N(t)}$ is injective and $F(L \times N(t))$ is a PL-submanifold of dimension 2 in $S \times [\varepsilon, 3\varepsilon]$. By the compactness, there are finite points $0 = t_0 < t_1 < \cdots < t_n = \varepsilon$ such that $[t_i, t_{i+1}] \subset N(t)$ for some $t \in [0, \varepsilon]$ with the above property. Since $F(L \times \{t_i\})$ is deformed into $F(L \times \{t_{i+1}\})$ by a finite number of disk moves for each i, $L = f^{-1} \circ F(L \times \{0\})$ is transformed into $L^* = f^{-1} \circ F(L \times \{\varepsilon\})$ by elementary moves. Let $h : (S^3, L) \to (S^3, L')$ be a PL-homeomorphism given by (2) such that $B' \subset S^3 \setminus (L \cup L')$ is fixed by h. By the above argument, L and L^* are equivalent by element moves or disk removes.

(3) $\Rightarrow$ (4) Exercise.

(4) $\Rightarrow$ (1) is clear. $\square$

For $Y = \mathbb{R}^3$ (noncompact 3-manifold), define L and L' in $\mathbb{R}^3$ to be *ambient isotopic with a compact support* if there is an h_t of $\mathbb{R}^3$ such that $h_0 = Id_{\mathbb{R}^3}$ and $h_1(L) = L'$, and a compact subset $X \subset \mathbb{R}^3$ with $h_t = Id_{\mathbb{R}^3 \setminus X}$ for all $t \in [0, 1]$.

Theorem 1.4.4. *For PL-links L and L' in $\mathbb{R}^3$ with regular presentations with respect to the plane $\mathbb{R}^2$, the following statements are equivalent :*

(1) L and L' are equivalent in the sense of Definition 1.1.3.

(2) L and L' are equivalent in the sense that there exists a PL-homeomorphism $h : \mathbb{R}^3 \to \mathbb{R}^3$ such that $h(L) = L'$, and $h : \mathbb{R}^3 \to \mathbb{R}^3, h|_L : L \to L'$ are orientation-preserving.

(3) L and L' are equivalent in the sense of Definition 1.2.4.

(4) L and L' are regular isotopic as in Definition 1.4.2.

(5) L and L' are ambient isotopic with a compact support.

Proof: Note that S^3 is a 1-point compactification of $\mathbb{R}^3$. Let $\overline{f} : (S^3, L) \rightarrow (S^3, l')$ be the homeomorphism obtained by a 1-point compactification of a homeomorphism $f : (\mathbb{R}^3, L) \rightarrow (\mathbb{R}^3, L')$. Thus $(1) \Rightarrow (2) \Rightarrow (3)$ follows by Theorem 1.4.3.

   $(3) \Rightarrow (4)$ If L' is obtained by a finite sequence of elementary moves on L, we triangulate all disks we removed. So L can be transformed into L' by a finite sequence of disk moves on 2-simplices. We can always adjust the projection plane $\mathbb{R}^2$ so that the regular diagrams of L and L' with respect to the new projection plane $\mathbb{R}^2$ are identical with the original ones and the link resulting from the disk move of each 2-simplex is regularly projected into the new projection plane $\mathbb{R}^2$.

Fig. 1.6   An elementary move of the disk

   $(4) \Rightarrow (5) \Rightarrow (1)$ is clear. $\qquad\qquad\qquad\qquad\qquad\qquad\qquad\square$

Exercise 1.4.5. Show that each of the Reidemeister moves can be obtained by one or several elementary moves.

Exercise 1.4.6. Show that the knots in Figure 1.7 are isotopic to each other.

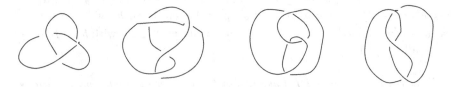

Fig. 1.7   Isotopic equivalent trefoil knots

1.5   Crossing number and knot tabulation

Note that each double point of the regular projection $p(L)$ in Definition 1.2.2 is also called a crossing. We form a basis of $\mathbb{R}^3$ by a unit vector on the normal line and the basis of the projection plane $\mathbb{R}^2$. Then the height function of $p^{-1}(x)$ is the coordinate of the unit vector on the normal line. The set $C_p(L) = \{x \in p(L) : o(x) = 2\}$ is a finite set of crossings for a regular projection p. For each $x \in C_p(L), p^{-1}(x)$ consists of two points. Let $x^+ \in p^{-1}(x)$ be a point with bigger height value, and x^- be a point of smaller height value. So x^+ is called *overcrossing*, and x^- is called *undercrossing*. Similarly, a short line segment of x^+ is called the overpass of x, and a short line segment of x^- is called the underpass of x.

Definition 1.5.1. The crossing number of L is

$$c(L) = \min_p\{|C_p(L)| : p \text{ is a regular projection of } L\},$$

where $|C_p(L)|$ is the cardinality of the crossings $C_p(L)$.

A link L with $c(L) = 0$ must be an unlink or the unknot. The crossing number of a knot is the smallest number of crossings of any regular projection presentation of the knot. It is a knot invariant.

Exercise 1.5.2. If $c(K) = 1$ or $c(K) = 2$ for knots, then K must be unknotted.

The trefoil knot has a crossing number three $c(3_1) = 3$ and the figure-eight knot has a crossing number four $c(4_1) = 4$. But there are two knots with crossing number five $c(5_1) = c(5_2) = 5$. It is not able to distinguish the two different knots. The number of knots with a fixed crossing number increases rapidly as the crossing number increases.

Exercise 1.5.3. For $n \in \mathbb{N}$, show that there are only finitely many link (knot) types L whose $c(L) \leq n$.

A single knot (link) may have infinitely many regular projections, but only finitely many diagrams of the knot (link) have a minimal number of crossings. Tait was the first person to draw a knot table with respect to the crossing number. The knot (link) tabulation helps us to enrich the examples and counterexamples, and serves a good general testing ground for intuition (see Section 2.6).

For any links L_1 and L_2, it is clear that $c(L_1 \# L_2) \leq c(L_1) + c(L_2)$. It is surprisingly hard to resolve the over 100-years difficult problem in knot theory to prove that

$$c(L_1 \# L_2) = c(L_1) + c(L_2).$$

In fact, it is not easy to show that $c(L_1 \# L_2) \geq c(L_i)$ for $i = 1, 2$. The important step to break this problem is due to the discovery of Jones polynomial in Chapter 5 and proofs of Tait's conjectures for alternating links and alternating projections of alternating links.

At each crossing of a regular projection of a link L, two strands come in and two strands go out following the orientation on the link. The Seifert's algorithm is to remove the crossing from the crossing and to reconnect the strands repeatedly until all crossings are removed. The result of this algorithm is to have a set of disjoint union of topological circles (called Seifert circles). The disks bounded by these Seifert circles are called Seifert disks. We then attach a half twisted band to two Seifert disks along their boundaries of each twisted band to the projection plane matches with the crossing of the original diagram (for example see [Diao, 2004, Figure 1 and Figure 2]). The result is a closed surface with L' as its boundary and is called the Seifert surface of L, where L' is equivalent to L. The genus of this Seifert surface is given by

$$ g = \frac{c(L) + 2 - s - \mu(L)}{2}, $$

where s is the number of Seifert circles produced by the Seifert's algorithm. Schubert proved that for any link $L_i (i = 1, 2)$ the following identity holds.

$$ g(L_1 \# L_2) = g(L_1) + g(L_2). $$

Let $s(L)$ be the minimum number of Seifert circles over all possible regular projections of links that are equivalent to L. Yamada [1987] showed that $s(L) = b(L)$, where $b(L)$ is the braid index of L (see also Chapter 3).

Definition 1.5.4. The deficiency of a link L is defined by

$$ d(L) = c(L) - (b(L) - 1) - 2g(L) - (\mu(L) - 1). $$

Example 1.5.5. Let $K = 3_1$ be the trefoil knot. We have $b(3_1) = 2$. By applying the Seifert algorithm to a minimum projection of the trefoil, we get two Seifert circles $s = 2$, and $g = (3 + 2 - 2 - 1)/2 = 1$. Thus $g(3_1) = 1$ and

$$ d(3_1) = 3 - (b(3_1) - 1) - 2g(3_1) - (\mu(3_1) - 1) = 3 - (2 - 1) - 2 - 0 = 0. $$

Exercise 1.5.6. For the figure eight knot 4_1, show that (i) $b(4_1) = 3$, (ii) $s(4_1) = 3$, (iii) $g(4_1) = 1$ and (iv) $d(4_1) = 0$.

Let M_n be a special family of Montesinos knots $(3, 3, \cdots, 3, 2)$ with n-times 3 and an alternating projection. Its braid index $b(M_n) \geq n+1$. Since M_n is an alternating knot and there is a reduced alternating diagram, we have $c(M_n) = 3n+2$ by results in Section 5.8 from Tait's conjectures. There are $n + 1$ Seifert circles so that $s(M_n) \leq n + 1$. We have $n + 1 \leq b(M_n) = s(M_n) \leq n + 1$. By using the Seifert surface, the genus of the Montesinos knot

$$g(M_n) = \frac{c(M_n) + 2 - s(M_n) - \mu(M_n)}{2}$$
$$= \frac{3n + 2 + 2 - (n + 1) - 1}{2}$$
$$= n + 1.$$

So $d(M_n) = 0$.

Proposition 1.5.7. $d(L) \geq 0$ *for any link L.*

Proof: We have the genus of a link L to be the minimum genus over the genera of all possible Seifert surfaces of links that are equivalent to L, and

$$g(L) \leq g = \frac{c(L) + 2 - s - \mu(L)}{2}$$
$$\leq \frac{c(L) + 2 - s(L) - \mu(L)}{2}$$
$$= \frac{c(L) + 2 - b(L) - \mu(L)}{2}$$
$$= \frac{c(L) - (b(L) - 1) - (\mu(L) - 1)}{2},$$

where the first equality follows from the basic genus identity, the second from the definition of $s(L)$ to be the minimum of the Seifert circles, the third from Yamada's [1987] result. Thus

$$c(L) - (b(L) - 1) - (\mu(L) - 1) - 2g(L) = d(L) \geq 0.$$

$\square$

Note that the deficiency of a link L is given by [Diao, 2004, Definition 2.5]. Also $d(5_2) = d(6_1) = d(6_2) = 1$. We have the following results from [Diao, 2004].

Theorem 1.5.8. *We have the following relations.*
(1) $c(L_1 \# L_2) \geq (c(L_1) - d(L_1)) + (c(L_2) - d(L_2))$.
(2) If $d(L_1) = d(L_2) = 0$, then $d(L_1 \# L_2) = 0$ and $c(L_1 \# L_2) = c(L_1) + c(L_2)$.

Proof: By Proposition 1.5.7, we have

$$d(L_1 \# L_2) = c(L_1 \# L_2) - (b(L_1 \# L_2) - 1) - 2g(L_1 \# L_2) - (\mu(L_1 \# L_2) - 1) \geq 0,$$

$$
\begin{aligned}
c(L_1 \# L_2) &\geq (b(L_1 \# L_2) - 1) + 2g(L_1 \# L_2) + (\mu(L_1 \# L_2) - 1) \\
&= (b(L_1) - 1) + (b(L_2) - 1) + 2(g(L_1) + g(L_2)) \\
&\quad + (\mu(L_1) - 1) + (\mu(L_2) - 1) \\
&= (c(L_1) - d(L_1)) + (c(L_2) - d(L_2))
\end{aligned}
$$

where the first identity follows from a result of Birman and Menasco [1991] that $b(L_1 \# L_2) - 1 = (b(L_1) - 1) + (b(L_2) - 1)$, a result of [Schubert, 1956] that $g(L_1 \# L_2) = g(L_1) + g(L_2)$ and the additivity $\mu(L_1 \# L_2) - 1 = (\mu(L_1) - 1) + (\mu(L_2) - 1)$, the last from Definition 1.5.4.

If $d(L_1) = d(L_2) = 0$, then $c(L_1 \# L_2) \geq c(L_1) + c(L_2)$ by (1). Therefore the additivity of the crossing number follows $c(L_1 \# L_2) = c(L_1) + c(L_2)$. The result $d(L_1 \# L_2) = d(L_1) + d(L_2)$ follows from the definition and the proof in the above. $\qquad \square$

Exercise 1.5.9. (1) If $d(L_i) = 0$, then prove that $c(L_1 \# L_2) \geq c(L_i)$ for $i = 1, 2$.

(2) If $d(L_1) = 0$ and L_2 is a nontrivial knot, show that $c(L_1 \# L_2) \geq c(L_1) + 3$.

(3) If $d(L_1) = 0$ and L_2 is a nontrivial link, show that $c(L_1 \# L_2) \geq c(L_1) + 2(\mu(L_2) - 1)$.

1.6 Brief history on the knot tabulation

A Greek physician Heraklas during the first century A.D. who was a possible associate of Heliodorus, wrote an essay on surgical stitching methods which was included in Oribasius of Pergamum's Medical Collections. In his essay, Heraklas described step-by-step instructions to tie orthopedic stitches in eighteen ways. The endless knot appears in Tibetan Buddhism, while the Borromean rings have made repeated appearances in different cultures, often symbolizing unity.

The Scottish mathematical physicist Sir William Thomson (Lord Kelvin) suggested that atoms were knotted vertices in the aether. In 1867, Thomson wrote [1867]:

> *A full mathematical investigation of the mutual action between two vortex rings of any given velocities passing one another in any two lines, so directed that they never come nearer to one*

> *another than a large multiple of the diameter of either, is a per-*
> *fect mathematical problem; and the novelty of the circumstances*
> *contemplated presents difficulties of an exciting character.*

Thomson viewed space as being filled with a perfect fluid, the aether. The vortex atoms were then knotted tubes of aether which retained their form despite being distorted by Helmholtz's theory. With a understanding of knots, one could unravel the secrets of the atom and of matter itself. It was this motivation for Peter Guthrie Tait (1831–1901) to investigate knots and their tabulation. Diagrams of knots and links drawn by Tait were included in Thomson's second paper on vortex atoms, which appeared in 1869. Tait's method is to enumerate all possible diagrams up to a given crossing number and to group those diagrams that represent the same knot type. Gauss and his student Listing studied knots with a scheme of encoding knot diagrams around 1848. Tait invented his own way to encode knots diagram. The encoding scheme is quite subtle, as explained in [Hoste, Thistleethwaite and Weeks, 1998].

James Clerk Maxwell, a Scottish mathematical physicist, was interested in knots because of electromagnetic considerations, and in a letter to Tait written on the fourth of December 1867 he rediscovered an integral formula counting the linking number of two closed curves which Gauss had discovered, but had not published in 1833. In September 1868, Maxwell wrote several manuscripts which study knots and links. He wrote:

> *Let any system of closed curves in space be given and let them*
> *be supposed capable of having their forms changed in any con-*
> *tinuous manner, provided that no two curves or branches of a*
> *curve ever pass through the same point of space, we propose to*
> *investigate the necessary relations between the positions of the*
> *curves and the degree of complication of the different curves of*
> *the system.*

Maxwell was the first to set out the classification problem of knots and links, and he considered the two-dimensional projections to devise a way of coding the diagram. Maxwell noted that the region could be eliminated by uncoiling the curve, and also defined the "Reidemeister moves" with no mathematical rigor. Maxwell's manuscripts were published more than 100 years after they were written.

To encode a diagram, choose a base point that is indicated by a black dot at the overpass labeled 1, and an orientation of the knot, see Figure 1.8. Traveling from this base point labeled 1 in the given direction, label points on the knot curve lying directly above or below crossings with consecutive integers $1, 2, 3, 4, \cdots$. Each crossing thus receives two integers as labels, one even and one odd, and this defines a one-to-one correspondence between the set of odd labels and the set of even labels. The overcrossing and undercrossing structure is captured by associating a minus sign to each even integer which is the label of an overpass.

A diagram is alternating (over-under-over-under) if and only if all even integers have the same sign. Note that the odd numbers have been written in their natural increasing order $1, 3, 5, \cdots, 2n - 1$ for crossing number n. Therefore the code for the diagram is just the sequence of even (signed) integers. The standard code for a diagram is a sequence which is minimal over all choices of starting point and direction with respect to a suitable ordering of sequences. See an example in Figure 1.8.

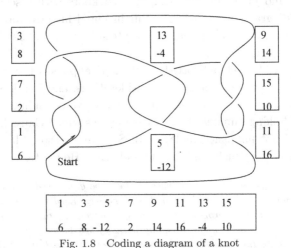

Fig. 1.8   Coding a diagram of a knot

In 1876, Tait decided to embark on a classification of plane closed curves, writing in a report to the British Association for the Advancement of Science:

> *The development of this subject promises absolutely endless work - but work of a very interesting and useful kind - because it is intimately connected with the theory of knots, which (especially as applied in Sir W. Thomson's Theory of Vortex Atoms) is likely soon to become an important branch of mathematics.*

He was considering alternating knots, namely those which when traversing the projection on 2-dimensional space the crossings go alternately over and under.

By 1877, Tait had classified all knots with seven crossings successfully by his mathematical and geometric intuition. He knew that what was really required was a knot invariant. He wrote:

> ... *though I have grouped together many widely different but equivalent forms, I cannot be* absolutely *certain that all those groups are essentially different one from another.*

Thomas P. Kirkman, another knot theorist, sent Tait his results on knot projections with up to nine crossings in May 1884 but he did not solve the equivalence problem among those knots. Tait solved the equivalence problem within a few weeks. Tait seemed to know how to tell whether two knots were equivalent without rigorous methods, and produced a remarkably correct table. When Kirkman sent him all knot projections with 10 crossings in January 1885 again, Tait found all equivalent knots with 10 crossings. The tables were printed in September 1885 and again they are completely correct. It is important to notice that Tait had no theorems from topology to enable him to distinguish different knots. This is the task of knot invariants we are going to study in the next sections.

Tait set forth several conjectures concerning alternating knots, none of which were resolved until the Jones polynomial in 1984. (For figures of Tait's seven crossing knots, see [Hoste, Thistleethwaite and Weeks, 1998]).
Tait Conjecture I: Reduced alternating diagrams have minimal crossing number.
Tait Conjecture II : Any two reduced alternating diagrams of a given knot have equal writhe.
Tait Conjecture III : Any two reduced alternating diagrams of a given knot are related by a sequence of flypes (see Figure 1.9).

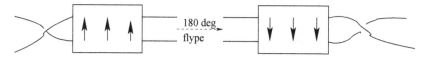

Fig. 1.9   Flype move

A flype move is obtained by rotating the tangle by 180-degree. Since flypes preserve writhe, Tait Conjecture III implies Tait Conjecture II. Tait defined the "beknottedness" (now known as the writhe number) and believed that this was a knot invariant. In fact it is not, but it is an invariant for alternating knots as a consequence of Tait Conjecture II.

Tait Conjecture I has been proved by several authors with different methods, but all proofs use properties of Jones polynomial or the Kauffman two-variable polynomial. See [Kauffman, 1987a; Murasugi, 1987; Thistlethwaite, 1987, 1988]. Tait Conjecture III is solved by Menasco and Thistlethwaite [1993], their methods are mostly geometric, and are related essentially on properties of the Jones polynomial. We present proofs of Tait Conjecture I and II from the method of Murasugi [1987] in §4.8.

The era of rigorous knot theory began in the early part of 1900. In 1914, Max Dehn was able to distinguish the right-handed and left-handed trefoil knots by the development of topology. In 1927, J. Alexander and G. Briggs were able to distinguish all the tabulated knots through nine crossings with a small exception, by using the first homology groups of branched cyclic covers.

Kurt Reidemeister in 1932 completely classified all knots up to nine crossings, by using the linking numbers of branch curves in irregular covers associated to homomorphisms of the knot group onto dihedral groups. Schubert [1949] proved that every knot can be uniquely decomposed, up to permutation, as a connected sum of prime knots. This reduces the work of knot tabulation to only prime knot tabulation.

Hoste, Thistlethwaite and Weeks [1998] explained how all the alternating diagrams with a given number of crossings are generated, how the ones corresponding to composite knots are eliminated, and finally how duplicates are eliminated. From alternating diagrams, all the non-alternating diagrams with a given number of crossings are then generated, and the (more difficult) process of elimination is repeated. For all prime knots up to and including those with 16 crossings, download from http://www.math.utk.edu/~morwen for the computer programs Knotscape and SnapPea through Linux system and Sun system.

In 1919, young J. Alexander wrote a letter to O. Veblen stating:

> When looking over Tait on Knots among other things, he really does not get very far. He merely writes down all the plane projections of knots with a limited number of crossings, tries out a few transformations that he happen to think of and assumes without proof that if he is unable to reduce one knot to another with a reasonable number of tries, the two are distinct. His invariant, the generalization of the Gaussian invariant ... for links is an invariant merely of the particular projection of the knot that you are dealing with, - the very thing I kept running

up against in trying to get an integral that would apply. The same is true of his "Beknottednes".

Max Dehn [1987] and Poul Heegaard formulated the subject of the Knot Theory in 1907 by proposing lattice knots and the precise definition of the lattice equivalence relation among lattice knots. Later Reidemeister and Alexander studied more general polygonal knots in a space with equivalent knots related by a sequence of Reidemeister moves.

Wolfgang Haken in 1961 discovered an algorithm to decide whether or not a knot is nontrivial, and Friedhelm Waldhausen [1970] completed Haken's program in 1970. In the late 1970s William Thurston's hyperbolization theorem introduced the theory of hyperbolic 3-manifolds into knot theory and made it of prime importance and received a Fields Medal in 1982 largely due to this work (Grigori Perelman [2002] sketched a proof of the full geometrization conjecture in 2003 using Ricci flow with surgery, and was awarded the Fields Medal. He declined to accept the award or to appear at the congress. He stated "I am not interested in money or fame; I do not want to be on display like an animal in a zoo").

In 1984, Vaughan Jones [1987] discovered the Jones polynomial (see Chapter 4) and was awarded the Fields Medal in 1990 for this work. Edward Witten [1989] proposed a topological quantum field theoretical approach to the Jones polynomial (Witten also received the Fields Medal in 1990 partly for this work), and described other topological invariants of 3-manifolds related to the Jones polynomial. John Horton Conway [1970] analyzed a procedure for unknotting knots (Conway notation). Victor A. Vassiliev and Mikhail N. Goussarov in early 1990s, discovered the finite type invariants of knots which encompass the Jones polynomial and its generalizations. Kashaev [1997] observed that the asymptotic behavior of a certain state sum of knots gives the hyperbolic volume of the complement of knots as verified for knots $4_1, 5_2$ and 6_1 (Murakami and Murakami [2001] identified the state sum of knots with the colored Jones polynomial, thus the asymptotic behavior of the colored Jones polynomial gives the hyperbolic volume of the knot as the Volume Conjecture.) Maxim Kontsevich [1993] defined a Kontsevich integral to get a universal Vassiliev invariant with certain algebraic structures, and was awarded the Fields Medal in 1998.

In 1995, the author first introduced the symplectic Floer homology for knots as an extension of Casson-Lin invariant, where the Euler number of the defined symplectic Floer homology of a knot is the Casson-Lin invariant of the knot (see [Li, 1997]). This is parallel to the instanton Floer homology defined by Floer [1989] which extends the Casson invariant of the integral homology 3-spheres and relates to 4-dimensional Donaldson invariants. In 2003, Peter Ozsváth and Zoltán Szabó and independently Jacob Rasmussen defined knot Floer homology through the Heegaard decompositions and holomorphic disks. Mikhail Khovanov in 2000 used the Khovanov bracket (the analogue of the Kauffman bracket in the construction of the Jones polynomial) to construct the normalized complex such that the Khovanov homology of the link as the homology of this normalized complex is an invariant of the link with the graded Euler characteristic of the Jones polynomial. Dror Bar-Natan developed a computer program to calculate the Khovanov homology of any knot in 2006.

Chapter 2

Braids and links

The earliest article on the braid notion can be traced to Hurwitz [1891] and Fricke and Klein [1897]. But Artin [1925] gave the first rigorous definition of the braid. His paper in 1925 contains the fundamental isomorphism between braids and braid automorphisms by which braids are classified. The proof, though, was not satisfying. In his second paper [Artin, 1947], he set rigorous definitions and proofs including the normal form of a braid. Artin invented the braid as a mathematical model to be used in the textile industry. Braid has various applications in mathematics (complex polynomials, knots and links, configuration spaces, representations of functions in n-variables) and in physics (classical mechanics, statistical physics and quantum field theory).

Braids arise naturally from a collection of n connected strings in three-dimensional space and they are crucial to the study of links and knots by Markov's theorems and moves. Starting with the Jones invariant of links through the braid constructions, a large number of link invariants has been followed along the similar line by using the braids. This relates to the fundamental group of the configuration space of n points on the plane as the definition of the braid group, where the configuration space is a $K(\pi, 1)$-space and carries a natural geometric model of a classifying space of the braid group. Arnold [1970] interpreted this configuration space as the space of polynomials over a complex variable of degree n without multiple roots.

2.1 Braid definition

Let Σ be a manifold of dimension 2, and Σ^n be the n-fold product space. Let $F\Sigma^n$ be the subspace of Σ^n defined by

$$F\Sigma^n = \{(z_1, \cdots, z_n) \in \Sigma^n | z_i \neq z_j, \text{if } i \neq j\}.$$

Definition 2.1.1. z and z' in $F\Sigma^n$ are equivalent if there is a permutation in Σ_n such that $z' = \sigma \cdot z$, where Σ_n is the permutation group of n-letters.

Under the equivalent relation, we have $B\Sigma^n$ the space of the equivalent classes of $F\Sigma^n$. Hence there is a natural projection $p : F\Sigma^n \to B\Sigma^n$ by $p(z) = [z]$. Fixing a base point $z_0 \in F\Sigma^n$, there is an induced map

$$\pi_1(p) : \pi_1(F\Sigma^n, z_0) \to \pi_1(B\Sigma^n, [z_0])$$

on fundamental groups. The group $\pi_1(F\Sigma^n, z_0)$ is called the *pure braid group* of Σ and $\pi_1(B\Sigma^n, [z_0])$ is the *full braid group* of Σ.

Lemma 2.1.2. *The natural projection $p : F\Sigma^n \to B\Sigma^n$ is a regular covering space projection. The group of covering transformations is the full symmetric group Σ_n.*

Proof: It is easy to check that the symmetric group Σ_n acts on $F\Sigma^n$ transitively and fixed-point free. Therefore one has a canonical isomorphism $\pi_1(B\Sigma^n, [z_0])/\pi_1(F\Sigma^n, z_0) \cong \Sigma_n$ or the short exact sequence

$$1 \to \pi_1(F\Sigma^n, z_0) \to \pi_1(B\Sigma^n, [z_0]) \to \Sigma_n \to 1.$$

$\square$

By the short exact sequence, we can study the pure braid group $\pi_1(F\Sigma^n, z_0)$ for understanding the full braid group $\pi_1(B\Sigma^n, [z_0])$. Let $F_m = \{s_1, \cdots, s_m\} \subset \Sigma$ be a set of fixed distinguished points. Define $F_m\Sigma^n = F(\Sigma \setminus F_m)^n$. The topological type of $F_m\Sigma^n$ is independent of the choice of $s_1, \cdots, s_m$ in Σ. Similarly, $B_m\Sigma^n = B(\Sigma \setminus F_m)^n$. Let $\pi_n^r : F_m\Sigma^n \to F_m\Sigma^r$ be the projection induced from $p_1 : \Sigma^n = (\prod_{i=1}^r \Sigma) \times (\prod_{i=r+1}^n \Sigma) = \Sigma^r \times \Sigma^{n-r} \to \Sigma^r$.

Theorem 2.1.3. *[Fadell and Neuwirth, 1962] The projection π_n^r is a fibration with fiber $F_{m+r}\Sigma^{n-r}$.*

Proof: For a base point $(z_1^0, \cdots, z_r^0) \in F_m\Sigma^r$, we have $(z_1^0, \cdots, z_r^0) \in F(\Sigma \setminus F_m)^r$. For any $i \neq j$, $z_i^0 \neq z_j^0$ and z_i^0 not in F_m. The fiber $(\pi_n^r)^{-1}(z_1^0, \cdots, z_r^0)$

$$= \{(z_1^0, \cdots, z_r^0, y_{r+1}, \cdots, y_n) : (z_1^0, \cdots, y_n \in F(\Sigma \setminus F_m)^n\}.$$

Let $F_{m+r} = \{s_1, \cdots, s_m, z_1^0, \cdots, z_r^0\} \subset \Sigma$. Then we have $F_{m+r}\Sigma^{n-r} = F(\Sigma \setminus F_{m+r})^{n-r}$ which can be identified with the set of $(y_{r+1}, \cdots, y_n)$ all distinct in $\Sigma \setminus F_{m+r}$. So there is a homeomorphism $h : F_{m+r}\Sigma^{n-r} \to (\pi_n^r)^{-1}(z_1^0, \cdots, z_r^0)$ by

$$h(y_{r+1}, \cdots, y_n) = (z_1^0, \cdots, z_r^0, y_{r+1}, \cdots, y_n).$$

Take $(z_1^0, \cdots, z_r^0) \in F_m\Sigma^r$ and add other m distinct points F_m to be F_{m+r}. Pick a homeomorphism $f : \Sigma \to \Sigma$ such that $f(F_m) = F_m$ and $f(F_{m+r} \setminus F_m) = \{z_1^0, \cdots, z_r^0\}$. Let U be a neighborhood of $(z_1^0, \cdots, z_r^0)$ in $\Sigma \setminus F_m$. Define $\theta : U \times \overline{U} \to \overline{U}$ by $\theta(z, w) = \theta_z(w)$ with the property that $\theta_z : \overline{U} \to \overline{U}$ is a homeomorphism fixed on $\partial \overline{U}$ and $\theta_z(z) = (z_1^0, \cdots, z_r^0)$. Extend θ to $\Theta : U \times \Sigma \to \Sigma$ as $\Theta|_{U \times \overline{U}} = \theta$ and $\Theta : U \times (\Sigma \setminus U) \to (\Sigma \setminus U)$ as the identity map. Since θ and id agree on a closed set $\partial \overline{U}$, the extended map Θ is continuous and homeomorphism for all $z \in U$. The local product structure $\phi_U : (\pi_n^r)^{-1}(U) \to U \times F_{m+r}\Sigma^{n-r}$ is given by

$$(z, y_{r+1}, \cdots, y_n) \to (z, \Theta_z(y_{r+1}), \cdots, \Theta_z(y_n))$$

$$\to (z, f^{-1} \circ \Theta_z(y_{r+1}), \cdots, f^{-1} \circ \Theta_z(y_n)).$$

Since $(y_{r+1}, \cdots, y_n) \in \Sigma \setminus F_{m+r}$ and Θ_z, f are homeomorphisms, $f^{-1} \circ \Theta_z(y_{r+1}), \cdots, f^{-1} \circ \Theta_z(y_n)$ are all distinct. If $f^{-1} \circ \Theta_z(y_i) = z_j^0$ for $i \in \{r+1, \cdots, n\}$ and $j \in \{1, \cdots, r\}$, then $\Theta_z(y_i) \in F_{m+r} \setminus F_m \subset \Sigma \setminus F_m$. Thus $\Theta_z(y_i) = y_i \in F_{m+r} \setminus F_m$ contradicts with $(y_{r+1}, \cdots, y_n) \in \Sigma \setminus F_{m+r}$. Therefore $f^{-1} \circ \Theta_z(y_i) \in \Sigma \setminus F_{m+r}$. The map ϕ_U has an inverse given by

$$\phi_U^{-1}(z, z_{r+1}, \cdots, z_n) = (z, \Theta_z^{-1} \circ f(z_{r+1}), \cdots, \Theta_z^{-1} \circ f(z_n)).$$

$\square$

Note that $F_m\Sigma = F(\Sigma \setminus F_m) = \Sigma \setminus F_m$. For any $n \in \mathbb{N}$, $\pi_n^1 : F_m\Sigma^n \to F_m\Sigma = \Sigma \setminus F_m$ is a fibration with fiber $F_{m+1}\Sigma^{n-1}$. Thus there is a homotopy exact sequence

$$\cdots \to \pi_{i+1}(F_m\Sigma) \to \pi_i(F_{m+1}\Sigma^{n-1}) \to \pi_i(F_m\Sigma^n) \overset{\pi_i(\pi_n^1)}{\to} \pi_i(F_m\Sigma) \to \cdots. \tag{2.1}$$

Example 2.1.4. If $\pi_2(F_m\Sigma) = \pi_3(F_m\Sigma) = 1$ for all $m \geq 0$, then we have, by (2.1),

$$1 \to \pi_2(F_{m+1}\Sigma^{n-1}) \overset{\cong}{\to} \pi_2(F_m\Sigma^n) \to 1.$$

Hence

$$\pi_2(F\Sigma^n) \cong \pi_2(F_1\Sigma^{n-1})$$
$$\cong \pi_2(F_2\Sigma^{n-2}) \cdots$$
$$\cong \pi_2(F_{n-1}\Sigma) = \{1\}.$$

Example 2.1.5. Let $\pi_n^{n-1} : F\Sigma^n \to F\Sigma^{n-1}$. Let $F_{n-1}\Sigma = \Sigma \setminus F_{n-1}$. Pick $(z_1^0, \cdots, z_{n-1}^0)$ to be a base point of $\pi_1(F\Sigma^n)$ and $F_{n-1} = \{z_1^0, \cdots, z_{n-1}^0\}$. Let $j : F_{n-1}\Sigma \to F\Sigma^n$ be the inclusion given by $j(z_n) = (z_1^0, \cdots, z_{n-1}^0, z_n)$ for $z_n \in \Sigma \setminus F_{n-1}$. Then the fibration is

$$F_{n-1}\Sigma \overset{j}{\hookrightarrow} F\Sigma^n \overset{\pi_n^{n-1}}{\to} F\Sigma^{n-1}.$$

There is an exact sequence of homotopy groups

$$\cdots \to \pi_{i+1}(F\Sigma^{n-1}) \to \pi_i(F_{n-1}\Sigma) \overset{\pi_i(j)}{\to} \pi_i(F\Sigma^n) \overset{\pi_i(\pi_n^{n-1})}{\to} \pi_i(F\Sigma^{n-1}) \to \cdots.$$

If $\pi_i(F_m\Sigma) = \{1\}$ for all $m \geq 0$ and $i = 0, 2, 3$, then $\pi_2(F\Sigma^n) = \{1\}$ by the previous example. So $\pi_2(F\Sigma^{n-1}) = \{1\}$. We also have the following short exact sequence

$$1 \to \pi_1(F_{n-1}\Sigma) \overset{\pi_1(j)}{\to} \pi_1(F\Sigma^n) \overset{\pi_1(\pi_n^{n-1})}{\to} \pi_1(F\Sigma^{n-1}) \to \pi_0(F_{n-1}\Sigma) = \{1\}.$$

Let H_g be a handlebody of genus g. The braid group B_n^g on n strings in H_g was studied by Sossinsky [1992]. Let M be the complex plane $\mathbb{C}$ and $Q_g = \{z_1^0, \cdots, z_g^0\}$ be g distinct points. The interior of H_g can be interpreted as the direct product of the complex plane $\mathbb{C}$ without g points $\mathbb{C} \setminus Q_g$ and an open interval $(-1, 2)$, $IntH_g = (\mathbb{C} \setminus Q_g) \times (-1, 2)$. The braids in B_n^g can be identified as those lying between the planes with coordinates $z = 0$ and $z = 1$ and joining the points $(g + 1, 0), \cdots, (g + n, 0)$. Thus B_n^g can be considered as the subgroup of the classical braid group on $g + n$ strings whose braids leave the first g strings unbraided.

Let $\tau_k, k = 1, 2, \cdots, g$ be the following braids

$$\tau_k = \sigma_g' \sigma_{g-1}' \cdots \sigma_{k+1}' (\sigma_k')^2 (\sigma_{k+1}')^{-1} \cdots (\sigma_{g-1}')^{-1} (\sigma_g')^{-1},$$

where σ_i' represents standard generator of B_n^g. The elements τ_k ($k = 1, 2, \cdots, g$) generate a free subgroup F_g of the braid group B_{g+n}. Shift $\sigma_i = \sigma_{g+i}', i = 1, \cdots, n - 1$. Then the braid group B_n^g is generated by τ_k and σ_i with the following relations:

$$\sigma_i \sigma_j = \sigma_j \sigma_i \quad \text{for } |i - j| > 1$$

$$\sigma_i \sigma_{i+1} \sigma_i = \sigma_{i+1} \sigma_i \sigma_{i+1},$$

$$\tau_k \sigma_i = \sigma_i \tau_k \quad \text{for } k \geq 1, i \geq 2$$

$$\tau_k \sigma_1 \tau_k \sigma_1 = \sigma_1 \tau_k \sigma_1 \tau_k \quad \text{for } 1 \leq k \leq g$$

$$\tau_k \sigma_1^{-1} \tau_{k+l} \sigma_1 = \sigma_1^{-1} \tau_{k+l} \sigma_1 \sigma_k \quad \text{for } 1 \leq k \leq g - 1, 1 \leq l \leq g - k.$$

2.2 Artin's classical braid group $\Sigma = \mathbb{R}^2$

The classical braid group introduced by Artin [1947] is the full braid group $\pi_1(B(\mathbb{R}^2)^n)$.

(a) Geometric Description: Let $z^0 = (z_1^0, \cdots, z_n^0) \in F(\mathbb{R}^2)^n$ be a base point which is a collection of n distinct points in $\mathbb{R}^2$ plane. Let $[z^0] \in B(\mathbb{R}^2)^n$ be its equivalent class. If $f \in \pi_1(B(\mathbb{R}^2)^n, [z^0]) = \pi_1(B(\mathbb{R}^2)^n)$ is represented by a loop $f : [0,1] \to B(\mathbb{R}^2)^n$ with $f(0) = f(1) = [z^0]$ which lifts uniquely to a path $F : [0,1] \to F(\mathbb{R}^2)^n$ with $F(0) = z^0$, then the curve $F(t) = (f_1(t), \cdots, f_n(t)) \in F(\mathbb{R}^2)^n$ for all $t \in [0,1]$. Hence $f_i(t) \neq f_j(t)$ for $i \neq j$. View $(f_i(t), t) \in \mathbb{R}^2 \times [0,1]$ is an arc in $\mathbb{R}^3$. So the image $A_i = \{(f_i(t), t) : t \in [0,1]\}$ is called the i-th braid string, and $F(1)$ is a permutation of $F(0)$ as an equivalence class $[z^0]$. A braid in n-string is a set of disjoint arcs in $\mathbb{R}^3$ with equivalent end points.

(b) Group Operation: (i) The product of two braids b_1 and b_2 is obtained by putting end to end as the product operation in the fundamental group, $b_1 \star b_2 = [f_1 \star f_2]$. Here f_1 and f_2 are representatives of b_1 and b_2 in $\pi_1(B(\mathbb{R}^2)^n)$ with $f_1 \star f_2 : [0,1] \to B(\mathbb{R}^2)^n$:

$$f_1 \star f_2 = \begin{cases} f_1(2t) & 0 \leq t \leq \frac{1}{2} \\ f_2(2t-1) & \frac{1}{2} \leq t \leq 1 \end{cases}$$

(ii) The operation $\star$ is associative. The proof is same as that where the fundamental group operation is associative.

(iii) There exists a unique element identity e as straight vertical strings $e = (z_1^0, \cdots, z_n^0)$ for all $t \in [0,1]$.

(iv) Every braid b has a unique inverse b^{-1} as the mirror image of b.

(c) Presentation of the braid group: Any braid can be presented as the product of the elements $\sigma_i^{\pm 1}$ for $i = 1, 2, \cdots, n-1$ in $\pi_1(B(\mathbb{R}^2)^n)$. See Figure 2.1.

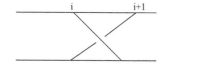
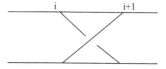

Fig. 2.1 σ_i and σ_i^{-1}

Thus elements $\sigma_1, \cdots, \sigma_{n-1}$ generate the group $\pi_1(B(\mathbb{R}^2)^n)$. Relations are arising from the equivalence class of any braid.

(i) The projection of strings cannot be tangent to each other. The transformation to create tangency yields the relation $\sigma_i \sigma_i^{-1} = 1$ in Figure 2.2 which trivially holds in any group.

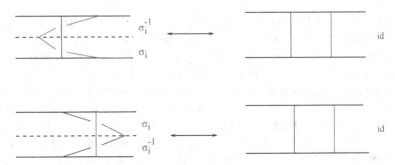

Fig. 2.2   Trivial relations $\sigma_i \sigma_i^{-1} = 1 = \sigma_i^{-1} \sigma_i$

(ii) The following transformation yields the relation (called braid relation) $\sigma_i \sigma_{i+1} \sigma_i = \sigma_{i+1} \sigma_i \sigma_{i+1}$.

Fig. 2.3 $\sigma_i \sigma_{i+1} \sigma_i^{-1} = \sigma_{i+1}^{-1} \sigma_i \sigma_{i+1}$

Fig. 2.4 $\sigma_i^{-1} \sigma_{i+1} \sigma_i = \sigma_{i+1} \sigma_i \sigma_{i+1}^{-1}$

There are two relations $\sigma_i \sigma_{i+1} \sigma_i^{-1} = \sigma_{i+1}^{-1} \sigma_i \sigma_{i+1}$ and $\sigma_i^{-1} \sigma_{i+1} \sigma_i =$

$\sigma_{i+1}\sigma_i\sigma_{i+1}^{-1}$ corresponding to Figures 2.3 and 2.4. But they are equivalent to the first one.

(iii) All the crossings occur at different heights. This requirement yields relation $\sigma_i\sigma_j = \sigma_j\sigma_i$ whenever $|i - j| \geq 2$.

Fig. 2.5 $\sigma_i\sigma_j = \sigma_j\sigma_i$ whenever $|i - j| \geq 2$

2.3 Artin's theorem on presentation of the full braid group

Let $\tilde{\sigma}_i \in \pi_1(B(\mathbb{R}^2)^n)$ with a representative given by the path

$$((1,0),\cdots,(i-1,0),(i+t,-\sqrt{t-t^2}),(i+1-t,\sqrt{t-t^2}),(i+2,0),\cdots,(n,0))$$

with a base point $z^0 = ((1,0),\cdots,(n,0))$ and $p(z^0) = \tilde{z}^0$, where $p : F(\mathbb{R}^2)^n \to B(\mathbb{R}^2)^n$ is the natural projection. Let B_n be the group with presentation given by

$$\langle \sigma_1,\cdots,\sigma_{n-1} \mid \begin{matrix} \sigma_i\sigma_j = \sigma_j\sigma_i & 1 \leq i,j \leq n-1, |i-j| \geq 2 \\ \sigma_i\sigma_{i+1}\sigma_i = \sigma_{i+1}\sigma_i\sigma_{i+1} & 1 \leq i \leq n-2 \end{matrix} \rangle. \quad (2.2)$$

Thus there is a natural homomorphism $l : B_n \to \pi_1(B(\mathbb{R}^2)^n)$ defined by $l(\sigma_i) = \tilde{\sigma}_i$ for $i = 1, 2, \cdots, n - 1$. For any element β in $\pi_1(B(\mathbb{R}^2)^n)$, there is a representative given by a loop $\tilde{b} : [0,1] \to B(\mathbb{R}^2)^n$. Let b be its unique lift of $\tilde{b}$ where $b : ([0,1],0) \to (F(\mathbb{R}^2)^n, z^0)$ with $b(0) = z^0$ and $b(1) \in \ker p$ as an element of symmetric group Σ_n. Let $\tilde{v} : \pi_1(B(\mathbb{R}^2)^n, \tilde{z}^0) \to \Sigma_n$ be a map defined by $\tilde{v}(\tilde{b}) = \begin{pmatrix} b_1(0) & \cdots & b_n(0) \\ b_1(1) & \cdots & b_n(1) \end{pmatrix} \in \Sigma_n$. Note that

$$\ker \tilde{v} = \pi_1(F(\mathbb{R}^2)^n, z^0).$$

Let v be the composition of $\tilde{v}$ and l, $v = \tilde{v} \circ l$:

$$B_n \xrightarrow{l} \pi_1(B(\mathbb{R}^2)^n, \tilde{z}^0) \xrightarrow{\tilde{v}} \Sigma_n, \quad \sigma_i \mapsto \tilde{\sigma}_i \mapsto (i, i+1), 1 \leq i \leq n-1.$$

Denote $\ker v = P_n$ (also called the pure braid group).

Lemma 2.3.1. *The group P_n has a presentation with generators in Figure 2.6*

$$\{A_{ij} = \sigma_{j-1}\sigma_{j-2}\cdots\sigma_{i+1}\sigma_i^2\sigma_{i+1}^{-1}\cdots\sigma_{j-2}^{-1}\sigma_{j-1}^{-1} : 1 \leq i,j \leq n\},$$

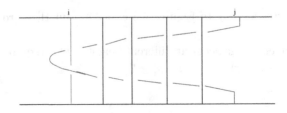

Fig. 2.6 A_{ij} braid

and relations given by the following

$$A_{rs}A_{ij}A_{rs}^{-1} = \begin{cases} A_{ij} & \text{if } r < s < i < j \\ & \text{or } i < r < s < j \\ A_{rj}A_{ij}A_{rj}^{-1} & \text{if } s = i \\ A_{ij}A_{sj}A_{ij}^2 A_{sj}^{-1} & \text{if } r = i < j < s \\ A_{rj}A_{sj}A_{rj}^{-1}A_{sj}^{-1}A_{ij}A_{rj}A_{sj}A_{rj}^{-1}A_{sj}^{-1} & \text{if } r < i < s < j \end{cases}$$

Proof: Note that $[B_n : P_n] = n!$. Choose any one of the cosets representatives of $n!$ word in B_n whose image v range over all of Σ_n. A set of right coset representatives are the collection of $\{\prod_{j=2}^{n} M_{j,k_j} : j \geq k_j \geq 1\}$, where

$$M_{j,i} = \begin{cases} \sigma_{j-1}\sigma_{j-2}\cdots\sigma_i & \text{if } j \neq i \\ 1 & \text{if } j = i \end{cases}$$

For the case of $n = 3$, we have a Schreier set

$$M_{22}M_{33} = 1 \quad M_{22}M_{32} = \sigma_2 \quad M_{22}M_{31} = \sigma_2\sigma_1$$
$$M_{21}M_{33} = \sigma_1 \quad M_{21}M_{32}\sigma_1\sigma_2 \quad M_{21}M_{31} = \sigma_1\sigma_2\sigma_1$$

Any initial segment of a coset representative is again a coset representative. Hence one may apply the Schreier-Reidemeister method [Lyndon and Schupp, 1977, p. 102] to obtain a group presentation for P_n.

There is an alternative approach given by W. -L. Chow [1948]. Let D_n be the subgroup of B_n which has the permutation fixing the letter n (the last letter). A Schreier set of coset representatives for D_n in B_n are the set $\{M_{n,k_n} : n \geq k_n \geq 1\}$. Let U_n be the normal subgroup generated by $\{A_{jn} : 1 \leq j \leq n - 1\}$. By the Schreier-Reidemeister method, $D_n = U_n \propto \overline{D}_n$ (a semi-direct product), where $\overline{D}_n$ is the subgroup of B_n which is generated by $\sigma_1, \cdots, \sigma_{n-2}$. So $\overline{D}_n = B_{n-1}$ and $D_{n-1} \subset B_{n-1}$ is the subgroup of B_{n-1} which has the permutation fixing the letter $n - 1$. Thus $D_{n-1} = U_{n-1} \propto \overline{D}_{n-1}$ with $\overline{D}_{n-1} = B_{n-2}$. Inductively, the element of B_n is a product of

$$u_2 M_{2,k_2} u_3 M_{3,k_3} \cdots u_n M_{n,k_n},$$

with $u_i \in U_i (2 \leq i \leq n)$. Each M_{j,k_j} is commutative with u_s if $j < s$. So any element can be represented as $u_2 u_3 \cdots u_n M_{2,k_2} M_{3,k_3} \cdots M_{n,k_n}$ uniquely [Chow, 1948, Theorem I]. □

Lemma 2.3.1 was proved first by Markov [1945]. Moreover Markov showed that P_n^k / P_n^{k-1} is a free group with free generators $A_{i(k+1)}$ for $1 \leq i \leq k$. Assume that $G_n (n \geq 0, G_0 = \{e\})$ is a system of subgroups of $AutF_n$ such that G_{n+m} as a subgroup of $AutF_{n+m}$ contains the image of $G_n \times G_m$ under μ for any n and m. Then we have a system of homomorphism $\mu_{n,m}(G) : G_n \times G_m \to G_{n+m}$. These pairings are strictly associative.

Exercise 2.3.2. (1) Verify that $\mu_{l,n+m} \circ (Id \times \mu_{n,m}) = \mu_{l+n,m} \circ (\mu_{l,n} \times Id)$.

(2) If the symmetric group Σ_n is contained in G_n as canonically embedded in $AutF_n$ for any $n \geq 0$, then a strict monodical (tensor) category can be defined in the usual way. That is the objects $\{0, 1, \cdots, n, \cdots\}$ of G correspond to the integers from 0 to infinity, and the morphisms are defined by $Mor(k, l) = G_k$ if $k = l$, $\emptyset$ otherwise. The symmetry in the category G generated by the symmetric groups induces a symmetry in the category G.

(3) Suppose that a system of normal subgroups $N_m \lhd G_m$ for any m such that N_{n+m} as a subgroup $AutG_{n+m}$ contains $N_n \times N_n$ for all n and m. Let H_m be the system of quotient groups $H_m = G_m/N_m$, and H be the corresponding category. Show that H is permeative, and the functor $G \to H$ induced by the epimorphisms $G_m \to H_m$ is a morphism of permutative categories.

Lemma 2.3.3. *Let $x = \sigma_1 \sigma_2 \cdots \sigma_{n-1}$ in B_n. We have the following identities:*

(a) $x\sigma_i = \sigma_{i+1}x$, for $1 \leq i \leq n-2$,
(b) $\sigma_1 x^i = x^i \sigma_{n-i+1}$, for $1 < i \leq n-1$,
(c) $x^i \sigma_1 = \sigma_{i+1} x^i$, for $1 \leq i \leq n-2$.

Proof: The identities follow from the relations in (2.2).

$$x\sigma_1 = \sigma_1 \sigma_2 \cdots \sigma_{n-1} \sigma_1 = \sigma_1 \sigma_2 \sigma_1 \sigma_3 \cdots \sigma_{n-1}$$

$$= \sigma_2 \sigma_1 \sigma_2 \sigma_3 \cdots \sigma_{n-1} = \sigma_2 x$$

Similarly, we have for $1 \leq i \leq n-2$,

$$x\sigma_i = \sigma_1 \sigma_2 \cdots \sigma_{n-1} \sigma_i = \sigma_1 \cdots \sigma_i \sigma_{i+1} \sigma_i \sigma_{i+2} \cdots \sigma_{n-2}$$

$$= \sigma_1 \cdots \sigma_{i+1} \sigma_i \sigma_{i+1} \sigma_{i+2} \cdots \sigma_{n-2}$$

$$= \sigma_{i+1} \sigma_1 \cdots \sigma_{n-2} = \sigma_{i+1} x.$$

We show (b) by using identity (a).

$$\sigma_1 x^2 = \sigma_1(x\sigma_1)\sigma_2\cdots\sigma_{n-1}$$

$$= \sigma_1(\sigma_2 x)\sigma_2\cdots\sigma_{n-1} \qquad\qquad \text{by (a) for } i=1 \text{ case}$$

$$= \sigma_1\sigma_2\cdots\sigma_{n-1}x\sigma_{n-1} \qquad\qquad \text{by (a) for } i=2,\cdots,n-2 \text{ cases}$$

$$= x^2\sigma_{n-1}.$$

One can inductively derive the formula (b) for i. Suppose (b) is true for $i-1$ power of x. Then one gets

$$\sigma_1 x^i = (\sigma_1 x^{i-1})x = x^{i-1}\sigma_{n-(i-1)+1}x$$

$$= x^{i-1}\sigma_{n-i+2}\sigma_1\sigma_2\cdots\sigma_{n-1}$$

$$= x^{i-1}\sigma_1\sigma_2\cdots\sigma_{n-i+2}\sigma_{n-i+1}\sigma_{n-i+2}\cdots\sigma_{n-1}$$

$$= x^{i-1}\sigma_1\sigma_2\cdots\sigma_{n-i+1}\sigma_{n-i+2}\sigma_{n-i+1}\sigma_{n-i+3}\cdots\sigma_{n-1}$$

$$= x^{i-1}\sigma_1\cdots\sigma_{n-1}\sigma_{n-i+1} = x^i\sigma_{n-i+1},$$

for all $1 < i \le n-1$. (c) follows from applying (a) i-times. $\qquad\square$

Lemma 2.3.4. *The element* $x^n = (\sigma_1\sigma_2\cdots\sigma_{n-1})^n$ *belongs to the center of* B_n.

Proof: All we need to show is that x^n commutes with each generator σ_i in B_n. For $2 \le i \le n-1$,

$$x^n\sigma_i = x^{i-1}x^{n-i+1}\sigma_i$$

$$= x^{i-1}\sigma_1 x^{n-i+1} \qquad\qquad \text{by Lemma 2.3.3 (b)}$$

$$= x^{i-1}\sigma_1 x^{n-i+1}$$

$$= \sigma_{(i-1)+1}x^{i-1}x^{n-i+1} \qquad\qquad \text{by Lemma 2.3.3 (c)}$$

$$= \sigma_i x^n.$$

Now for $i=1$, we have to choose j such that $1 \le j \le n-2$. So

$$x^n\sigma_1 = x^{n-j}x^j\sigma_1$$

$$= x^{n-j}\sigma_{j+1}x^j \qquad\qquad \text{by Lemma 2.3.3 (c)}$$

$$= \sigma_1 x^{n-j}x^j \qquad\qquad \text{by Lemma 2.3.3 (b)}$$

$$= \sigma_1 x^n.$$

Hence we get x^n in the center of the braid group B_n. $\qquad\square$

In fact, the center of B_n is generated by x^n [Chow, 1948, Theorem III].

Exercise 2.3.5. Verify the identity

$$(\sigma_2\sigma_3\cdots\sigma_{n-1})(\sigma_2\sigma_3\cdots\sigma_{n-2})\cdots(\sigma_2\sigma_3)\sigma_2$$

$$= \sigma_{n-1}(\sigma_{n-2}\sigma_{n-1})\cdots(\sigma_2\sigma_3\cdots\sigma_{n-1}).$$

Lemma 2.3.6. *If* $l|_{P_n} : P_n \to \pi_1(F(\mathbb{R}^2)^n, z^0)$ *is isomorphic, then* $l : B_n \to \pi_1(B(\mathbb{R}^2)^n, \tilde{z}^0)$ *is isomorphic.*

Proof: Note that $(i, i+1)$ generates Σ_n. Hence $\{v(\sigma_i) : 1 \leq i \leq n-1\}$ generates the full image Σ_n. We obtain the epimorphism of v. We have a commutative diagram

$$
\begin{array}{ccccccccc}
1 & \longrightarrow & \ker v = P_n & \longrightarrow & B_n & \xrightarrow{v} & \Sigma_n & \longrightarrow & 1 \\
& \downarrow{id} & & \downarrow{l_n = l|_{P_n}} & & \downarrow{l} & & \downarrow{id} & & \downarrow{id} \\
1 & \longrightarrow & \ker \tilde{v} = \pi_1(F(\mathbb{R}^2)^n) & \longrightarrow & \pi_1(B(\mathbb{R}^2)^n) & \xrightarrow{\tilde{v}} & \Sigma_n & \longrightarrow & 1,
\end{array}
$$
$$(2.3)$$

with exact rows. The identity is isomorphic. Hence $l_n = l|_{P_n}$ isomorphism implies l isomorphism by the Five Lemma. □

Theorem 2.3.7. *[Artin, 1947] There is a natural isomorphism* $\pi_1(B(\mathbb{R}^2)^n) \cong B_n$, *and thus the full braid group* $\pi_1(B(\mathbb{R}^2)^n)$ *admits a presentation as in (2.2).*

Proof: We prove it by Principle of Mathematical Induction on n. When $n = 1$, we have $P_1 = \{1\}$ and $\pi_1(F(\mathbb{R}^2)^n) = \{1\}$, so l_1 is isomorphic. Hence the result is true for $n = 1$ by Lemma 2.3.6. Assume that the isomorphism is true for $n-1$. Note that $P_{n-1} = \{A_{ij} : 1 \leq i < j \leq n-1\}$ as a subgroup of P_n. The natural projection $\epsilon : P_n \to P_{n-1}$ is defined by

$$
\epsilon(A_{ij}) = \begin{cases} A_{ij} & \text{if } 1 \leq i < j \leq n-1, \\ 1 & \text{if } 1 \leq i < n, \, j = n. \end{cases}
$$

Then elements $A_{1n}, A_{2n}, \cdots, A_{n-1n}$ are generators of the normal subgroup $\ker \epsilon$. Using the map $\pi_n^{n-1} : \pi_1(F(\mathbb{R}^2)^n) \to \pi_1(F(\mathbb{R}^2)^{n-1})$, we have

$$
\begin{array}{ccccccccc}
1 & \longrightarrow & U_n = \ker \epsilon & \longrightarrow & P_n & \xrightarrow{\epsilon} & P_{n-1} & \longrightarrow & 1 \\
& \downarrow{id} & & \downarrow{l_n|_{U_n}} & & \downarrow{l_n} & & \downarrow{l_{n-1}} & & \downarrow{id} \\
1 & \longrightarrow & \pi_1(F_{n-1}\mathbb{R}^2) & \longrightarrow & \pi_1(F(\mathbb{R}^2)^n) & \xrightarrow{\pi_1(\pi_n^{n-1})} & \pi_1(F(\mathbb{R}^2)^{n-1}) & \longrightarrow & 1
\end{array}
$$
$$(2.4)$$

where the rows are exact. Note that $l_n|_{U_n} = \{l_n(A_{jn}) : 1 \leq j \leq n-1\}$ is a free basis of $\pi_1(F_{n-1}\mathbb{R}^2)$, where $l_n(A_{jn})$ is an element representing a loop based at z_n^0 which encircles z_j^0 once and separates it from $z_1^0, \cdots, z_{j-1}^0, z_{j+1}^0, \cdots, z_{n-1}^0$. By the Hopfian property of free groups, U_n must be free and $l_n|_{U_n}$ is isomorphic for all n. Therefore l_n is isomorphic by

Induction hypothesis (l_{n-1} isomorphism from Lemma 2.3.6 and (2.3)) and Five Lemma from the commutative diagram (2.4). Then l is isomorphic by Lemma 2.3.6. By the principle of mathematical induction, we conclude that $\pi_1(B(\mathbb{R}^2)^n) \cong B_n$ is true for all n.   □

2.4   Normal form of braid elements

From the proof of Theorem 2.3.7, the group P_n is a semi-product of U_n and P_{n-1}. We will identify the two groups B_n and $\pi_1(B(\mathbb{R}^2)^n)$ through the identification l, similarly we identify P_n with $\pi_1(F(\mathbb{R}^2)^n)$ through the identification l_n. The coset representatives for P_n in B_n which are defined in the following Proposition 2.4.1 will be called the permutation braids.

Proposition 2.4.1. *Every element $b \in B_n$ can be represented uniquely as the normal form*

$$b = b_2 b_3 \cdots b_n \pi(b),$$

where $\pi(b)$ is a permutation braid in $B_n = \cup_{\pi(b)} P_n \cdot \pi(b)$, and each $b_j \in U_j$ as in Theorem 2.3.7.

Proof: From the coset representatives $\pi(b)$, we have $b = \bar{b}_n \cdot \pi(b)$ for some $\bar{b}_n \in P_n$. Since P_n is a semi-product of U_n and P_{n-1}, $\bar{b}_n = \bar{b}_{n-1} \cdot b_n$ for $\bar{b}_{n-1} \in P_{n-1}$ and $b_n \in U_n$ with $n > 2$. Applying inductively, we obtain

$$b = \bar{b}_2 b_3 \cdots b_n \pi(b),$$

where $b_i \in U_i$ ($i = 3, 4, \cdots, n$) and $\bar{b}_2 \in P_2 = U_2$. Let $b_2 = \bar{b}_2$. This provides the expression of $b \in B_n$.

Since each b_i lies in a free group on known free generators, it is possible to algorithmically calculate standard representatives for $b_2, b_3, \cdots, b_n$, and $\pi(b)$ in the given generators for $\pi(b)$. Thus the uniqueness of each b_i($i = 2, 3, \cdots, n$) follows inductively from the semi-product of U_n and P_{n-1}.   □

In fact, taking the representative of $b \in B_n$ in the form of Proposition 2.4.1 is called *combing the braid* b. Each $b_i \in U_i$ is a product of the free generators $A_{1i}, A_{2i}, \cdots, A_{i-1,i}$.

2.5   Braid action on a free group

Let $F_n = \langle x_1, \cdots, x_n \rangle$ be a free group of rank n. Define a representation $\rho_F : B_n \to Aut F_n$ by $\rho_F(\sigma_i)(1 \le i \le n-1)$, where $\rho_F(\sigma_i) : F_n \to F_n$ is given by

$$\rho_F(\sigma_i): \quad x_i \mapsto x_i x_{i+1} x_i^{-1}, \quad x_{i+1} \mapsto x_i, \quad x_j \mapsto x_j \quad (j \ne i, i+1).$$

Note that the automorphism $\rho_F(\sigma_i)$ has its inverse $\rho_F(\sigma_i)^{-1}$ given by

$$\rho_F(\sigma_i)^{-1}: \quad x_i \mapsto x_{i+1}, \quad x_{i+1} \mapsto x_{i+1}^{-1} x_i x_{i+1}, \quad x_j \mapsto x_j \quad (j \neq i, i+1).$$

This provides another important approach to the braid group as a subgroup of the automorphism group of F_n. One can also assign the elements in the symmetric group Σ_n into $Aut F_n$ by $\xi_i(x_i) = x_{i+1}, \xi_i(x_{i+1}) = x_i, \xi_i(x_j) = x_j$ for $j \neq i, i+1$.

Proposition 2.5.1. *The representation* $\rho_F : B_n \to Aut F_n$ *is faithful .*

Proof: We first show that ρ_F is a representation which amounts to checking relations, and then prove that $\ker \rho_F = \{1\}$ to conclude that ρ_F is a faithful representation .

(1) Check ρ_F is a well-defined representation. Verify $\rho_F(\sigma_i \sigma_j) = \rho_F(\sigma_j \sigma_i)$ whenever $|i - j| \geq 2$ for $1 \leq i, j \leq n - 1$.

$$\rho_F(\sigma_i \sigma_j): \begin{cases} x_i \overset{\rho_F(\sigma_j)}{\mapsto} x_i \overset{\rho_F(\sigma_i)}{\mapsto} x_i x_{i+1} x_i^{-1} \\ x_{i+1} \overset{\rho_F(\sigma_j)}{\mapsto} x_{i+1} \overset{\rho_F(\sigma_i)}{\mapsto} x_i \\ x_j \overset{\rho_F(\sigma_j)}{\mapsto} x_j x_{j+1} x_j^{-1} \overset{\rho_F(\sigma_i)}{\mapsto} x_j x_{j+1} x_j^{-1} \end{cases}$$

$$\rho_F(\sigma_j \sigma_i): \begin{cases} x_i \overset{\rho_F(\sigma_i)}{\mapsto} x_i x_{i+1} x_i^{-1} \overset{\rho_F(\sigma_j)}{\mapsto} x_i x_{i+1} x_i^{-1} \\ x_{i+1} \overset{\rho_F(\sigma_i)}{\mapsto} x_i \overset{\rho_F(\sigma_j)}{\mapsto} x_i \\ x_j \overset{\rho_F(\sigma_i)}{\mapsto} x_j \overset{\rho_F(\sigma_j)}{\mapsto} x_j x_{j+1} x_j^{-1} \end{cases}$$

Verify $\rho_F(\sigma_i \sigma_{i+1} \sigma_i) = \rho_F(\sigma_{i+1} \sigma_i \sigma_{i+1})$.

$$\rho_F(\sigma_i \sigma_{i+1} \sigma_i): \begin{cases} x_i \overset{\rho_F(\sigma_i)}{\mapsto} x_i x_{i+1} x_i^{-1} \overset{\rho_F(\sigma_{i+1})}{\mapsto} x_i x_{i+1} x_{i+2} x_{i+1}^{-1} x_i^{-1} \\ \overset{\rho_F(\sigma_i)}{\mapsto} (x_i x_{i+1} x_i^{-1}) x_i x_{i+2} x_i^{-1} (x_i x_{i+1} x_i^{-1})^{-1} \\ = x_i x_{i+1} x_{i+2} x_{i+1}^{-1} x_i^{-1} \\ x_{i+1} \overset{\rho_F(\sigma_i)}{\mapsto} x_i \overset{\rho_F(\sigma_{i+1})}{\mapsto} x_i \overset{\rho_F(\sigma_i)}{\mapsto} x_i x_{i+1} x_i^{-1} \\ x_j \overset{\rho_F(\sigma_i)}{\mapsto} x_j \overset{\rho_F(\sigma_{i+1})}{\mapsto} x_j \overset{\rho_F(\sigma_i)}{\mapsto} x_j \text{ if } j \neq i, i+1, i+2 \\ x_{i+2} \overset{\rho_F(\sigma_i)}{\mapsto} x_{i+2} \overset{\rho_F(\sigma_{i+1})}{\mapsto} x_{i+1} \overset{\rho_F(\sigma_i)}{\mapsto} x_i \text{ if } j = i+2 \end{cases}$$

$$\rho_F(\sigma_{i+1} \sigma_i \sigma_{i+1}): \begin{cases} x_i \overset{\rho_F(\sigma_{i+1})}{\mapsto} x_i \overset{\rho_F(\sigma_i)}{\mapsto} x_i x_{i+1} x_i^{-1} \\ \overset{\rho_F(\sigma_{i+1})}{\mapsto} x_i (x_{i+1} x_{i+2} x_{i+1}^{-1}) x_i^{-1} \\ x_{i+1} \overset{\rho_F(\sigma_{i+1})}{\mapsto} x_{i+1} x_{i+2} x_{i+1}^{-1} \overset{\rho_F(\sigma_i)}{\mapsto} x_i x_{i+2} x_i^{-1} \\ \overset{\rho_F(\sigma_{i+1})}{\mapsto} x_i x_{i+1} x_i^{-1} \\ x_j \overset{\rho_F(\sigma_{i+1})}{\mapsto} x_j \overset{\rho_F(\sigma_i)}{\mapsto} x_j \overset{\rho_F(\sigma_{i+1})}{\mapsto} x_j \text{ if } j \neq i, i+1, i+2 \\ x_{i+2} \overset{\rho_F(\sigma_{i+1})}{\mapsto} x_{i+1} \overset{\rho_F(\sigma_i)}{\mapsto} x_i \overset{\rho_F(\sigma_{i+1})}{\mapsto} x_i \text{ if } j = i+2 \end{cases}$$

Hence $\rho_F : B_n \to Aut F_n$ is a group homomorphism.

(2) The homomorphism $\rho_F(\sigma_i) : F_n \to F_n$ satisfies

$$\rho_F(\sigma_i)[a, b] = [\rho_F(\sigma_i)(a), \rho_F(\sigma_i)(b)],$$

and $\rho_F(\sigma_i) : [F_n, F_n] \to [F_n, F_n]$. Thus $\rho_F(\sigma_i)$ induces $\overline{\rho}_F : F_n/[F_n, F_n] \to F_n/[F_n, F_n]$. Let $\overline{\rho}_F : B_n \to Aut(F_n/[F_n, F_n])$ be defined by $\overline{\rho}_F(\sigma_i)(x_i) = x_i x_{i+1} x_i^{-1} = [x_i, x_{i+1}] \cdot x_{i+1} = \overline{x}_{i+1} \in F_n/[F_n, F_n]$ as an equivalence class, $\overline{\rho}_F(\sigma_i)(x_{i+1}) = \overline{x}_i$ and $\overline{\rho}_F(\sigma_i)(x_j) = \overline{x}_j$ for $j \neq i, i+1$. Therefore the induced map $\overline{\rho}_F : B_n \to Aut(F_n/[F_n, F_n])$ sends $\sigma_i \mapsto (i, i+1) \in \Sigma_n$. We have, two short exact sequence in rows,

$$
\begin{array}{ccccc}
P_n & \longrightarrow & B_n & \longrightarrow & \Sigma_n \\
\downarrow & & \downarrow{\scriptstyle id} & & \downarrow{\scriptstyle \cong} \\
\ker \overline{\rho}_F & \longrightarrow & B_n & \xrightarrow{\overline{\rho}_F} & Aut(F_n/[F_n, F_n]).
\end{array}
$$

Hence $P_n \cong \ker \overline{\rho}_F$ by the Five Lemma. Now with short exact rows,

$$
\begin{array}{ccccc}
P_n & \longrightarrow & B_n & \longrightarrow & \Sigma_n \\
\downarrow{\scriptstyle \rho_F|_{P_n}} & & \downarrow{\scriptstyle \rho_F} & & \downarrow{\scriptstyle \cong} \\
\ker \rho & \longrightarrow & Aut F_n & \xrightarrow{\rho} & Aut(F_n/[F_n, F_n]),
\end{array}
$$

Hence the representation ρ_F is *monomorphism* ($\ker \rho_F = \{Id\}$) if $\rho_F|_{P_n}$ is monomorphism by the Five Lemma. The result will follow the monomorphism of $\rho_F|_{P_n}$.

(3) $\ker \rho_F|_{P_n}$ is trivial. Note that the representation $\rho_F|_{P_n}$ arises in a natural way by the previous structure on $P_{n+1} = P_n \propto U_{n+1}$, where U_{n+1} is a free normal subgroup of rank n in P_{n+1}. Hence P_n acts by conjugation as a group of automorphisms of U_{n+1}. Define $f : U_{n+1} \to F_n$ by $f(A_{j,n+1}) = x_j (j = 1, \cdots, n)$. The conjugate action makes the following diagram commutative:

$$
\begin{array}{ccc}
P_n & \xrightarrow{c} & Aut U_{n+1} \\
\downarrow{\scriptstyle id} & & \downarrow{\scriptstyle Aut(f)} \\
P_n & \xrightarrow{\rho_F|_{P_n}} & Aut F_n.
\end{array}
$$

Note that f is isomorphic, so is $Aut(f)$. The kernel $\ker \rho_F|_{P_n} = \ker c|_{P_n}$ is the subgroup of elements in P_n which commute with U_{n+1}.

We prove it by contradiction. Assume $\ker \rho_F|_{P_n}$ is not trivial. There is an element $b(\neq 1) \in \ker \rho_F|_{P_n}$ such that $b = b_2 \cdots b_i$ with i being the largest integer such that $\beta_i \neq 1$ but $b_{i+1} = b_{i+2} = \cdots = b_n = 1$ by

Proposition 2.4.1. Since b is in the kernel of c via the isomorphism identification, b commutes with U_{n+1}. Therefore b commutes with the generators $\{A_{i,n+1} : i = 1, 2, \cdots, n\}$, where

$$A_{i,n+1} = \sigma_n \sigma_{n-1} \cdots \sigma_{i+1} \sigma_i^2 \sigma_{i+1}^{-1} \cdots \sigma_{n_1}^{-1} \sigma_n^{-1}.$$

Each $b_j (2 \leq j \leq i)$ belongs to the free group $U_j \subset P_{n+1}$ which is freely generated by $A_{1j}, \cdots, A_{j-1,j}$. Hence b_j depends only on $\sigma_1, \cdots, \sigma_{j-1}$. Thus β_j commutes with $A_{i,n+1}$ whenever $j \leq i-1$. Therefore b commutes with $A_{i,n+1}$:

$$b_i A_{i,n+1} b_i^{-1} = A_{i,n+1}.$$

Let $\pi = \sigma_n \sigma_{n-1} \cdots \sigma_i$. Then we have $A_{i,n+1} = \pi \sigma_i^2 \pi^{-1}$ and $A_{n,n+1} = \sigma_n^2$.

$$A_{i,n+1} = \sigma_n \sigma_{n-1} \cdots \sigma_{i+1} \sigma_i^2 \sigma_{i+1}^{-1} \cdots \sigma_n^{-1} = \sigma_i^{-1} \cdots \sigma_{n-1}^{-1} \sigma_n^2 \sigma_{n-1} \cdots \sigma_i.$$

Fig. 2.7 $A_{i,n+1}$ braid

$$\begin{aligned}
\pi A_{i,n+1} \pi^{-1} &= \pi \sigma_n \sigma_{n-1} \cdots \sigma_{i+1} \sigma_i^2 \sigma_{i+1}^{-1} \cdots \sigma_n^{-1} \pi^{-1} \\
&= \pi \sigma_i^{-1} \cdots \sigma_{n-1}^{-1} \sigma_n^2 \sigma_{n-1} \cdots \sigma_i \pi^{-1} \\
&= \pi \pi^{-1} \sigma_n^2 \pi \pi^{-1} \\
&= \sigma_n^2 = A_{n,n+1}
\end{aligned}$$

Similarly, we obtain the relations, for $k < i$,

$$\pi A_{k,i} \pi^{-1} = \sigma_n \cdots \sigma_i \sigma_{i-1} \cdots \sigma_{k+1} \sigma_k^2 \sigma_{k+1}^{-1} \cdots \sigma_{i-1}^{-1} \sigma_i^{-1} \cdots \sigma_n^{-1} = A_{k,n+1}. \tag{2.5}$$

So $b_i A_{i,n+1} b_i^{-1} = A_{i,n+1}$. Equivalently

$$(\pi b_i \pi^{-1})(\pi A_{i,n+1} \pi^{-1})(\pi b_i^{-1} \pi^{-1}) = \pi A_{i,n+1} \pi^{-1} = A_{n,n+1}.$$

$$(\pi b_i \pi^{-1})(\pi A_{n,n+1} \pi^{-1})(\pi b_i^{-1} \pi^{-1}) = A_{k,n+1}.$$

Thus by (2.5),

$$(\pi b_i \pi^{-1}) = \pi(\prod_{i'} A_{i'i}^{\varepsilon_i})\pi^{-1}$$
$$= \prod_{i'} (\pi A_{i'i}^{\varepsilon_i} \pi^{-1})$$
$$= \prod_{i'} A_{i',n+1}^{\varepsilon_i} \in U_{n+1}.$$

The first equality follows from $b_i \in U_i$ which is generated by $\{A_{i'i}\}(i' < i)$, the last equality from (2.5). So $\pi b_i \pi^{-1}$ lies in a free group U_{n+1}. Two elements commute in a free group, they must be powers of the same element in the free group. Since $A_{n,n+1}$ is a generator of U_{n+1} and $\pi b_i \pi^{-1}$ commutes with $A_{n,n+1}$, there exists an integer l such that

$$\pi b_i \pi^{-1} = A_{n,n+1}^l.$$

We get $b_i = \pi^{-1} A_{n,n+1}^l \pi = A_{i,n+1}^l$. Recall $b_i \in U_i$ (generated by $A_{1,i}, \cdots, A_{i-1,i}$). The only way $A_{i,n+1}^l \in U_i$ is the case $l = 0$ and $b_i = 1$. Hence we obtain a contradiction with the choice of b_i. $b \in \ker \rho_F|_{P_n}$ must be the identity element. Thus $\rho_F|_{P_n}$ is a faithful representation, so is ρ_F. $\square$

The free group F_n is the fundamental group of the disk D_n (the standard disk in R^2 punctured at n distinct points), and a generator x_i corresponding to a loop around the i-th point. In this way, the B_n is the mapping class group of D_n with fixed boundary and acts on the fundamental group of D_n by the representation ρ_F. There is a canonical embedding

$$AutF_n \times AutF_m \to AutF_{n+m}$$

that is compatible with the standard pairing of the braid groups

$$\mu : B_n \times B_m \to B_{n+m}.$$

Thus $\mu(\sigma_i', e) = \sigma_i, \mu(e, \sigma_j'') = \sigma_{n+j}$ for $\sigma_i' \in B_n$ and $\sigma_j'' \in B_m$ with $1 \le i \le n-1$ and $1 \le j \le m-1$.

2.6   Alexander theorem

Definition 2.6.1. The closure of a braid $\beta \in B_n$ is to identify the initial points and end points of each of the braid strings. Denoted by $\bar{\beta}$.

The following fundamental result of Alexander is related the link by a closure of braids.

Theorem 2.6.2. *There is a surjective map* $A : \cup_{n=2}^{\infty} B_n \to \mathcal{L}(\mathbb{R}^3)$ *defined by* $\beta \mapsto \bar{\beta}$.

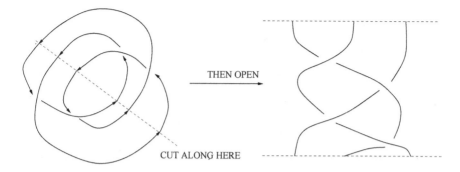

Fig. 2.8 · Switch a knot to a closure of a braid

Proof: Call a link winding around a point O if every edge (as PL link) of L is seen from O as positive (oriented from right to left). See Figure 2.8.

Hence every link winding around a point can be obtained by closing a braid. In order to prove the surjectivity of the Alexander map A, we need to show that any link is isotopic to one which is winding around the point.

The Alexander trick is to replace any negative edge of L by two positive sides of a triangle with a new vertex c behind the point O. Repeat applications of the Alexander trick, until no negative edge remains, then the new link (isotopic equivalent to the previous one) has the desired property. Hence it can be given by an image of A.

(a) AB has no crossing points. The application of the Alexander trick is straightforward. By moving c perpendicularly upward to the ABO plane such that $ABC \cap L = \emptyset$.

(b) AB has an overpass point (underpass). The application of the Alexander trick is to move c perpendicularly upward (downward) to the ABO plane such that $ABC \cap L = \emptyset$.

(c) If the negative edge has several crossing points, then we subdivide the edge into smaller edges with only one crossing point and apply (b) several times. $\square$

Exercise 2.6.3. A is not injective. Note that $\sigma_1 \neq \sigma_1^{-1}$ in B_n for $n \geq 2$. Show that $A(\sigma_1) = A(\sigma_1^{-1})$.

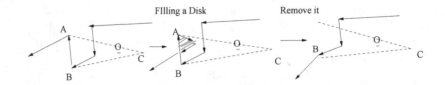

Fig. 2.9   Alexander trick

The following theorem describes the preimage of the Alexander map A.

Theorem 2.6.4. *[Markov, 1945; Birman, 1974] The preimage $A^{-1}(L)$ consists of any finite sequence of two Markov moves of $L = \overline{\beta}$.*

Markov move I: $\beta \to b\beta b^{-1}$ for $b, \beta \in B_n$.

Markov move II: $\beta \to \beta \sigma_n^{\pm 1}$ (as imbedding $B_n \hookrightarrow B_{n+1}$) (see Figure 2.10).

Proof: The two Markov moves of β do not effect the link as closure of the braid $\beta \in B_n$.

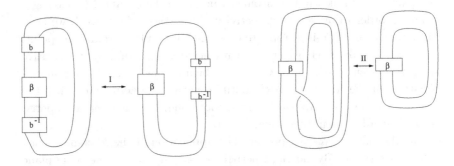

Fig. 2.10   Markov moves of type I and II

Hence any finite sequence of their combinations does not change the link type as well. The difficult part of Theorem 2.6.4 is that those operations generate the preimage of A for any link $L = \overline{\beta}$. We refer to [Birman, 1974, pp. 48–67] for the detailed proof.   □

Remark 2.6.5. The first published proof of the Markov theorem is due to Joan Birman [1974]. Markov presented an oral exposition of his theorem and its proof, but never published it.

Definition 2.6.6. (1) A link $L = \overline{\alpha}$ is said to have braid index n if it can be represented by a braid word $\alpha \in B_n$, but cannot be represented by a braid word in B_{n-1}.

(2) A link $L = \overline{\beta}$ is a pure link of multiplicity n if it is represented by an element β in the pure braid group P_n, but cannot be represented by an element in P_{n-1}.

If $\alpha \in B_n$, then $\overline{\alpha\sigma_n}$ and $\overline{\alpha\sigma_n^{-1}}$ are equivalent links by two Markov move II. But the words $\alpha\sigma_n$ and $\alpha\sigma_n^{-1}$ are not conjugate to each other in B_{n+1} since the exponent sum $e(\alpha\sigma_n) - e(\alpha\sigma_n^{-1}) = 2 \neq 0$, where the exponent sum is an invariant of conjugate class in B_{n+1}. Thus there are two non-conjugate braids which are equivalent. Is the same true for the fixed string index ? It would be extremely difficult to determine the braid index of an arbitrary link. But it is tractable for the pure links of multiplicity n.

Theorem 2.6.7. *Every pure link of multiplicity n has braid index n.*

Proof: Let $\beta \in P_n$ be a pure braid representative of the pure link $\overline{\beta}$. The link group $\pi_1(S^3 \backslash \overline{\beta})$ is generated by at least n elements due to n components of the pure link $\overline{\beta}$. If $\overline{\beta}$ is equivalent to $\overline{\alpha}$ for $\alpha \in B_m$ and $m < n$, then, by Artin's theorem (Theorem 2.8.1), one can obtain a link group presentation of $\pi_1(S^3 \backslash \overline{\beta})$ with m generators, which is impossible. Hence $\overline{\beta}$ has braid index n. □

Murasugi and Thomas (1972) constructed an example that $\beta_1 = \sigma_1^3\sigma_2^5\sigma_3^7$ and $\beta_2 = \sigma_1^5\sigma_2^3\sigma_3^7 \in B_4$ such that both have braid index 4 and they are equivalent, but β_1 is not conjugate to β_2 in B_4.

$$\overline{\beta_1} = \overline{\sigma_1^3\sigma_2^5\sigma_3^7} \sim \overline{\sigma_1^3 \# \sigma_1^5 \# \sigma_1^7} \sim \overline{\sigma_1^5 \# \sigma_1^3 \# \sigma_1^7} \sim \overline{\sigma_1^5\sigma_2^3\sigma_3^7} = \overline{\beta_2}.$$

A question is whether a braid word $\beta \in B_{n+1}$ can be conjugate to an element of the form $\alpha_1\sigma_n\alpha_2\sigma_n^{-1}$ with $\alpha_1, \alpha_2 \in B_n$. Note that $\gamma = \alpha_1\sigma_n^{-1}\alpha_2\sigma_n$ has the equivalent relation $\overline{\beta} \sim \overline{\gamma}$, and γ and β may not be necessarily conjugate to each other in B_{n+1}.

Exercise 2.6.8. Let $\beta = \sigma_3\sigma_1^2\sigma_2^4\sigma_3^{-1}\sigma_2^2$ and $\gamma = \sigma_3^{-1}\sigma_1^2\sigma_2^4\sigma_3\sigma_2^2$ be two pure 4-braids.

(i) Show that $\overline{\beta} \sim \overline{\gamma}$ (the links $\overline{\beta}$ and $\overline{\gamma}$ are equivalent), by drawing the closures of those braid elements.

(ii) Show that β and γ are not conjugate to each other in B_3 under the homomorphism defined by mapping $\sigma_1 \mapsto \sigma_1, \sigma_2 \mapsto \sigma_2, \sigma_3 \mapsto \sigma_1$.

(iii) Show that β and γ are not conjugate to each other in B_4.

2.7   Characterization of braid representation

From Proposition 2.5.1, we can identify B_n as a subgroup of $AutF_n$. Hence we identify $B_n \cong \rho_F(B_n) \subset AutF_n$. For each element $b \in B_n$, we have the automorphism

$$\rho_F(b) = \rho_F(\sigma_{i_1})^{\varepsilon_1} \cdots \rho_F(\sigma_{i_k})^{\varepsilon_k},$$

where $b = \sigma_{i_1}^{\varepsilon_1} \cdots \sigma_{i_k}^{\varepsilon_k}$ with $i_1, \cdots, i_k \in \{1, 2, \cdots, n-1\}$. Each $\rho_F(\sigma_i)$ acts as a conjugation of x_{i+1} and x_i, and also fixes other generators of F_n. So $\rho_F(\sigma_i)(x_j) = A_j x_{\mu_j} A_j^{-1}$, where $A_j = 1$ if $j \neq i$, $A_j = x_j$ if $j = i$; and $\mu_j = j$ if $j \neq i, i+1$, $\mu_j = i$ if $j = i+1$ and $\mu_j = i+1$ if $j = i$. Thus for each generator σ_i, we can express the faithful representation as a conjugation. Hence $\rho_F(\sigma_i^{\varepsilon_i}) = \rho_F(\sigma_i)^{\varepsilon_i}$ is also a conjugation of some element with more complicated relations on A_j and μ_j. Note that

$$
\begin{aligned}
\rho_F(\sigma_i)(x_1 \cdots x_n) &= \rho_F(\sigma_i)(x_1) \cdots \rho_F(\sigma_i)(x_n) \\
&= x_1 \cdots \rho_F(\sigma_i)(x_i)\rho_F(\sigma_i)(x_{i+1})x_{i+2} \cdots x_n \\
&= x_1 \cdots x_{i-1}x_ix_{i+1}x_i^{-1}x_ix_{i+2} \cdots x_n \\
&= x_1 \cdots x_n.
\end{aligned}
$$

For any $\sigma_i^{\varepsilon_i}$, we also have $\rho_F(\sigma_i^{\varepsilon_i})(x_1 \cdots x_n) = (x_1 \cdots x_n)$. Now we find one typical relation for $b \in B_n$,

$$\rho_F(b)(x_1 \cdots x_n) = x_1 \cdots x_n. \tag{2.6}$$

One natural question arises as to whether the conjugate relation and (2.6) characterize the image of the faithful representation of the braid group. The following Artin theorem offers the affirmative answer to such a question.

Lemma 2.7.1. *Let ρ be an endomorphism of F_n with the properties (1) $\rho(x_i) = A_i x_{\mu_i} A_i^{-1}$ $(1 \leq i \leq n)$ and (2)*

$$A_1 x_{\mu_1} A_1^{-1} \cdot A_2 x_{\mu_2} A_2^{-1} \cdots A_n x_{\mu_n} A_n^{-1} = x_1 \cdot x_2 \cdots x_n.$$

Then there must exist some $v \in \{1, 2, \cdots, n-1\}$ such that either (a) $x_{\mu_v} A_v^{-1}$ is absorbed by A_{v+1}, i.e., $A_{v+1} = A_v x_{\mu_v}^{-1} \tilde{A}_{v+1}$; or (b) A_v^{-1} absorbs $A_{v+1} x_{\mu_{v+1}}$.

Proof: If one of $A_v x_{\mu_v} A_v^{-1}$ is completely absorbed by the other terms in the free cancelations which reduce to the identity, then x_{μ_v} is absorbed by a letter of the left of $x_{\mu_{v+1}}$. Hence (a) follows. If x_{μ_v} is absorbed by a letter of A_{v-1}^{-1}, then (b) follows. If x_{μ_v} is absorbed by a letter in A_{v+1}, then (a) is true. Same on the left side of x_{μ_v}. Since x_{μ_v} cannot be absorbed by $x_{\mu_{v-1}}, x_{\mu_{v+1}}$, then all possible cases reduce to either (a) or (b).

It remains that no subscript v with the property $A_v x_{\mu_v} A_v^{-1}$ is completely absorbed. After all free reduction, $A_v x_{\mu_v} A_v^{-1}$ can be reduced to a word R_v. The identity (2) in the lemma becomes

$$R_1 \cdot R_2 \cdots R_n = x_1 \cdot x_2 \cdots x_n.$$

That is $R_i = x_i$ for each $i = 1, 2, \cdots, n$. Therefore $A_1 x_{\mu_1} A_1^{-1} = x_1$. If $A_1 x_{\mu_1} A_1^{-1} \neq x_1$, then

$$A_1 x_{\mu_1} A_1^{-1} = x_1 (x_1^{-1} A_1 x_{\mu_1} A_1^{-1} x_1) x_1^{-1} = x_1 (\tilde{A}_1 x_{\mu_1} \tilde{A}_1^{-1} x_1^{-1}).$$

So the term $\tilde{A}_1 x_{\mu_1} \tilde{A}_1^{-1} x_1^{-1}$ must be completely absorbed. Either x_{μ_1} is absorbed by A_2 ((a) is true) or by a letter x_{μ_2} ((b) is true). If $A_1 x_{\mu_1} A_1^{-1} = x_1$ is true, then we have

$$R_2 \cdots R_n = x_2 \cdots x_n.$$

Look at $A_2 x_{\mu_2} A_2^{-1}$ by the same argument. Thus the result follows. $\qquad\square$

Lemma 2.7.2. *Let $l(\rho)$ be the length of the endomorphism of ρ to be the sum of the letter lengths of the words $A_i x_{\mu_i} A_i^{-1}$. If Lemma 2.7.1 (a) is true for the endomorphism ρ, then $l(\rho \circ \rho_F(\sigma_v)) < l(\rho)$. If Lemma 2.7.1 (b) is true for ρ, then $l(\rho \circ \rho_F(\sigma_v^{-1})) < l(\rho)$.*

Proof: If Lemma 2.7.1 (a) is true for the endomorphism ρ, then the action ρ on x_v and x_{v+1} are given by

$$x_v \xrightarrow{\rho} A_v x_{\mu_v} A_v^{-1}$$
$$x_{v+1} \xrightarrow{\rho} A_{v+1} x_{\mu_{v+1}} A_{v+1}^{-1} = A_v x_{\mu_v}^{-1} \tilde{A}_{v+1} x_{\mu_{v+1}} \tilde{A}_{v+1}^{-1} x_{\mu_v} A_v^{-1}.$$

Then the endomorphism $\rho \circ \rho_F(\sigma_v)$ action is given by

$$x_v \xrightarrow{\rho_F(\sigma_v)} x_v x_{v+1} x_v$$
$$\xrightarrow{\rho} (A_v x_{\mu_v} A_v^{-1})(A_v x_{\mu_v}^{-1} \tilde{A}_{v+1} x_{\mu_{v+1}} \tilde{A}_{v+1}^{-1} x_{\mu_v} A_v^{-1})(A_v x_{\mu_v}^{-1} A_v^{-1})$$
$$= A_v \tilde{A}_{v+1} x_{\mu_{v+1}} A_{v+1}^{-1} A_v^{-1}$$
$$x_{v+1} \xrightarrow{\rho_F(\sigma_v)} x_v \xrightarrow{\rho} A_v x_{\mu_v} A_v^{-1}$$
$$x_j \xrightarrow{\rho_F(\sigma_v)} x_j \xrightarrow{\rho} \rho(x_j), \quad j \neq v, v+1.$$

Now we can compute the length of the new endomorphism

$$l(\rho) = \sum_{i=1}^{n} l(A_i x_{\mu_i} A_i^{-1})$$

$$= \sum_{i \neq v, v+1} l(A_i x_{\mu_i} A_i^{-1}) + l(\rho(x_v)) + l(\rho(x_{v+1}))$$

$$= \sum_{i \neq v, v+1} l(A_i x_{\mu_i} A_i^{-1}) + l(\rho \circ \rho_F(\sigma_v))(x_{v+1})$$

$$+ l(\rho \circ \rho_F(\sigma_i))(x_v) + l(x_{\mu_v}^{-1}) + l(x_{\mu_v}^{1})$$

$$= \sum_{i \neq v, v+1} l(\rho \circ \rho_F(\sigma_v))(x_i) + l(\rho \circ \rho_F(\sigma_i))(x_v)$$

$$+ l(\rho \circ \rho_F(\sigma_v))(x_{v+1}) + 2$$

$$= \sum_{i=1}^{n} l(\rho \circ \rho_F(\sigma_v))(x_i) + 2$$

$$= l(\rho \circ \rho_F(\sigma_v)) + 2.$$

Exercise 2.7.3. Prove that if Lemma 2.7.1 (b) is true for ρ, then

$$l(\rho \circ \rho_F(\sigma_v^{-1})) < l(\rho).$$

Hence the result follows. □

Theorem 2.7.4. *[Artin, 1947] An endomorphism ρ of F_n is in $\rho_F(B_n) \subset$ AutF$_n$ (i.e., there exists a $b \in B_n$ such that $\rho = \rho_F(b)$) if and only if the endomorphism ρ satisfies*

$$\rho(x_i) = A_i x_{\mu_i} A_i^{-1}, \quad 1 \leq i \leq n, \tag{2.7}$$

$$\rho(x_1 \cdots x_n) = x_1 \cdots x_n, \tag{2.8}$$

where $\mu \in \Sigma_n$ is a permutation of $(1, 2, \cdots, n)$ and $A_i = A_i(x_1, x_2, \cdots, x_n)$ is a word in F_n.

Proof: $\Longrightarrow$ It follows from what we discussed above.

$\Longleftarrow$: If every automorphism of F_n with (2.7) and (2.8) is generated by $\{\rho_F(\sigma_i) : 1 \leq i \leq n-1\}$, then there is a $b \in B_n$ such that $\rho = \rho_F(b)$. From (2.7) and (2.8), we obtain

$$A_1 x_{\mu_1} A_1^{-1} \cdot A_2 x_{\mu_2} A_2^{-1} \cdots A_n x_{\mu_n} A_n^{-1} = x_1 \cdot x_2 \cdots x_n. \tag{2.9}$$

In order for an endomorphism ρ to have such a relation (2.8), there must be some cancelation as Lemma 2.7.1. We show the result by induction on the length of the endomorphism. One can represent every automorphism of F_n with (2.7) and (2.8) by another endomorphism $\rho \circ \rho_F(\sigma_v^{\pm 1})$. Hence by Lemma 2.7.2, the endomorphism $\rho \circ \rho_F(\sigma_v^{\pm 1})$ can be represented by a power product of $\rho_F(\sigma_i)(1 \le i \le n-1)$ by the induction of the length since $l(\rho \circ \rho_F(\sigma_v^{\pm 1})) < l(\rho)$. Hence ρ can be represented by a power product of $\rho_F(\sigma_i)$ as well. By Proposition 2.5.1, there exists an element $b \in B_n$ such that $\rho = \rho_F(B)$. $\qquad\square$

Theorem 2.7.5. *Let $M_n = Aut(\pi_1(D^2 \setminus Q_n)) \cap Homeo(D^2 \setminus Q_n, Id_{\partial D^2})$, where Q_n is the set of n distinct points in the interior of 2-disk D^2 and homeomorphisms are fixed to be identity on the boundary of the 2-disk. Then $M_n = \rho_F(B_n) \cong B_n$.*

Proof: Let $x_1, x_2, \cdots, x_n$ be a basis of $\pi_1(D^2 \setminus Q_n)$ generated by a simple loop enclosing points $q_1, q_2, \cdots, q_n \in Q_n$. Then any homeomorphism of $D^2 \setminus Q_n$ with restriction on $\mathbf{D^2}$ to be the identity map satisfies (2.7) and (2.8) in Theorem 2.7.4. Hence $M_n \subset \rho_F(B_n) \cong B_n$.

Identify B_n with its image of $\rho_F : B_n \to AutF_n$. The action of σ_i $(\rho_F(\sigma_i))$ on F_n can be realized by a homeomorphism on $D^2 \setminus Q_n$ with interchanging q_i, q_{i+1} and fixing the outside of a little disk which includes q_i and q_{i+1}. So $\rho_F(\sigma_i)(x_j) = x_j$ for $j \neq i, i+1$. Therefore we obtain $B_n \cong \rho_F(B_n) \subset M_n$. $\qquad\square$

2.8 Link group via the braid presentation

Every link L has a braid representative $\beta \in B_n$ such that $L = \overline{\beta}$ by the Alexander theorem (Theorem 2.6.2). Every braid $\beta \in B_n$ can be identified with an automorphism $\rho_F(\beta) : F_n \to F_n$ with prescribed action by Proposition 2.5.1. Hence every link group in S^3 can be presented in terms of braids. The following theorem shows that every link group admits a presentation related to Theorem 2.7.4. Every abstract group defined in Theorem 2.8.1 is the fundamental group of the complement of a link in S^3, since every endomorphism $\rho : F_n \to F_n$ defined in Theorem 2.7.4 is a braid automorphism $\rho_F(\beta)$. By the faithful representation ρ_F, ρ can be realized by a geometric braid. Hence Theorem 2.8.1 gives the link group of $\overline{\beta}$ in S^3.

Theorem 2.8.1. *Let $\beta \in B_n$ and ρ_β acts on F_n by Theorem 2.7.4. Let $\overline{\beta}$ be the link determined by closing the braid in S^3. Then the link group*

$\pi_1(S^3 \setminus \overline{\beta})$ *admits the presentation:*

   generators: $y_1, y_2, \cdots, y_n,$

    relations: $y_i = A_i(y_1, \cdots, y_n) y_{\mu_i} A_i^{-1}(y_1, \cdots, y_n), \quad 1 \le i \le n - 1.$

Proof: Using Theorem 2.7.5 to identify the braid element with the homeomorphism of $D^2 \setminus Q_n$, we have the action $\beta \in B_n$ on $F_n = \pi_1(D^2 \setminus Q_n) = \langle x_1, \cdots, x_n \rangle$ given by Proposition 2.5.1. Using $\beta : D^2 \setminus Q_n \to D^2 \setminus Q_n$ to represent the $\beta \in B_n$ through the identification, we have $\{(D^2 \setminus Q_n) \times I\} / \sim$ the quotient space $(z, 1) \sim (\beta(z), 0)$. This is the complement of a link in the solid torus $T = \{D^2 \times I\} / \sim$. Denote $\{(D^2 \setminus Q_n) \times I\} / \sim = T \setminus \overline{\beta}$. It fibers over S^1 with fiber $D^2 \setminus Q_n$. Then the homotopy exact sequence shows that

$$1 = \pi_2(S^1) \to \pi_1(D^2 \setminus Q_n) \to \pi_1(T \setminus \overline{\beta}) \to \pi_1(S^1) \to \pi_0(D^2 \setminus Q_n) = 1.$$

Let $y_1, \cdots, y_n$ be the image of $x_1, \cdots, x_n$ under the natural embedding of

$$\pi_1(D^2 \setminus Q_n) = \pi_1((D^2 \setminus Q_n) \times \{0\}) \to \pi_1(T \setminus \overline{\beta}).$$

Let $t \in \pi_1(T \setminus \overline{\beta})$ be $z \times I$, $z \in \partial D^2$, oriented from $\partial D^2 \times \{1\}$ to $\partial D^2 \times \{0\}$. ($t$ is a longitude on the boundary of the solid torus T.) The element t is also a lift of a generator of $\pi_1(S^1)$ to $\pi_1(T \setminus \overline{\beta})$. Then the short exact sequence gives a centralization of $\pi_1(T \setminus \overline{\beta})$ and defines a presentation for $\pi_1(T \setminus \overline{\beta})$:

   generators: $y_1, y_2, \cdots, y_n, t,$

    relations: $t y_i t^{-1} = A_i(y_1, \cdots, y_n) y_{\mu_i} A_i^{-1}(y_1, \cdots, y_n), \quad 1 \le i \le n.$

Let m be a meridian on the boundary of $T \setminus \overline{\beta}$, and (M, L) be the meridian and longitude pair on $\partial(S^3 \setminus T)$. Then we have $S^3 \setminus \overline{\beta} = (T \setminus \overline{\beta}) \cup_{\partial T} (S^3 \setminus T)$ which by Seifert-Van Kampen's theorem gives the fundamental groups

$$
\begin{array}{ccc}
\pi_1(\partial T) & \longrightarrow & \pi_1(T \setminus \overline{\beta}) \\
\downarrow & & \downarrow \\
\pi_1(S^3 \setminus T) & \longrightarrow & \pi_1(S^3 \setminus \overline{\beta}).
\end{array}
$$

Thus the link group $\pi_1(S^3 \setminus \overline{\beta})$ has a presentation:

generators: $y_1, y_2, \cdots, y_n, t, m, M, L,$

   relations: $t y_i t^{-1} = A_i(y_1, \cdots, y_n) y_{\mu_i} A_i^{-1}(y_1, \cdots, y_n), \quad 1 \le i \le n,$

                 $M = 1, t = M, m = L, m = y_1 y_2 \cdots y_n.$

A simple sequence of Tietze transformations shows that

$$\pi_1(S^3 \setminus \overline{\beta}) \cong \langle y_1, y_2, \cdots, y_n \mid y_i = A_i y_{\mu_i} A_i^{-1}, 1 \le i \le n - 1 \rangle.$$

The relation $y_n = A_n(y_1, \cdots, y_n) y_{\mu_n} A_n^{-1}(y_1, \cdots, y_n)$ is omitted since it follows from (2.9) for y_i's. $\qquad \square$

2.9 Braid groups are linear

A group is said to be linear if it is isomorphic to a subgroup of $GL(m, K)$ for some natural number m and some field K. Krammer [2000] proved that there is a faithful B_4 representation which was studied by Lawrence previously. Shortly thereafter, Bigelow [2001a] found a proof that the same representation is faithful for all n by a beautiful topological argument. Here, we present Krammer's proof from [Krammer, 2002].

Let $s_{ij} = s(i, j) \in \Sigma_n$ be the permutation (called a reflection) interchanging i with j and fixing the rest. Denote $s_i = s_{i(i+1)}$ and $S = \{s_1, s_2, \cdots, s_{n-1}\}$. The pair (Σ_n, S) is a Coxeter system of type A_{n-1}, and Ref is the set of reflections in the symmetric group Σ_n. Define $l : \Sigma_n \to \mathbb{Z}_{\geq 0}$ to be the length function with respect to S, i.e., $l(x)$ is the smallest integer k such that there exists $s_{i_1}, \cdots, s_{i_k} \in S$ with $x = s_{i_1} \cdots s_{i_k}$.

The braid group B_n admits a presentation with generators $\{rx : x \in S_n\}$, where S_n is the symmetric group on $\{1, 2, \cdots, n\}$, and relations $r(xy) = (rx)(ry)$ whenever $l(xy) = l(x) + l(y)$. Denote $\Omega = \text{Image} \{r : S_n \to B_n\}$ by viewing rx as an element of B_n. There exists a well-known homomorphism $B_n \to S_n$ defined by $rx \mapsto x$ ($x \in S_n$), identifying $r(s_i)$ with σ_i in the Artin presentation of the braid group. The element $\Delta = r(w_0)$ is the half-twist, for the unique longest element $w_0 \in S_n$ with $w_0(i) = n + 1 - i$.

Denote B_n^+ (including identity element, as positive braids) as the submonoid of B_n generated by Ω, and the length function $l : B_n^+ \to \mathbb{Z}_{\geq 0}$ is the unique mooned homomorphism with $l(rx) = l(x)$ for all $x \in S_n$. Let Ω_k be the set of elements of Ω of length k. Define an ordering on B_n^+ by $x \leq y \iff y \in xB_n^+$. Restriction of this ordering yields an ordering on Ω, and thereby on S_n. The ordering on S_n is equivalent to the weak Bruhat ordering with the smallest ordering element 1 and the largest ordering element w_0.

For any positive braid $x \in B_n^+$, there is a greatest ordering element in the set $\{y \in \Omega : y \leq x\}$, denoted by $LF(x)$ for the leftmost factor. A sequence $\{x_1, \cdots, x_k\} \in \Omega_k$ is said to be (left) greedy if $LF(x_i x_{i+1}) = x_i$ for all $i = 1, \cdots, k-1$. For any positive braid $x \in B_n^+$, there exists a unique greedy sequence $\{x_1, \cdots, x_k\}$ (the left greedy form for x) with $x_1 \cdots x_k = x$ for $x_k \neq 1$. An important identity follows

$$LF(xy) = LF(xLF(y)), \quad x, y \in B_n^+. \tag{2.10}$$

Proposition 2.9.1. *Let B_n act on a set U.*

(i) If the inclusion $xC_y \subset C_{LF(xy)}$ holds for all pairs $(x, y) \in \Omega_1 \times \Omega$

with $C_x \subset U$ for $x \in \Omega$, then it holds for all pairs in $B_n^+ \times \Omega$;

(ii) If C_x is nonempty and pairwise disjoint and Property (i) holds, then the B_n-action on U is faithful.

Proof: (i) The result follows from the induction on the length $l(x)$. If $l(x) \le 1$, then there is nothing to prove. Let $l(x) > 1$ with $x = uv$ for $u, v \in B_n^+ \setminus \{1\}$. Then

$$xC_y = (uv)C_y = u(vC_y) \subset uC_{LF(vy)} \subset C_{LF(uLF(vy))} = C_{LF(uvy)} = C_{LF(xy)}.$$

Two inclusions in the above follow from the induction principle, the middle equality follows from (2.10). Hence, the result follows from the induction steps.

(ii) Let $Sym(U)$ be the group of permutations of U, and let $\pi : B_n \to Sym(U)$ be the action. Simply denote xu for the action $(\pi x)u$ for $x \in B_n$ and $u \in U$. For any $z \in B_n$, there are $x, y \in B_n^+$ such that $z = xy^{-1}$. If we show that for any $x, y \in B_n^+$ with $\pi(x) = \pi(y)$, then $x = y$. Hence, the result follows from the induction on $l(x) + l(y)$.

Suppose $x, y \in B_n^+$ with $\pi(x) = \pi(y)$. If $l(x) + l(y) = 0$, then $x = 1, y = 1$. So $x = y$. Consider the induction on $l(x) + l(y)$ with the case $l(x) + l(y) > 0$. By the assumption, C_1 is nonempty. For any $u \in C_1$, by (i), we have $xu \in xC_1 \subset C_{LF(x)}$ and similarly $yu \in C_{LF(y)}$. By $\pi(x) = \pi(y)$, we have $\pi(x)u = xu = yu = \pi(y)u$. Therefore, $xu = yu \in C_{LF(x)} \cap C_{LF(y)}$. By the pairwise disjoint assumption, $LF(x) = LF(y)$. Define $x', y' \in B_n^+$ with $x = LF(x)x', y = LF(y)y'$ and $LF(x) \neq 1$ otherwise $x = y = 1$ which contradicts with $l(x) + l(y) > 0$. We obtain $l(x') + l(y') < l(x) + l(y)$. The induction assumption yields $x' = y'$, and hence $x = LF(x)x' = LF(y)y' = y$. The result follows.   □

Let R be a commutative ring with two invertible elements q and t. Let V be the free R-module with basis $\{x_{s_{ij}} : s_{ij} \in \Sigma_n\}$. Thus dim $V = \frac{n(n-1)}{2}$. Denote x_{ij} for $x_{s_{ij}}$ for $1 \le i < j \le n$. Let V^* be the dual module of V and let $\langle \cdot, \cdot \rangle : V^* \times V \to R$ be the natural pairing. Denote x_{ij}^* for $x_{s_{ij}}^*$ be the dual basis for all $s_{ij} \in \Sigma_n$. Let $End(V)$ act on V^* on the right by

$$\langle uA, v \rangle = \langle u, Av \rangle, \quad A \in End(V), \quad (u, v) \in V^* \times V.$$

Definition 2.9.2. Define a representation $\rho : B_n \to GL(V)$(action of $GL(V)$ on V from the left, and denote $(\rho x)(v)$ simply by xv for $x \in B_n$

and $v \in V$) by

$$\sigma_k x_{k(k+1)} = tq^2 x_{k(k+1)}$$
$$\sigma_k x_{ik} = (1-q)x_{ik} + qx_{i(k+1)}, \quad i < k,$$
$$\sigma_k x_{i(k+1)} = x_{ik} + tq^{k-i+1}(q-1)x_{k(k+1)}, \quad i < k,$$
$$\sigma_k x_{kj} = tq(q-1)x_{k(k+1)} + qx_{(k+1)j}, \quad k+1 < j,$$
$$\sigma_k x_{(k+1)j} = x_{kj} + (1-q)x_{(k+1)j}, \quad k+1 < j,$$
$$\sigma_k x_{ij} = x_{ij}, \quad i < j < k, \quad \text{or} \quad k+1 < i < j,$$
$$\sigma_k x_{ij} = x_{ij} + tq^{k-i}(q-1)^2 x_{k(k+1)}, \quad i < k < k+1 < j.$$

Proposition 2.9.3. *The defined* $\rho : B_n \to GL(V)$ *map indeed satisfies (i)* $\rho(\sigma\sigma') = \rho(\sigma)\rho(\sigma')$ *(homomorphism), (ii)* $\rho(\sigma_i\sigma_{i+1}\sigma_i) = \rho(\sigma_{i+1})\rho(\sigma_i)\rho(\sigma_{i+1})$, *(iii)* $\rho(\sigma_i\sigma_j) = \rho(\sigma_j)\rho(\sigma_i)$ *for* $|i - j| > 1$, *and (iv)* $\rho(\sigma_k)$ *is invertible and* $\rho^{-1}(\sigma_k) = \rho(\sigma_k)$.

Exercise 2.9.4. Proof Proposition 2.9.3. Note that (i), (ii) and (iii) show that the map ρ defines a braid group homomorphism. (iv) follows from previous properties. This is a straightforward verification.

For each element $x \in B_n$, $\rho(x)$ can be identified with its matrix with respect to the basis $\{x_{ij} : 1 \leq i < j \leq n\}$ of the free R-module V. For $x \in B_n^+$, the entries of the matrix of $\rho(x)$ are in $\mathbb{Z}[q, q^{-1}, t]$, since B_n^+ is generated by $\Omega_1 = \{\sigma_1, \sigma_2, \cdots, \sigma_{n-1}\}$. Let $R = \mathbb{R}[t, t^{-1}]$ be the one-variable Laurent polynomial ring. For $q \in \mathbb{R} \subset R$ with $0 < q < 1$, define

$$V_1 = \bigoplus_{s_{ij}} \mathbb{R}[t] \cdot x_{s_{ij}} \subset V = \bigoplus_{s_{ij}} \mathbb{R}[t, t^{-1}] \cdot x_{s_{ij}}, \quad 1 \leq i < j \leq n.$$

We have $B_n^+ V_1 \subset V_1$, and all entries of $\rho(\sigma_k)$ are in $\{0, 1, q, 1-q\} + t\mathbb{Z}[q, q^{-1}, t] = [0, 1] + t\mathbb{R}[t]$ for a real number $q \in (0, 1)$. We set, for $A \subset \{s_{ij} : 1 \leq i < j \leq n\}$,

$$V_2 = \bigoplus_{s_{ij}} ([0, 1] + t\mathbb{R}[t]) \cdot x_{s_{ij}} = \left(\bigoplus_{s_{ij}} \mathbb{R}_{\geq 0} \cdot x_{s_{ij}} \right) \oplus tV_1$$

$$= \{v \in V : \langle x_{ij}^*, v \rangle \in \mathbb{R}_{\geq 0} + t\mathbb{R}[t] \text{ for all } s_{ij}\},$$

$$D_A = \{v \in V_2 : \langle x_{ij}^*, v \rangle \in t\mathbb{R}[t] \text{ for all } s_{ij} \iff s \in A\}.$$

Thus, we have $B_n^+ V_2 \subset V_2$ and V_2 is the disjoint union of the D_A. Let $X \in B_n^+$ and $A \subset \{s_{ij} : 1 \leq i < j \leq n\}$. There is a unique $B \subset \{s_{ij} :$

$1 \leq i < j \leq n\}$ with $xD_A \subset D_B$, denoted by $B = xA$. Let 2^{Ref} be the power set of $\{s_{ij} : 1 \leq i < j \leq n\}$ (the set of reflections in S_n). The map $B_n^+ \times 2^{\text{Ref}} \to 2^{\text{Ref}}$ defined by $(x, A) \mapsto xA$, defines an action of B_n^+ on 2^{Ref}. The explicit formula for the B_n^+-action on 2^{Ref} is given by the following (see [Krammer, 2002, Lemma 4.1]). The $\sigma_k A$ is the set of those $s_{ij}, 1 \leq i < j \leq n$ and

$$\begin{cases} \text{true statement,} & i = k, j = k+1 \\ \{s_{ik}, s_{i(k+1)}\} \subset A, & i < k, j = k \\ s_{ik} \in A & i < k, j = k+1 \\ s_{(k+1)j} \in A, & i = k, j > k+1 \\ \{s_{(k+1)j}, s_{kj}\} \subset A, & i = k+1, j > k+1 \\ s_{ij} \in A & \{i, j\} \cap \{k, k+1\} = \emptyset. \end{cases}$$

This follows directly from the σ_k-action on V^* and self-adjoint action property.

Definition 2.9.5. (i) Define a map $L : S_n \to 2^{\text{Ref}}$ by

$$L(x) = \{s_{ij} : 1 \leq i < j \leq n, x^{-1}i > x^{-1}j\}.$$

(ii) A set $A \subset 2^{\text{Ref}}$ is called a half-permutation if whenever $1 \leq i < j < k \leq n$,

$$s_{ij}, s_{jk} \in A \Longrightarrow s_{ik} \in A.$$

Denote by HP for the set of half-permutations.

(iii) Let GB (Greatest Braid) denote the map

$$GB = rL^{-1}Pro : HP \to \Omega,$$

where there is a greatest $B \in L(S_n)$ with $B \subset A$ and defined to be $B = Pro(A)$.

Denote the length $l(x) = |L(x)|$. For $x, y \in S_n$,

$$x \leq xy \Longleftrightarrow l(xy) \leq l(x) + l(y)$$

$$\Longleftrightarrow L(xy) = L(x) \cup xL(y)x^{-1} \Longleftrightarrow L(x) \subset L(xy).$$

The image of L is denoted by $L(S_n)$. We may identify S_n with the image $L(S_n)$ due to the injective map L. Every element of the image set $L(S_n)$ is a half-permutation. There is a bisection from the set of partial ordering $<_0$ on $\{1, 2, \cdots, n\}$ with $(i <_0 j) \Longrightarrow (i < j)$ to HP which takes $<_0$ to

$\{s_{ij} : i <_0 j\}$. A subset $A \subset \text{Ref}$ is in $L(S_n)$ if and only both A and $\text{Ref} - A$ are half-permutations.

Proposition 2.9.6. *(i) If $x \in B_n^+, A \in HP$, then $xA \in HP$.*

(ii) For every half-permutation A, there is a greatest with respect to inclusion $B \in L(S_n)$ with $B \subset A$. Hence, $Pro(A)$ is well-defined in Definition 2.9.5 (iii).

(iii) The map $GB : HP \to \Omega$ is B_n^+-equivariant, i.e., $GB(xA) = LF(xy)$ holds for $x \in B_n^+, y = GB(A)$ and $A \in HP$.

Proof: (i) Suppose $x \in \Omega_1$ (say $x = \sigma_k$), and $1 \le p < q < r \le n$. The result follows if this statement $(s(p, q), s(q, r) \in xA \Rightarrow s(p, r) \in xA)$ is true. This amounts to verifying five cases which are consequences of Proposition 2.9.3.

Case 1 ($\{p, q, r\} \cap \{k, k+1\} = \emptyset$): $s(p, q) \in xA \Longleftrightarrow s(p, q) \in A$, so are $s(q, r) \in A$ and $s(p, r) \in A$;

Case 2 ($r = k$): $s(p, q) \in xA \Longleftrightarrow s(p, q) \in A, s(q, r) \in xA \Longleftrightarrow s(q, r) \in A, s(q, k+1) \in A$, so is for $s(p, r)$;

Case 3 ($q < k, r = k+1$): $s(p, q) \in xA \Longleftrightarrow s(p, q) \in A, s(q, r) \in xA \Longleftrightarrow s(q, k) \in A$, so is $s(p, r)$;

Case 4 ($q = k, r > k+1$): $s(p, q) \in xA \Longleftrightarrow s(p, q) \in A$ and $(p, k+1) \in A$, and $s(q, r) \in xA \Longleftrightarrow s(k+1, r) \in A$, so $s(p, q), s(p, k+1), s(k+1, r) \in A$ implies $s(p, k+1), s(k+1, r) \in A$, thus, $s(p, r) \in A (\Longleftrightarrow s(p, r) \in xA)$;

Case 5 ($q = k, r = k+1$): $s(p, q) \in xA \Longleftrightarrow s(p, q) \in A$ and $s(p, r) \in A$ to have $s(q, r) \in xA$ true and $s(p, r) \in xA \Longleftrightarrow s(p, q) \in A$.

(ii) It is the same to prove that $P = \{y \in S_n : L(y) \subset A\}$ contains a greatest element. The ordering on P is generated by $x \le xs$ whenever true with $x, xs \in P, s \in S_n$. Let $x, xs, xt \in P$ with $x \le xs, x \le xt, s, t \in S_n (s \ne t)$. Since P is finite and has a smallest element, it is sufficient to show that there exists $y \in P$ with $xs, xt \le y$. Let $m_{st} \in \{2, 3\}$ be the order of st, and $y = xst$ if $m_{st} = 2$; $y = xsts$ if $m_{st} = 3$.

Case 1 ($m_{st} = 2$): $L(y) = L(xs) \cup L(xt) \subset A$ whence $y \in P$;

Case 2 ($m_{st} = 3$): Define $C = x^{-1}(A - L(x))x$, for any $u \in S_n$ with $x \le xu$, $xu \in P \Longleftrightarrow L(xu) \subset A \Longleftrightarrow L(x) \coprod xL(u)x^{-1} \subset A \Longleftrightarrow xL(u)x^{-1} \subset A \backslash L(x) \Longleftrightarrow L(u) \subset x^{-1}(A \backslash L(x))x = C$, thus for any $u \in S_n$ with $x \le xu$, one has

$$xu \in P \Longleftrightarrow L(u) \subset C,$$

where $\coprod$ is the disjoint union. For $u = s, t$, we get $s, t \in C$, i.e., $s(k, k+1) = s_k, s(k+1, k+2) = s_{k+1} \in C$.

By [Krammer, 2002, Lemma 5.3], if $A \in HP, x \in S_n, L(x) \subset A$ and $B = x^{-1}(A \setminus L(x))x$, then $B \in HP$. We have $C \in HP$ and conclude that $s(k, k+2) \in C$. Therefore, $L(sts) = \{s(k, k+1), s(k+1, k+2), s(k, k+2)\} \subset C$. By the above equivalent relation, $sts \in P$ and $xstd = y \in P$. This is done in this case.

(iii) Let $x = ru, y = rv$ for $u, v \in S_n$ and $L(u) = \{u\}$. We have (a) $Pro(ruA) = B \in S_n$ the greatest B with $L(u) \subset B \subset (ruA)$ by (ii) and $\{u\}(L(u) \subset L(S_n)) \subset (ruA)$, (b) (ruA) is the greatest $C \in HP$ with $\{u\} \subset C \subset \{u\} \cup uAu$ [Krammer, 2002, Lemma 5.5]. Thus, $Pro(ruA)$ is the greatest $B \in L(S_n)$ with $L(u) \subset B \subset (ruA)$. Set $B = L(uw)$ for $w \in S_n$. Hence, $u \leq uw$. we get $B \subset (ruA)$

$$\iff L(uw) \subset \{u\} \cup uAu \iff \{u\} \coprod uL(w)u \subset \{u\}uAu$$

$$\iff L(w) \subset A \iff (\text{by (ii)}) \quad L(w) \subset Pro(A) = L(v)$$

$$w \leq v \iff uw \leq r^{-1}LF(xy),$$

where the last equivalence follows from the $u \leq uw$. The greatest B which satisfies these properties is given by $uw = r^{-1}LF(xy)$. This shows that $Pro(ruA) = Lr^{-1}LF(xy)$, and

$$GB(xA) = rL^{-1}Pro(ruA) = LF(xy).$$

$\square$

Proposition 2.9.6 is a combination of lemmata by Krammer [2002, Lemma 4.2, Lemma/Definition 4.3 and Lemma 4.4]. Now we are ready for the main theorem on the linearity of the braid group.

Theorem 2.9.7. *The representation $\rho : B_n \to GL(V)$ given in Definition 2.9.3 is faithful.*

Proof: This amounts to applying Proposition 2.9.1. We just verify that (a) $xC_y \subset C_{LF(xy)}$ holds for all pairs $(x, y) \in B_n^+ \times \Omega$, and (b) C_x is nonempty and disjoint.

(a) Recall $C_y = \cup\{D_A : A \in HP, GB(A) = y\}$. We need to show that for $A \in HP, GB(A) = y$, $xD_A \subset C_{LF(xy)}$ holds. We have $xD_A \subset D_{xA}$ by the definition of the B_n^+-action on 2^{Ref}, $D_{xA} \subset C_{GB(xA)}$ follows from $xA \in HP$ by Proposition 2.9.6 (i) and that $Pro(xA)$ is well-defined by Proposition 2.9.6 (ii) and so are $GB(xA)$ and $C_{GB(xA)}$, and the definition of C_z. We get $C_{GB(xA)} = C_{LF(xy)}$ by Proposition 2.9.6 (iii). Therefore $xC_y = \cup xD_A \subset C_{LF(xy)}$.

(b) The set C_x is nonempty since $D_{L(r^{-1}x)} \subset C_x$ and $D_{L(r^{-1}x)}$ is not empty. That C_x are disjoint is trivial.

Hence, the result follows from Proposition 2.9.1. $\square$

2.10   Singular braids

The development of Vassiliev invariants lead mathematicians to consider singular knots and singular links which are the immersed circles with a finite number of transversal self-intersections (double points). Let $\mathcal{L}^{\text{sing}}(R^3)$ be the space of singular links. Via the Alexander theorem principle (Theorem 2.6.2), there is a corresponding notion which we briefly introduce in this section, the so-called singular braid.

Singular braids form the singular braid monoid (denoted by SB_n) which is generated by $\sigma_1, \cdots, \sigma_{n-1}$ from presentation of B_n and additional singular generator $\tau_1, \cdots, \tau_{n-1}$ (which do not have inverses).

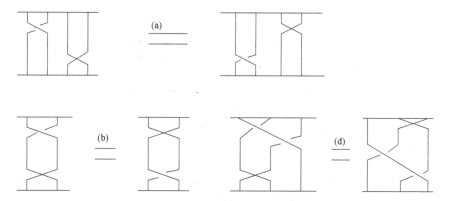

Fig. 2.11   Relations (a), (b) and (d) in the presentation of SB_n

Proposition 2.10.1. *The monoid SB_n is generated by $\{\sigma_i, \tau_i : 1 \leq i \leq n-1\}$ with the following relations:*

(a) $[\sigma_i, \sigma_j] = [\sigma_i, \tau_j] = [\tau_i, \tau_j] = 1$ *for* $|i - j| \geq 2$,

(b) $[\sigma_i, \tau_i] = 1$ *for* $1 \leq i \leq n - 1r$,

(c) $\sigma_i \sigma_{i+1} \sigma_i = \sigma_{i+1} \sigma_i \sigma_{i+1}$ *for* $1 \leq i \leq n - 2$,

(d) $\sigma_i \sigma_j \tau_i = \tau_j \sigma_i \sigma_j$ *for* $|i - j| = 1$.

Proof: It is easy to see that σ_i's and τ_i's generate SB_n from the definition of singular knots and singular links. It suffices to verify those four relations. Let $\tilde{\beta}_0$ and $\tilde{\beta}_1$ be two singular braids which represents the same element in the monoid SB_n. Let $\tilde{\beta}_t$ be the one-parameter family of singular braids such that it connects $\tilde{\beta}_0$ and $\tilde{\beta}_1$. The fact that the $\tilde{\beta}_t$ has no triple points implies that there is a well-defined order of singularities along each braid strand which is preserved during the isotopy. The isotopy gives a bijective mapping between the transverse self-intersections of $\tilde{\beta}_0$ and $\tilde{\beta}_1$. Generically, we can divide the t-interval $[0,1]$ in such a way that only the changes in the singular braid diagram $\tilde{\beta}_t$ occurs:

(1) Two transverse self-intersections in the singular braid projection interchange their t-level, as relations (a) and (b);
(2) A triple point in the singular braid projection is created momentarily as a "free" strand crosses a self-intersection point or a crossing point, as the relation (d);
(3) New crossing points are created or destroyed via the Reidemeister move of type II.

All possible cases of (1) are given by relations (a) and (b). Note that the mirror image of (σ_i, τ_i) is (σ_i^{-1}, τ_i). Thus all possible cases of (2) are given by relations (c) and (d). All possible cases of (3) are given by the relation $\sigma_i \sigma_i^{-1} = \sigma_i^{-1} \sigma_i = 1$ from creating and/or destroying a pair of crossing points at specific t-intervals. Outside these specific t-intervals, the singular braids diagrams can be modified without any extra relation. Thus we obtain a presentation for the singular braid monoid SB_n (see [Baez, 1992; Birman, 1993] also).   □

The monoid SB_n is also called *the Baez-Birman monoid, or singular braid monoid*. There is an obvious homomorphism $j_n : B_n \to SB_n$ from the braid group to the singular braid monoid. Fenn, Keyman and Rourke [1998] proved that the singular braid monoid can be embedded in a group (the singular braid group) $SB_n \to SG_n$. Hence τ_i is invertible in SG_n for $1 \le i \le n-1$, and all the relations of SB_n in Proposition 2.10.1 remain valid.

Theorem 2.10.2. *Let $A : \cup_{n=2}^{\infty} B_n \to \mathcal{L}(R^3)$ be the Alexander surjective map. Then there is a natural extension of A to a surjective map $\overline{A} : \cup_{n=2}^{\infty} SB_n \to \mathcal{L}^{sing}(R^3)$.*

Proof: Note that $B_n \subset SB_n$ for every n. The closure operation of a braid can be naturally extended to an operation for a singular braid. The singular link $\tilde{L} = \overline{\tilde{\beta}}$ is a closed singular braid if there is an axis A in R^3 such that if the representative of a singular link $\tilde{L}$ is parametrized by cylindrical coordinate (a, θ) relative to the A-axis, the polar angle function restricted to $\tilde{L}$ is monotonically increasing.

After isotopy $\tilde{L}$ in $R^3 \setminus A$, we can have the singular link $\tilde{L}$ defined by a diagram in (a, θ)-plane. By a further isotopy one can choose a neighborhood

$N(s_k) \in \tilde{L}$ at each transverse self-intersection point s_k such that the polar angle function restricted to $\cup_{k=1}^m N(s_k)$ is monotonically increasing. Then following exactly the method as in Theorem 2.6.2 to obtain piecewise linear family of arcs in $\tilde{L} \backslash \cup_{k=1}^m N(s_k)$. Thus there is a braid element corresponding to it. With the transverse self-intersection points on the braids, we obtain a closed singular braid for $\tilde{L}$. $\qquad \square$.

Let $\tilde{\sigma}_i$ be the image of the natural map $\epsilon : B_n \to \mathbb{C}B_n$, where $\mathbb{C}B_n$ is a group algebra of B_n over $\mathbb{C}$. There is a well-defined monoid homomorphism $\overline{\epsilon} : SB_n \to \mathbb{C}B_n$ which is an extension of ϵ such that

$$\overline{\epsilon}(\tau_i) = \epsilon(\sigma_i) - \epsilon(\sigma_i^{-1}) = \tilde{\sigma}_i - \tilde{\sigma}_i^{-1}.$$

This will allow us to extend every finite-dimensional representation $\rho_n : B_n \to GL_n(K)$ for some algebra K (see §4.6 for example) to $\overline{\rho}_n : SB_n \to GL_n(K)$ by $\overline{\rho}_n(\tau_i) = \rho_n(\sigma_i) - \rho_n(\sigma_i^{-1})$. It would be interesting to extend various representations of B_n in [Birman, 1974] to the ones of SB_n.

2.11 (Co)Homology of braid groups and loop spaces

Arnold [1970] studied the cohomology $H^*(B_n, \mathbb{Z})$ of the braid group B_n by identifying $K(B_n, 1) = B(\mathbb{R}^2, n)$ the space of complex polynomials of degree n without multiple roots and with leading coefficient equal to one $P_n(t) = t^n + z_1 t^{n-1} + \cdots + z_{n-1} t + z_n$.

Let $p : \mathbb{C}_\lambda^n \to \mathbb{C}_z^n$ be the Viete map from $\mathbb{C}^n$ (domain denoted by $\mathbb{C}_\lambda^n$) into $\mathbb{C}^n$ (range denoted by $\mathbb{C}_z^n$) sending a point $\lambda = (\lambda_1, \cdots, \lambda_n) \in \mathbb{C}_\lambda^n$ to the polynomial $P_n(t)$ that has roots $\lambda_1, \cdots, \lambda_n$ counting with multiplicity. The orbit space $Orb(\mathbb{C}^n, \Sigma_n)$ of the canonical action of the symmetric group Σ_n on $\mathbb{C}^n$ is isomorphic to $\mathbb{C}_\lambda^n$ under the map defined by the elementary symmetric functions in the variables $\lambda_1, \cdots, \lambda_n$. Define

$$z_k(\lambda) = (-1)^k \sum_{i_1 < \cdots < i_k} \lambda_{i_1} \cdots \lambda_{i_k}, \quad 1 \le k \le n,$$

to be the standard basis of the symmetric polynomials. Hence, $\mathbb{C}_\lambda^n \cong Orb(\mathbb{C}^n, \Sigma_n)$. Let $Orb^{reg}(\mathbb{C}^n, \Sigma_n)$ be the subspace of $Orb(\mathbb{C}^n, \Sigma_n)$ consisting of regular orbits of this symmetric action. Throughout the Viete map p, it induces a homeomorphism

$$p : Orb^{reg}(\mathbb{C}^n, \Sigma_n) \to C_{(z)}^n \backslash \Delta,$$

where Δ is the discriminant surface of the polynomial $P_n(t) = t^n + z_1 t^{n-1} + \cdots + z_{n-1} t + z_n$. The Δ is given by

$$\Delta = \{z \in \mathbb{C}^n : \Delta^n(z) = \prod_{i \neq j} (\lambda_i - \lambda_j) = (-1)^{\frac{n(n-1)}{2}} \prod_{i<j} (\lambda_i - \lambda_j)^2 = 0\}.$$

Arnold [1970] looked at the canonical embedding of $C^n_{(z)} \setminus \Delta$ into the sphere S^{2n}. The complement of the image of $C^n_{(z)} \setminus \Delta$ in S^{2n} can be identified with the one-point compactification Δ^* of Δ in S^{2n}. By the Alexander duality, we have

$$H^i(B_n; \mathbb{Z}) = H^i(C^n_{(z)} \setminus \Delta; \mathbb{Z}) \cong H_{2n-i}(S^{2n}, \Delta^*; \mathbb{Z}) \cong \tilde{H}_{2n-i+1}(\Delta^*; \mathbb{Z}),$$

where $\tilde{H}_*(X)$ is the reduced homology of the space X.

Theorem 2.11.1. *[Arnold, 1970]*

Finiteness Theorem *The cohomology groups of the braid group B_n are finite except $H^0(B_n, \mathbb{Z}) \cong \mathbb{Z}, H^1(B_n, \mathbb{Z}) \cong \mathbb{Z}$ for $n \geq 2$. Moreover, $H^(B_n; \mathbb{Z}) = 0$ for $i \geq n$.*

Repetition Theorem *All the cohomology groups of each braid group on an odd number of strings are the same as those of the braid group on the preceding even number of strings:*

$$H^i(B_{2n+1}; \mathbb{Z}) \cong H^i(B_n; \mathbb{Z}).$$

Stabilization Theorem *The cohomology group $H^i(B_n; \mathbb{Z})$ stabilizes as n increases:*

$$H^i(B_n; \mathbb{Z}) = H^i(B_{2i-2}; \mathbb{Z}), \quad n \geq 2i - 2.$$

Theorem 2.11.1 provides that the Euler characteristic of the braid group B_n is zero from the Finiteness Theorem. The isomorphism $H^1(B_n, \mathbb{Z}) \cong \mathbb{Z}$ follows from the face that the abelianization of B_n is $\mathbb{Z}$. Define a homomorphism deg:

$$\deg : B_n \to \mathbb{Z}, \quad \deg((\sigma_{i_1})^{m_1}(\sigma_{i_2})^{m_2} \cdots (\sigma_{i_k})^{m_k}) = \sum_j m_j.$$

The map deg takes each element of the braid group via the canonical generators to the sum of the exponents of the occurrence of all the generators. The kernel of deg is generated by the commutators. The result on $H^i(B_n; \mathbb{Z}) = 0$ for $i \geq n$ follows from the fact that the space $\mathbb{C}^n_{(z)} \setminus \Delta$ is a Stein manifold. Arnold [1970] also computed the groups $H^i(B_n; \mathbb{Z})$ for $n \leq 11$ and $i \leq 9$.

The study of the cohomology of the braid groups was continued by Fuks [1970], who calculated the cohomologies of the braid groups mod 2. Let us denote for simplicity by Γ_n the configuration space $B(\mathbb{R}^2, n) = B(\mathbb{C}, n) = F(\mathbb{C}, n)/\Sigma_n$. Let Γ_n^* be the one-point compactification of Γ_n. Poincaré duality yields

$$H^k(\Gamma_n; \mathbb{Z}/2) \cong \tilde{H}_{2n-k}(\Gamma_n^*; \mathbb{Z}/2).$$

To investigate the group $\tilde{H}_{2n-k}(\Gamma_n^*; \mathbb{Z}/2)$, some natural cellular decomposition of the space Γ_n^* is constructed. Using this decomposition, all the groups $H^i(B_n; \mathbb{Z}/2)$ are computed, the multiplicative structure of the ring $H^*(B_n; \mathbb{Z}/2)$ and connections with the cohomology $H^*(BO_n; \mathbb{Z}/2)$ are described. The Hopf algebra structure of the cohomologies of the infinite braid group $H^*(B_\infty; \mathbb{Z}/2)$ arising from the canonical pairing: $B_n \times B_m \to B_{n+m}$ is also considered in [Fuks, 1970]. Although the results are formulated there in the language of cohomology it is more convenient to translate them to homology. Then the main results will be the following:

Theorem 2.11.2. *[Fuks, 1970]*

(1) *The homology of the infinite braid group with coefficients in $\mathbb{Z}/2$ as a Hopf algebra is isomorphic to the polynomial algebra on infinitely many generators a_i ($i \geq 1$) with $\deg(a_i) = 2^i - 1$:*

$$H_*(B_\infty; \mathbb{Z}/2) \cong \mathbb{Z}/2[a_1, a_2, \cdots, a_i, \cdots]$$

with the coproduct given by the formula:

$$\Delta(a_i) = 1 \otimes a_i + a_i \otimes 1.$$

(2) *The canonical inclusion $B_n \to B_\infty$ induces a monomorphism in homology with coefficients in $\mathbb{Z}/2$. Its image is the sub-coalgebra of the polynomial algebra $\mathbb{Z}/2[a_1, a_2, \cdots, a_i, \cdots]$ with $\mathbb{Z}/2$-basis consisting of monomials*

$$a_1^{k_1} \cdots a_l^{k_l}, \quad \sum_i k_i 2^i \leq n.$$

(3) *The canonical homomorphism $B_n \to BO_n$ $(1 \leq n \leq \infty)$ induces a monomorphism (of Hopf algebras if $n = \infty$)*

$$H_*(B_n; \mathbb{Z}/2) \to H_*(BO_n; \mathbb{Z}/2).$$

The cohomology of classical braid groups studied in [Arnold, 1970] were introduced in connection with the problem of representing an algebraic function of several variables by the superposition of algebraic functions of fewer variables. Arnold proved three important theorems on the integral cohomology of the braid groups in Theorem 2.11.1, and also completely described the cohomology of the pure braid groups. Fuks [1970] continued the study of the cohomology of the classical braid groups and calculated the cohomology modulo 2.

Chapter 3

Knot and link invariants

3.1 Basic background

Definition 3.1.1. A knot invariant is a function $f : \mathcal{L}(Y^3) \to \mathcal{C}$ which assigns to each knot (link) K in Y^3 an object in a category $\mathcal{C}$ such that $f(K_1) = f(K_2)$ with the equality in the sense of $\mathcal{C}$ if K_1 is equivalent to K_2 in Y^3.

Remark 3.1.2. (1) Most knot invariants are for knots in $\mathbb{R}^3$ and S^3. Using Dehn surgery to relate knots in a 3-manifold Y^3, one can study the 3-manifold topology of Y.

(2) The category $\mathcal{C}$ can be just integers $\mathbb{Z}$ (numerical invariants), algebraic category (group, ring, module and polynomial invariants), geometric category (metric, curvature and spectrum invariants) and topological category (topological invariants).

(3) The function f can be a combination of various categories $\mathcal{C} = \prod_{\alpha \in A} \mathcal{C}_\alpha$.

Conjecture 3.1.3. *(Main Conjecture in Knot Theory in Y^3) There exists an injective function $f : \mathcal{L}(Y^3) \to \mathcal{C}_{Y^3}$ for some category $\mathcal{C}_{Y^3}$ related to the 3-manifold Y^3.*

This conjecture can be interpreted in many known conjectures and problems for specific invariants and the 3-manifold Y.

Remark 3.1.4. (1) If $Y^3 = S^3$ and $\mathcal{C}$ is a category of topological spaces with homeomorphisms as morphisms, then

$$f : \mathcal{L}(S^3) \to \mathcal{C}, \quad K \mapsto S^3 \setminus K.$$

The knot complement $S^3 \setminus K$ is an invariant up to homeomorphism. Hence the homeomorphism equivalent class of $S^3 \setminus K$ is a topological invariant.

Conjecture 3.1.3 can be read as if the knot complement is an injective map (see [Rolfsen, 1976, p 48]), i.e., if $S^3 \setminus K_1$ is homeomorphic to $S^3 \setminus K_2$, then there exists a homeomorphism $h : S^3 \to S^3$ such that $h(K_1) = K_2$ $((S^3, K_1) \cong (S^3, K_2))$. This is proved by Gordon and Luecke [1989]. But for $Y^3 \neq S^3$, the knot, in general, is not determined by its complement. Rong [1993] provided some examples related to Seifert spaces to show that the knot complement is not an injective function.

(2) For $Y^3 = S^3$ and $\mathcal{C}$ is the Laurent ring and f is the Jones polynomial, Conjecture 3.1.3 can be related to the very important problem of a nontrivial knot with trivial Jones polynomial.

One of the fundamental problems in knot theory is the classification problem of knots in S^3. This amounts to creating a complete representative list of knots in $\mathcal{L}(S^3)$. The knot tabulation in §2.5 is aimed at this problem.

There are several special classes of knots. A knot K is amphicheiral if K is equivalent to its mirror image K^* (the figure eight knot is amphicheiral, but torus knot $T_{p,q}$ with $p, q > 1$ is not). A knot is invertible if K is equivalent to $-K$ the opposite orientation (the left-hand trefoil knot is invertible). Characterizing amphicheiral, prime, invertible, slice and periodic knots is very attractive and popular in knot theory literature (see [Burde and Zieschang, 1985] and MathSciNet for more references).

In the following, we study some numerical knot invariants, algebraic invariants and Jones polynomials.

3.2 Unknotting and unknotting number

An embedding $f : D^2 \to Y^3$ is called flat in the topological sense if f extends to an embedding $\overline{f} : U_{D^2} \to Y^3$, where U_{D^2} is a neighborhood of D^2 in $\mathbb{R}^3$.

Proposition 3.2.1. *A knot K in S^3 is equivalent to the unknot (or trivial knot S^1) if and only if K is the boundary of a flat disk D^2 in S^3.*

Proof: If K is equivalent to the unknot, then there exists an isotopy f_t such that $f_1(K) = S^1 \subset S^3$. Hence the flat disk D^2 is the standard one bounded by the equator.

If K is the boundary of a flat disk $D^2 \subset S^3$, then there exists an embedding $f : U_{D^2} \to S^3$ such that U_{D^2} is a neighborhood of D^2 in $\mathbb{R}^3$ and $f(\partial D^2) = K$. So there is a closed smaller neighborhood $V_{D^2} \subset U_{D^2}$ in $\mathbb{R}^3$ such that there is a bicollar on the boundary $\partial V_{D^2} \subset U_{D^2}$ and V_{D^2}

is a 3-ball. Let $g : \mathbb{R}^3 \to S^3$ be the inverse of the stereographic projection with $g(\partial \mathbf{D^2}) = \mathbf{S^1} \hookrightarrow \mathbf{S^3}$. Thus $g(V_{D^2})$ is a 3-ball in S^3 with bicollared boundary. Then $S^3 \setminus g(V_{D^2})$ and $S^3 \setminus f(V_{D^2})$ are 3-balls in S^3, and

$$f \circ g^{-1} : g(V_{D^2})(\subset S^3) \to f(V_{D^2})(\subset S^3)$$

is homeomorphic. By Alexander lemma, $f \circ g^{-1}$ can be extended to a homeomorphism $h : S^3 \to S^3$ such that $h = f \circ g^{-1}$ on V_{D^2}. In particular, we have

$$\begin{aligned}
h(S^1) &= h(g(\partial D^2)) = h|_{V_{D^2}}(g(\partial D^2)) \\
&= f \circ g^{-1}(g(\partial D^2)) = f(\partial D^2) \\
&= K.
\end{aligned}$$

The homeomorphism h is the equivalence between the unknot $S^1 \subset S^3$ and K. $\qquad\square$

Recall that Alexander lemma: Any homeomorphism $\partial A \to \partial B$ extends to a homeomorphism $A \to B$ for A and B homeomorphic to the n-ball D^n.

Lemma 3.2.2. *Suppose that $f : D^2 \to M^3$ is a map of a disk with no-singularity on ∂D^2 ($f|_{\partial D^2}$ is injective). Then there exists an embedding $g : D^2 \to M^3$ such that $g(\partial D^2) = f(\partial D^2)$.*

Theorem 3.2.3. *A knot K in S^3 is the unknot if and only if $\pi_1(S^3 \setminus K) \cong \mathbb{Z}$ (infinite cyclic).*

Proof: If K is the unknot in S^3, then $(S^3, K) \cong (S^3, S^1)$ up to isotopy. Hence $\pi_1(S^3 \setminus K) = \pi_1(S^3 \setminus S^1)$. The latter group is generated by a meridian (a loop encircled S^1 once by a deformation retract), i.e. $\pi_1(S^3 \setminus S^1) \cong \mathbb{Z}\langle m \rangle$.

Assume K is a regular knot in S^3 with $\pi_1(S^3 \setminus K) \cong Z$. Then we need to show that K bounds a flat disk in the topological sense by Proposition 3.2.1. Consider a tubular neighborhood V of K with a preferred framing $V \cong S^1 \times D^2$, defining a meridian $x \times \partial D^2$ and a longitude $l \subset \partial V$ as a trivial homological class in $S^3 \setminus Int(V) \cong S^3 \setminus K$. Thus l is homotopically trivial in $S^3 \setminus Int(V)$. Thus the loop $l \subset \partial V \subset (S^3 \setminus Int(V))$ bounds a disk in $S^3 \setminus Int(V)$ by Dehn's Lemma (Lemma 3.2.2), i.e., $l = \partial D^2$. There is an annulus A in V connecting l and K ($\partial A = l \bigsqcup K$). The union $A \cup D^2$ of this annulus A and the disk D^2 bounded by l give the flat disk D^2 in S^3 bounded by K. $A \cup D^2$ with $\partial(A \cup D^2) = K$ is another flat disk. Hence K is unknotted by Proposition 3.2.1. $\qquad\square$

Exercise 3.2.4. (1) K is unknotted if and only if the inclusion homomorphism $\pi_1(\partial V) \to \pi_1(S^3 \setminus Int(V))$ is injective.

(2) For every unknotted $K \in S^3$, $\pi_1(S^3 \setminus K)$ contains a subgroup $\mathbb{Z} \oplus \mathbb{Z}$. (The image of the subgroup $\mathbb{Z} \oplus \mathbb{Z}$ in $\pi_1(S^3 \setminus K)$ is called the *peripheral subgroup* .)

Theorem 3.2.5. *[Waldhausen, 1968] Two knots K_1 and K_2 in S^3 with the equal peripheral subgroup (up to conjugate) if and only if there is an isomorphism $h : \pi_1(S^3 \setminus K_1) \to \pi_1(S^3 \setminus K_2)$ such that $h(m_1) = m_2$ and $h(l_1) = l_2$, where (m_i, l_i) is the pair of meridian and longitude of the knot $K_i(i = 1, 2)$.*

We cannot provide the proof of Theorem 3.2.5 since it depends on a fundamental theorem of Waldhausen on 3-manifolds. We refer interested readers to [Burde and Zieschang, 1985, p 38] for the explanation.

Definition 3.2.6. Let $K_j(j = 1, 2)$ be a knot in S^3. There is a neighborhood U of $p_j \in K_j$ in S^3 such that the pair $(U, U \cap K_j)$ is topologically equivalent to the canonical ball pair (B_j^3, B_j^1). By removing the resulting pairs (B_j^3, B_j^1) from (S^3, K_j) and sewing the resulting pairs by a homeomorphism $f : (\partial B_2^3, \partial B_2^1) \to (\partial B_1^3, \partial B_1^1)$, we form the pair connected sum $(S^3, K_1)\#(S^3, K_2) = (S^3, K_1) \cup_f (S^3, K_2)$.

Fig. 3.1   Composition of two knots

In the convention, we write the connected sum as $K = K_1 \# K_2$ for the connected sum of knots or the composite knot of K_1 and K_2, a nontrivial knot which is not composite is called a *prime* knot. If the knot K_j is represented by a closure of a braid element β_j, then the composite knot $K_1 \# K_2$ can be described as the following. If $K_1 = \overline{\beta_1}, \beta_1 \in B_n$ and $K_2 = \overline{\beta_2}, \beta_2 \in B_m$, then
$$K_1 \# K_2 = \overline{\beta_1 \Sigma^{n-1}(\beta_2)}, \qquad \beta_1 \Sigma^{n-1}(\beta_2) \in B_{n+m},$$
where Σ is the shift map on the inductive limit of the B_n's and $\Sigma(\sigma_i) = \sigma_{i+1}$. For example, the composite knot of the figure eight $K_1 = \overline{\sigma_1 \sigma_2^{-1} \sigma_1 \sigma_2^{-1}}$ and the trefoil $K_2 = \overline{\sigma_1^3}$ is $\overline{\beta}$, where $\beta = \sigma_1 \sigma_2^{-1} \sigma_1 \sigma_2^{-1} \Sigma^{3-1}(\sigma_1^3) = \sigma_1 \sigma_2^{-1} \sigma_1 \sigma_2^{-1} \sigma_3^3$ [Birman, 1974].

Theorem 3.2.7. *Suppose that a connected sum $K = K_1 \# K_2$ of two knots is trivial. Then both K_1 and K_2 are trivial.*

Proof: By the definition of the connected sum, we take $S^3 = B_1^3 \cup_{S^2} B_2^3$ such that $K_1 \subset B_1^3$ and $K_2 \subset B_2^3$ with the arc $K_1 \cap K_2 \subset S^2$. Hence we have

$$S^3 \setminus K = (B_1^3 \setminus K_1) \cup_{S^2 \setminus (K_1 \cap K_2)} (B_2^3 \setminus K_2).$$

Since $\pi_1(B_j^3 \setminus K_j) \cong \pi_1(S^3 \setminus K_j)$ for $j = 1, 2$, by Seifert-Van Kampen theorem we have

$$
\begin{array}{ccc}
\pi_1(S^2 \setminus (K_1 \cap K_2)) & \xrightarrow{i^2} & \pi_1(B_2^3 \setminus K_2) \\
\downarrow{\scriptstyle i_1} & & \downarrow{\scriptstyle i_2} \\
\pi_1(B_1^3 \setminus K_1) & \xrightarrow{i^1} & \pi_1(S^3 \setminus K).
\end{array}
$$

The inclusion i^j is injective for both $j = 1$ and $j = 2$. By Definition 3.2.6, one can see that i_j is also injective for both $j = 1$ and $j = 2$. Therefore $\pi_1(S^3 \setminus K)$ contains two subgroup of $\pi_1(B_j^3 \setminus K_j) \cong \pi_1(S^3 \setminus K_j)$. By the hypothesis, we have K unknotted and $\pi_1(S^3 \setminus K) \cong \mathbb{Z}$ by Theorem 3.2.3. Thus both $\pi_1(S^3 \setminus K_j)$ $(j = 1, 2)$ are Abelian and infinite cyclic. By Theorem 3.2.3, K_1 and K_2 are unknotted. $\qquad\square$

Remark 3.2.8. Theorem 3.2.7 is also called the non-cancelation theorem in [Rolfsen, 1976] which means that one cannot connect a nontrivial knot to obtain the unknot. There are several other methods to prove Theorem 3.2.7.

Theorem 3.2.9. *(i) Any two knots in S^3 have a well-defined composition.*
(ii) The connected sum $\#$ is associative:

$$(K_1 \# K_2) \# K_3 = K_1 \# (K_2 \# K_3).$$

(iii) For any knot K, $K \# O = K$, where O is the unknot.
(iv) The connected sum is commutative $K_1 \# K_2 = K_2 \# K_1$.

Exercise 3.2.10. Prove Theorem 3.2.9.

The unknot O in Theorem 3.2.9 represents a unit under the connected sum operation. Theorem 3.2.7 shows that for any nontrivial knot, there is no inverse knot under the connected sum operation. Hence the isotopy classes of knots in S^3 under the connected sum operation $(\mathcal{L}(S^3), \#)$ form a commutative monoid. The monoid $(\mathcal{L}(S^3), \#)$ has the unique prime decomposition property.

Theorem 3.2.11. *Every nontrivial knot K in S^3 is a finite product of prime knots and these factors are uniquely determined. That is $K = K_1 \# \cdots K_n$, where $K_i(1 \leq i \leq n)$ is prime and $K = K_1' \# \cdots K_m'$, then $n = m$ and $K_i = K_{\mu(i)}'$ for $\mu \in \Sigma_n$.*

Proof: See [Schubert, 1949; Burde and Zieschang, 1985, Chapter 7].   □

Definition 3.2.12. An unknotting number of a knot K is an integer $u(K)$ where there exists a projection of an equivalent knot presentation (isotopic knot diagram) such that changing n crossings turns the knot into the unknot and there is no projection such that changing k $(k < n)$ crossing would turn the knot K into the unknot.

The unknotting number $u(K)$ is the minimal number of crossings in a diagram of the knot K into the unknot, the minimum being taken over all possible diagrams of the knot K.

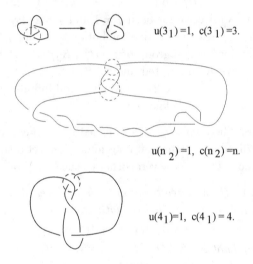

$u(3_1) = 1$, $c(3_1) = 3$.

$u(n_2) = 1$, $c(n_2) = n$.

$u(4_1) = 1$, $c(4_1) = 4$.

Fig. 3.2   Unknotting and crossing

Figure 3.2 shows several knots with unknotting number 1.

Exercise 3.2.13.

(1) Find other different knots with unknotting number 1.
(2) Prove that $u(K) \le c(K) - 2$, where $c(K)$ is the crossing number.

In general, computing $u(K)$ is very difficult. The knot 7_4 with symmetry projection easily shows that one crossing changing is not sufficient, but the difficulty is that there may be other different projections with one crossing changing taken into the unknot. From the given presentation, we may see that two crossing changing can take the knot 7_4 into the unknot. Hence

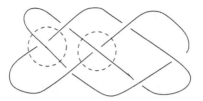

Fig. 3.3   The knot 7_4

$u(7_4) \leq 2$. The other direction is hard, and is given by Lickorish [1985]. Scharlemann [1985] proved that a knot K with $u(K) = 1$ is prime. This shows that any composite knot cannot be turned into the unknot by only one crossing change.

One may think the unknotting number of a knot is realized in a projection of the knot with minimal crossings. Bleiler [1984] and Nakaniski [1983] independently found a surprising example in 1983 that the projection with minimal crossings requires more crossing changes to the unknot [Gilbert and Porter, 1994, Theorem 1.3].

Conjecture 3.2.14. *(Unknotting Conjecture)* $u(K_1 \# K_2) = u(K_1) + u(K_2)$.

Milnor [1952] conjectured that the unknotting number of the torus knot of type (p, q) is $\frac{(p-1)(q-1)}{2}$. Kronheimer and Mrowka [1993] proved the Milnor unknotting conjecture and extend it to positive links.

The unknotting problem is the problem of algorithmically recognizing the unknot, given some presentation (knot diagrams) of a knot. A major unsolved challenge is to determine if the unknotting problem admits a polynomial time algorithm. Hass, Lagarias and Pippenger [1999] proved that the unknotting problem is in the complexity class NP by using normal surfaces in describing the Seifert surfaces of a given knot.

Murakami introduced a more accessible invariant, the algebraic unknotting number $u_a(K)$ which is defined to be the minimal number of crossing changes needed to turn K into a knot with Alexander polynomial equal to 1. Fogel [1994] showed that this definition is equivalent to Murakami's original definition in terms of certain operations on Seifert matrices. Borodzik and Friedl [2014] gave another characteristic of the algebraic unknotting number.

Let $X(K) = S^3 \setminus N(K)$ and the Blanchfield form

$$Bl(K) : H_1(X(K); \mathbb{Z}[t, t^{-1}]) \times H_1(X(K); \mathbb{Z}[t, t^{-1}]) \to \mathbf{Q}/\mathbb{Z}[t, t^{-1}],$$

defined by $(a, b) \mapsto \Phi(a)(b)$ for the map

$$\Phi : H_1(X(K); \mathbb{Z}[t, t^{-1}]) \to H_1(X(K), \partial X(K); \mathbb{Z}[t, t^{-1}])$$

$$\to \overline{H^2(X(K); \mathbb{Z}[t, t^{-1}])} \cong \overline{H^1((X(K); \mathbf{Q}(t)/\mathbb{Z}[t, t^{-1}])}$$

$$\to \overline{Hom_{\mathbb{Z}[t, t^{-1}]}(H_1(X(K); \mathbb{Z}[t, t^{-1}]), \mathbf{Q}(t)/\mathbb{Z}[t, t^{-1}])},$$

where the first map is the inclusion induced map, the second from Poincaré duality, the third from the long exact sequence in cohomology corresponding to the coefficients

$$0 \to \mathbb{Z}[t, t^{-1}] \to \mathbf{Q}(t) \to \mathbf{Q}(t)/\mathbb{Z}[t, t^{-1}] \to 0,$$

and the last one from the evaluation map. The Blanchfield form is well-know to be hermitian $Bl(K)(a, b) = \overline{Bl(K)(b, a)}$ and $Bl(K)(\mu_1 a, \mu_2 b) = \overline{\mu_1} Bl(K)(a, b)\mu_2$ for $\mu_i \in \mathbb{Z}[t, t^{-1}]$.

Given a hermitian $n \times n$-matrix A over $\mathbb{Z}[t, t^{-1}]$ with $\det(A) \neq 0$ define $\lambda(A)$ by the form

$$\mathbb{Z}[t, t^{-1}]^n/A\mathbb{Z}[t, t^{-1}]^n \times \mathbb{Z}[t, t^{-1}]^n/A\mathbb{Z}[t, t^{-1}]^n \to \mathbf{Q}(t)/\mathbb{Z}[t, t^{-1}],$$

with $(a, b) \mapsto \overline{a}^t A^{-1} b$ by viewing a and b column vectors in $\mathbb{Z}[t, t^{-1}]^n$. Borodzik and Friedl defined

$$n(K) = \min\{n : \text{there exists a hermitian } n \times n\text{-matrix } A \text{ over } \mathbb{Z}[t, t^{-1}]$$

such that $\lambda(A) \cong Bl(K)$ and $A(1)$ is diagonalizable over $\mathbb{Z}\}$.

Borodzik and Friedl [2014] showed that a matrix A exists, and hence $n(K)$ is defined. Moreover, $n(K) \leq \deg \Delta_K(t) + 1$, and $n(K)$ is the lower bound on the algebraic unknotting number. Therefore, $n(K) \leq u_a(K)$. Borodzik and Friedl [2014] proved that $n(K) \geq u_a(K)$, hence, $n(K) = u_a(K)$ for a knot $K \subset S^3$. The algebraic unknotting number (the minimal number of crossing needed to turn K into an Alexander polynomial one knot) is the minimal number of algebraic unknotting moves needed to change the Seifert matrix of K into the trivial matrix, and is the minimal second Betti number of a topological 4-manifold that strictly cobounds $M(K)$ (the 0-framed surgery along K), as well as the same number $n(K)$.

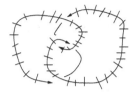 

Fig. 3.4 Bridges of the same knot

3.3 Bridge number and total curvature

By Proposition 1.2.3, any knot has finite crossings. Define an overpass to be a subarc of the knot that goes over at least one crossing but never goes under a crossing.

A maximal overpass is the overpass subarc which cannot be extended further. Then the bridge number of the projection is the number of maximal overpasses in the projection. Those maximal overpasses present the bridges over the rest of the arcs.

Definition 3.3.1. The bridge number $b(K)$ of the knot K is the least bridge number of all of the projections of the knot K.

Lemma 3.3.2. *A knot K has $b(K) = 1$ if and only if K is the unknot.*

Proof: For the unknot K, it is conventional to have $b(K) = 1$. If K is the knot with $b(K) = 1$, then there exists a regular presentation of the knot K with only one bridge and the least number of crossings n_K. If $n_K \leq 2$, then K is the unknot. Suppose $n_K \geq 3$. Except for the one maximal overpass (the one-bridge), there is no other arc overpassing. Hence there must be at least two adjacent crossings at which the same arc is the underpass. By the Reidemeister moves, we can reduce the number n_K by 2. Thus there is an isotopic presentation of K with $n_K - 2$ crossing without changing the bridge number. Hence K is either unknotted or one bridge knot with $n_K - 2$ crossings. The last one contradicts with the choice of n_K. The result follows. □

Theorem 3.3.3. *[Schubert, 1954]*
$$b(K_1 \# K_2) = b(K_1) + b(K_2) - 1.$$

Remark 3.3.4. If we define reduced bridge number $\bar{b}(K) = b(K) - 1$, then $\bar{b}(K_1 \# K_2) = \bar{b}(K_1) + \bar{b}(K_2)$. The reduced bridge number is additive under the connected sum.

Exercise 3.3.5. (1) Show that the trefoil knot has bridge number 2.

(2) Compute the number of maximal overpasses for $K_1 \# K_2$ in Figure 3.5.

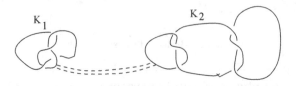

Fig. 3.5   Connected sum of K_1 and K_2

Corollary 3.3.6. *If $b(K) = 2$, then K is a prime knot.*

Proof: Suppose the contrary. So $K = K_1 \# K_2$ is a composite of two nontrivial knots. By Theorem 3.3.3, we have

$$b(K_1) + b(K_2) = b(K) + 1 = 3.$$

Thus $b(K_1) = 1$ or 2 since $b(K) \in \mathbb{N}$ for any knot. If $b(K_1) = 1$, then K_1 is the unknot by Lemma 3.3.2 which contradicts with the hypothesis of nontriviality of the knot K_1. If $b(K_1) = 2$, then $b(K_2) = 3 - b(K_1) = 1$ and K_2 is the unknot which contradicts with the hypothesis. Thus K cannot be a composite knot of two nontrivial knots. $\qquad\square$

From the proof of Corollary 3.3.6, we have that any composite knot K must have $b(K) \geq 3$.

Corollary 3.3.7. *The reduced bridge number $\bar{b} : (\mathcal{L}(S^3), \#) \to (\mathbb{N} \cup \{0\}, +)$ defines a monoid epimorphism with $\ker \bar{b} = \{O\}$, where O is the unknot.*

Proof: It follows from Theorem 3.3.3 and the definition of $\bar{b}(K)$. Since $\bar{b}(3_1) = 1$, for any $n \in \mathbb{N}$, we have a knot K with $\bar{b}(K) = n$. (K can be chosen as $\#_{i=1}^n 3_1$) Note that $\ker \bar{b}$ is equivalent to the knots with bridge number one. Hence $\ker \bar{b} = \{O\}$ follows from Lemma 3.3.2. $\bar{b}$ is not a monomorphism since $\bar{b}$ is a monoid homomorphism (not a group homomorphism). $\qquad\square$

Exercise 3.3.8. Compute the number of maximal overpasses for the following presentations of 4_1 (the figure-eight knot), and show that $b(4_1) = 2$. Hence $\bar{b}$ is not a monomorphism.

Fig. 3.6   Overpasses of the figure eight knot

Schubert [1956] classified knots and links with bridge number 2. Knots (links) that have bridge number 2 are called two-bridge knots (links). If we cut a two-bridge knot open along the projection plane, then we see two unknotted untangled arcs from the knot above the plane attached at 4-points (A, B, C, D) on the plane. These are two bridges from the definition. Below the projection plane, there are two unknotted untangled arcs. Each pair of arcs can be projected onto $\mathbb{R}^2$ without any crossing. Assume that one pair of arcs is projected onto straight line segments $w_1 = AB, w_2 = CD$. The other pair of arcs is projected onto two disjoint simple curves v_1 (from B to C) and v_2 (from D to A). The crossings or each arc v_i to meet w_i alternatively. The number of double points on AB and CD are even. Regularize the v_1 meets CD first, and $\alpha - 1$ double points on AB and CD. Then K is a knot if α is odd; K is a link if α is even and $\partial v_1 = \{A, B\}$ and $\partial v_2 = \{C, D\}$. See [Burde and Zieschang, 1985, Chapter 12] for the classification of 2-bridge links.

Let $c : [0, L] \to \mathbb{R}^3$ be a closed curve with $c(0) = c(L)$ and $c'(0) = c'(L)$, where L is the length of the curve c which is parametrized by the arc length. We may assume that $\|c'(s)\| = 1$ for $s \in [0, L]$. Let $c'(s)$ be the unit tangent vector. Then $c''(s)$ represents the rate at which the curvature deviates from being a straight line. The norm $\kappa(s) = \|c''(s)\|$ is called the curvature of c at s.

$$\kappa(c) = \int_0^L \kappa(s)ds = \int_0^L \|c''(s)\|ds$$

is called the total curvature of the curve c.

Exercise 3.3.9. Using the parametrization of trefoil knot given in §2.2, find the length and its total curvature of f.

Lemma 3.3.10. *Let c be a closed polygon with finite vertices $v_0, v_1, \cdots, v_m = v_0$. Let α_i be the angle between the vectors $v_{i+1} - v_i$ and*

$v_i - v_{i-1}$ *satisfying* $0 \leq \alpha_i \leq \pi$. *(α_i is called the exterior angle of the vertex v_i.) Then the total curvature of c is $\kappa(c) = \sum_{i=1}^{m} \alpha_i$.*

Proof: We take a sequence of smooth curves that approaches the polygon at v_i.

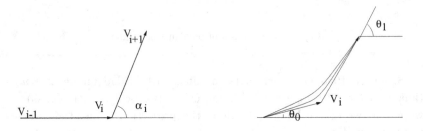

Fig. 3.7 Exterior angle

Hence $c^{'}(s) = (\cos\theta(s), \sin\theta(s))$ since $\|c^{'}(s)\| = 1$ and $\theta(a) = \theta_0$ and $\theta(b) = \theta_1$. We get $c^{''}(s) = \theta^{'}(s)(-\sin\theta(s), \cos\theta(s))$.

$$\kappa(c|_{[a,b]}) = \int_a^b \|c^{''}(s)\| ds$$

$$= \int_a^b |\theta^{'}(s)| ds$$

$$= \int_a^b \theta^{'}(s) ds \quad \theta(s) \text{ is concave up}$$

$$= \theta(b) - \theta(a) = \theta_1 - \theta_0 = \alpha_i$$

So $\lim_{a,b \to v_i} \kappa(c|_{[a,b]}) = \alpha_i$. Combining $\kappa(c|_{[v_i, v_{i+1}]}) = 0$ (a straight line), we have

$$\kappa(c) = \lim_{\varepsilon \to 0^+} \left(\sum_{i=1}^{m} \kappa(c|_{[v_i + \varepsilon, v_{i+1} - \varepsilon]}) + \sum_{i=1}^{m} \kappa(c|_{[v_i - \varepsilon, v_i + \varepsilon]}) \right)$$

$$= 0 + \sum_{i=1}^{m} \lim_{\varepsilon \to 0^+} \kappa(c|_{[v_i - \varepsilon, v_i + \varepsilon]})$$

$$= \sum_{i=1}^{m} \alpha_i.$$

$\square$

Now we define the total curvature of a smooth curve by

$$\kappa(c) = \sup_{P \subset c} \kappa(P), \tag{3.1}$$

where the sup is taken over all inscribed polygons P for c. Let c be a regular presentation of a knot, and let X be a unit normal vector of the project plane. Hence

$$h(s) = c(s) \cdot X = \|c(s)\|\|X\| \cos(\widehat{c(s)}, X) = \|c(s)\| \cos(\widehat{c(s)}, X)$$

is the height function of c. Let $\mu_X(c)$ be the number of local maxima of the height function $h(s)$. Then $b(c) = \min_X \mu_X(c)$ is the crookedness of a closed curve defined by Milnor [1952]. In [Schubert, 1956], this number is the same as the bridge number of the knot $b(K) = \inf_c b(c)$.

For the polygon P with vertices $v_0, v_1, \cdots, v_m = v_0$, the vertices yield m points $a_i = (v_i - v_{i-1})/\|v_i - v_{i-1}\|$ in S^2. Let Q_i be the shortest great circle arc connecting a_i and a_{i-1}. Denote $Q = \cup_{i=1}^m Q_i$ as closed curves in S^2. This curve is a spherical polygon which is called a spherical image of the polygon P.

For every point $a \in S^2$, let S_a^1 be the great circle of S^2 which has a pole at a. An edge $a_i a_{i-1}$ of Q crosses S_a^1 if and only if $a \cdot (a_{i+1} - a_i)$ and $a \cdot (a_i - a_{i-1})$ have opposite signs. So $a \cdot a_i$ is the local maximal or local minimal of $X \cdot a_i$ for $X = a$. If S_a^1 contains no vertex of Q (no edge Q_i is perpendicular to a), then the number of intersections $S_a^1 \cap Q$ is $2\mu_a(c)$.

$$\mu_a(c) = \sum_{i=1}^m \mu_a(c_i) = \sum_{i=1}^m \#(S_a^1 \cap Q_i) = \sum_{i=1}^m \mu_a(Q_i).$$

Let $M = \{a \in S^2 : \exists a_i \in S_a^1\}$. This is a set of poles with equators crossing vertices. Thus M is the union $\cup S_{a_i}^1$ of the finite collection of great circles with a_i on equators. In the component of $S^2 \setminus (\cup S_{a_i}^1)$, the value $2\mu_a(c)$ is constant. The integral $\int_{S^2} 2\mu_a(c)da$ is defined, where da is the area form of S^2.

Proposition 3.3.11. *The following identity holds.* $\int_{S^2} \mu_a(c)da = 2\kappa(c)$.

Proof: On S^2, the curve Q_i has length α_i, where α_i is the exterior angle at v_i of P. Let a_i be a pole with equator E_{a_i}. So E_{a_i} and $E_{a_{i-1}}$ form a "double lune" DL_i. If a is an interior point of DL_i, then $\#E_a \cap Q_i = 1$; if a is an exterior of DL_i, then $\#E_a \cap Q_i = 0$. The contribution of DL_i to the function $2\mu_a(c)$ is 1 if $a \in Int(DL_i)$ and 0 otherwise. So the function $2\mu_a(c)$ is a characteristic function:

$$2\mu_a(c) = \sum_{i=1}^m \chi_{DL_i}.$$

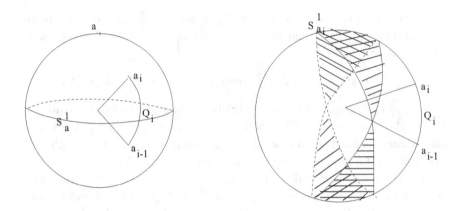

Fig. 3.8   Length α_i equals the exterior angle at a_i

Therefore we have

$$\int_{S^2} 2\mu_a(c)da = \int_{S^2} \sum_{i=1}^{m} \chi_{DL_i}da$$

$$= \sum_{i=1}^{m} \int_{S^2} \chi_{DL_i}da$$

$$= \sum_{i=1}^{m} \text{Area}(DL_i)$$

$$= \sum_{i=1}^{m} 4\alpha_i$$

$$= 4\kappa(c),$$

where $\text{Area}(DL_i) = \frac{\alpha_i}{\pi}\text{Area}(S^2) = \frac{\alpha_i \cdot 4\pi}{\pi}$. $\qquad\qquad\square$

Theorem 3.3.12. *[Milnor, 1952] (1) For any smooth closed curve c,*
$\int_{S^2} \mu_a(c)da = 2\kappa(c)$.
(2) $\kappa(c) \geq 2\pi b(c)$.

Proof: (1) follows from the refinement of c by a sequence of polygons $P_1 \subset P_2 \subset \cdots \subset c$, and

$$\int_{S^2} \mu_a(c)da = \lim_{P \subset c} \int_{S^2} \mu_a(P)da = \lim_{P \subset c} 2\kappa(P) = 2\kappa(c).$$

The second equality follows from Proposition 3.3.11 and the last one from the definition in (3.1).

(2) Since $b(c) = \min_X \mu_X(c)$, one has

$$2\kappa(c) = \int_{S^2} \mu_a(c)\,da \geq \int_{S^2} b(c)\,da = 4\pi b(c).$$

□

Recall that $\kappa(K) = \inf_c \kappa(c)$ and $b(K) = \sup_c b(c)$, where c is a closed curve with the same isotopy class of the knot K.

Theorem 3.3.13. *(Milnor Theorem [Milnor, 1952]) If $[c]$ is a simple type (there is a simple closed curve represented by the isotopy class $[c]$), then*

$$\kappa([c]) = 2\pi b([c]).$$

Corollary 3.3.14. *(1) (Fenchel's Theorem) For any closed curve c, $\kappa(c) \geq 2\pi$.*

(2) (Borsuk's conjecture) For any nontrivial knotted curve K, $\kappa(K) > 4\pi$.

Proof: (1) By Theorem 3.3.12, $\kappa(c) \geq 2\pi b(c)$, since every curve has an inscribed polygon P with $\kappa(P) = 2\pi$ or since $\mu_a(c) \geq 1$ for any closed curve c, we have $\kappa(c) \geq 2\pi$.

(2) By Lemma 3.3.2, $b(K) = 1$ if and only if K is the trivial knot. Then $b(K) \geq 2$ for any nontrivial knot and $\kappa(K) \geq 4\pi$. The total curvature of the knot type K cannot equal the curvature of its presented curve P. $(\kappa(K) = \inf \kappa(P))$ If there is an inscribed polygon P in K with $\kappa(P) \leq \kappa(K)(\leq \kappa(P'))$ for any P'). P cannot be planar. One can choose another $\overline{P}$ within the same type K, but $\kappa(\overline{P}) < \kappa(P)$ [Milnor, 1952, Corollary 1.2]. Thus $\kappa(\overline{P}) < \kappa(K)$ which contradicts with the definition of $\kappa(K)$. Hence

$$\kappa(K) > \kappa([c]) = 2\pi b([c]) \geq 4\pi.$$

□

3.4 Linking number and crossing number

Let $L = K_1 \cup K_2$ be two components of a link in S^3. We orient K_1 and K_2, and use the right-hand rule to define a sign ε at each crossing between these two components (see Figure 3.9). This is also called a local writhe number of the corresponding double point.

Definition 3.4.1. Let $C_p(L)$ be the set of crossings of K_1 and K_2. The linking number is defined to be

$$lk(L) = lk(K_1, K_2) = \frac{1}{2} \sum_{x \in C_p(L)} \varepsilon_x.$$

$$\varepsilon = +1 \qquad\qquad\qquad \varepsilon = -1$$

Fig. 3.9   A local writhe number of crossings

Example 3.4.2. We have a corresponding calculation for the linking number in Figure 3.10.

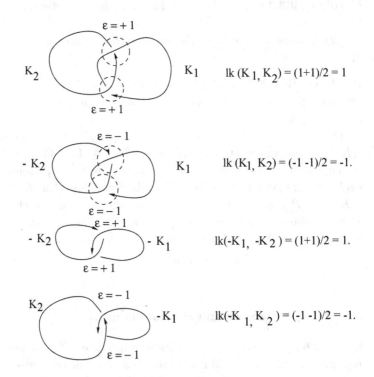

Fig. 3.10   Evaluation of the linking number

Example: This is the Whitehead link. Links can be linked even if their linking number is zero. See Figure 3.11.

Theorem 3.4.3. *If $L \sim L'$ are oriented isotopic links, then $lk(L) = lk(L')$.*

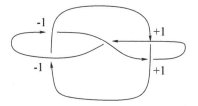

Fig. 3.11   Whitehead link $lk(K_1, K_2) = \frac{1}{2}(+1 + 1 - 1 - 1) = 0$

Proof: All we need to check is that the linking number is unchanged under the Reidemeister moves. Note that Reidemeister move type I (see Figure 3.12) has a crossing on a single component. So it does not effect the linking number. Reidemeister move type II (see Figure 3.12) has a pair of crossings with opposite linking number 1. So it adds up to a zero effect on the linking number.

Fig. 3.12   Reidemeister moves of type I and II

There are 4-different choices of the oriented pairs as in Figure 3.13. They all contribute zero to the linking number.

Fig. 3.13   Invariant under Reidemeister move of type II

Reidemeister move type III as in Figure 3.14 has zero contribution to the linking number. Note that $\varepsilon_{x_1} + \varepsilon_{x_2} = 0$ in Figure 3.14. Also we have $\varepsilon_{x_i} = -\varepsilon_{\overline{x}_i}$ $(i = 1, 2)$. Hence $\varepsilon_{\overline{x}_1} + \varepsilon_{\overline{x}_2} = 0$.

Similarly one can verify for the rest of the possible cases. Hence $lk(L)$ is a topological invariant. □

Fig. 3.14   Invariant under Reidemeister move of type III

Let L_+ and L_- be two component links that differ at one crossing with changing in Figure 3.15.

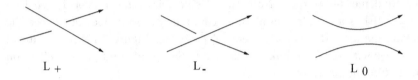

Fig. 3.15   Skein relation

Suppose that L_0 is obtained from L_+ and L_- by changing the crossing. Then we have

$$lk(L_+) = \frac{1}{2} \sum_{x \in C_p(L_+)} \varepsilon_x$$

$$= \frac{1}{2} \sum_{x \in C_p(L_+) \backslash x_0} \varepsilon_x + \frac{1}{2} \varepsilon_{x_0}(L_+)$$

$$= \frac{1}{2} \sum_{x \in C_p(L_-) \backslash x_0} \varepsilon_x + \frac{1}{2} \varepsilon_{x_0}(L_-) + l(L_0)$$

$$= lk(L_-) + l(L_0),$$

where $l(L_0) = \frac{1}{2}(\varepsilon_{x_0}(L_+) - \varepsilon_{x_0}(L_-))$. If $l(L_0) = 1$, then L_0 has one component (L_0 is a knot); if $l(L_0) = 0$, then L_0 has more than one component (link). Thus the relation

$$lk(L_+) - lk(L_-) = l(L_0)$$

gives the first simplest skein invariant under the crossing.

Definition 3.4.4. The crossing number of a knot K, $c(K)$, is the smallest number of crossings over all its regular projection.

In general we give a projection of the knot K and count the number of crossings n. Then we have $c(K) \leq n$. If all the knots with fewer crossings $(< n)$ are known and K is not in the known list, then we have $c(K) = n$. Determining the crossing number of a knot is quite difficult.

Conjecture 3.4.5. *The crossing number is additive under the connected sum,*

$$c(K_1 \# K_2) = c(K_1) + c(K_2).$$

Let $\gamma_i : S^1 \to \mathbb{R}^3$ be a smooth curve with $\gamma_i(0) = \gamma_i(L_i)(i = 1, 2)$ to represent the link L by two arc-length parametrized closed curves. Then

$$f : [0, L_1] \times [0, L_2] / \sim = T^2 \overset{\gamma_1 \times \gamma_2}{\to} \frac{\gamma_1(s_1) - \gamma_2(s_2)}{\|\gamma_1(s_1) - \gamma_2(s_2)\|} \in S^2$$

is the unit vector directed from $\gamma_2(s_2)$ to $\gamma_1(s_1)$. We oriented the S^2 by the inner normal and torus by $ds_1 \wedge ds_2$.

Lemma 3.4.6. $\deg f = lk(\gamma_1, \gamma_2)$ *[Rolfsen, 1976, pp. 133–135].*

Theorem 3.4.7. *(Gauss Theorem) The linking number of γ_1 and γ_2 in $\mathbb{R}^3$ is*

$$lk(\gamma_1, \gamma_2) = \frac{1}{4\pi} \int_0^{L_1} \int_0^{L_2} \frac{(d\gamma_1/ds_1, d\gamma_2/ds_2, \gamma_1 - \gamma_2)}{\|\gamma_1 - \gamma_2\|^3} ds_1 \wedge ds_2.$$

Proof: By definition of the degree, we have $\deg f = \frac{1}{4\pi} \int_{T^2} f^* \omega_{S^2}$, where ω_{S^2} is the area form given by the inner normal orientation. By the definition of f,

$$f^* \omega_{S^2}(a_1, a_2) = \omega_{S^2}(f_*(a_1), f_*(a_2)) = (f_*(a_1), f_*(a_2), -f),$$

where a_1 and a_2 are tangent to T^2 at (s_1, s_2) and the orientation of S^2 is given by the inner normal. Let $a_i = \frac{\partial}{\partial s_i}$. Then by the product rule

$$f_*(a_1) = \frac{\partial f}{\partial s_1} = \frac{d\gamma_1/ds_1}{\|\gamma_1 - \gamma_2\|} + (-\frac{(\gamma_1 - \gamma_2) \cdot d\gamma_1/ds_1}{\|\gamma_1 - \gamma_2\|^2}) \cdot f.$$

$$f_*(a_2) = \frac{-d\gamma_2/ds_2}{\|\gamma_1 - \gamma_2\|} + (\frac{(\gamma_1 - \gamma_2) \cdot d\gamma_2/ds_2}{\|\gamma_1 - \gamma_2\|^2}) \cdot f.$$

$$(f_*(\frac{\partial}{\partial s_1}), f_*(\frac{\partial}{\partial s_2}), -f) = (\frac{d\gamma_1/ds_1}{\|\gamma_1 - \gamma_2\|} + c(d\gamma_1/ds_1, f)f,$$

$$\frac{-d\gamma_2/ds_2}{\|\gamma_1 - \gamma_2\|} + c(d\gamma_2/ds_2, f)f, -f)$$

$$= (\frac{d\gamma_1/ds_1}{\|\gamma_1 - \gamma_2\|}, \frac{-d\gamma_2/ds_2}{\|\gamma_1 - \gamma_2\|}, -f)$$

$$= \frac{(d\gamma_1/ds_1, d\gamma_2/ds_2, \gamma_1(s_1) - \gamma_2(s_2))}{\|\gamma_1 - \gamma_2\|^3}.$$

The first equality is by the previous calculation, the second by the row-addition property of the determinant and the last by the scaler on rows. Hence

$$f^*\omega_{S^2} = (f_*(\frac{\partial}{\partial s_1}), f_*(\frac{\partial}{\partial s_2}), -f)ds_1 \wedge ds_2$$

$$= \frac{(d\gamma_1/ds_1, d\gamma_2/ds_2, \gamma_1(s_1) - \gamma_2(s_2))}{\|\gamma_1 - \gamma_2\|^3}ds_1 \wedge ds_2.$$

Thus $lk(\gamma_1, \gamma_2) = \deg f = \frac{1}{4\pi}\int_{T^2} f^*\omega_{S^2}$. $\qquad\square$

Let $\{g^t\}$ be the phase flow of the divergence free field ξ (div $\xi = 0$ in a simply connected closed $M \subset \mathbb{R}^3$), and let $\gamma_i(s_i) = g^{s_i} \cdot x_i$ be the trajectory of the point $x_i \in M$ $(i = 1, 2)$. Define the asymptotic linking number

$$\Lambda_\xi(\gamma_1, \gamma_2) =$$

$$\lim_{L_1, L_2 \to \infty} \frac{1}{4\pi L_1 L_2} \int_0^{L_1} \int_0^{L_2} \frac{(d\gamma_1/ds_1, d\gamma_2/ds_2, \gamma_1(s_1) - \gamma_2(s_2))}{\|\gamma_1 - \gamma_2\|^3}ds_1 \wedge ds_2.$$

Define a self-linking number of a field ξ

$$\lambda_\xi(\gamma_1, \gamma_2) = \int_{M \times M} \Lambda_\xi(\gamma_1, \gamma_2)\mu(x_1) \wedge \mu(x_2),$$

where $\mu(x_i)$ is the volume form on M.

Theorem 3.4.8. *(Arnold's Helicity Theorem) If ξ is a divergence free field on a simply connected manifold M with a volume form μ, then*

$$\lambda_\xi = \langle \xi, curl^{-1}\xi \rangle.$$

See [Arnold, 1973] for more details.

Definition 3.4.9. For two closed curves γ_1 and γ_2 in $\mathbb{R}^3$, the crossing number

$$c(\gamma_1, \gamma_2) = \frac{1}{4\pi} \int_0^{L_1} \int_0^{L_2} \frac{|(d\gamma_1/ds_1, d\gamma_2/ds_2, \gamma_1 - \gamma_2)|}{\|\gamma_1 - \gamma_2\|^3}ds_1 \wedge ds_2.$$

This is the integral of the absolute value of the Gauss integrand for linking number in Theorem 3.4.7. It is no longer a topological invariant under a curve isotopy. Parallelizing with asymptotic linking number, we can have the following definition.

Definition 3.4.10. The asymptotic crossing number of the field lines of a divergence free vector field ξ with γ in a simply connected manifold M^3 is

$$c_\xi(x, \gamma) = \lim_{L \to \infty} \sup \frac{1}{L} c(\Gamma_L(x), \gamma),$$

where $\Gamma_L(x)$ is the piece of the ξ-trajectory of $x \in M$ in the length $[0, L]$ and closed by a short path.

Note that $c_\xi(x, \gamma) \in L^1(M)$ and is well-defined (independent of the choice of the short paths). Hence the average crossing number

$$c_\xi(\gamma) = \int_M c_\xi(x, \gamma)\mu_x.$$

See [Freedman and He, 1991] for details on these numbers and [Lin and Wang, 1996] for other knot invariants expressed in similar integrals.

3.5 Knot group and Wirtinger presentation

The most important invariant of knots in S^3 is the fundamental group (also called *knot group*) of its complement in S^3. Let K be a knot in S^3. The space $S^3 \setminus K$ (also called *knot space*) with knot group $\pi_1(S^3 \setminus K)$ can be replaced by its homotopy equivalence $\overline{S^3 \setminus N(K)}$ (also called *knot exterior*), $\overline{\mathbb{R}^3 \setminus K}$ and $\mathbb{R}^3 \setminus K$, where $N(K)$ is a tubular neighborhood of K in S^3. First we prove that the knot space is a $K(\pi_1(S^3 \setminus K), 1)$-space. Recall that $K(G, n)$-space is the topological space with $\pi_n = G$ and $\pi_i = 0$ for $i \neq n$. These spaces serves the building blocks for homotopy theory.

Theorem 3.5.1. *(Papakyriakopoulos asphericity Theorem) The knot space* $S^3 \setminus K$ *is a* $K(\pi_1(S^3 \setminus K), 1)$-space. I.e., $\pi_i(S^3 \setminus K) = 0$ for $i \geq 2$.

Proof: We use the sphere theorem to prove $\pi_2(S^3 \setminus K) = 0$ first. Recall that the sphere theorem states that if M^3 is an orientable 3-manifold M^3 with $\pi_2 \neq 0$ then there is an embedding $F : S^2 \to M^3$ with homotopic nontriviality to present the nontrivial homotopy class. Suppose $\pi_2(S^3 \setminus K) \neq 0$. Then there exists a (PL) sphere which is noncontractible inside $S^3 \setminus K$. Hence there exists an embedding $F : S^2 \to S^3 \setminus K$ to represent the same class by the sphere theorem. Now S^3 splits the oriented 3-manifold into two pieces $X_1 \cup X_2$ by $F(S^2)$. We obtain $\overline{X}_i$ is 3-ball for both $i = 1$ and $i = 2$ by the PL Schönflies Theorem. Since K lies entirely on one side, say X_1, $F(S^2) \subset X_2$ lies on the other side. Then $F(S^2)$ is contractible in $X_2 \subset (S^3 \setminus K)$, so is in $S^3 \setminus K$. This contradicts with the homotopical nontriviality of $F(S^2)$. Therefore $\pi_2(S^3 \setminus K) = 0$.

Let $\widetilde{S^3 \setminus K}$ be the universal covering of $S^3 \setminus K$. Then $\pi_i(S^3 \setminus K) = \pi_i(\widetilde{S^3 \setminus K})$ for $i \geq 3$. For the open 3-manifold $S^3 \setminus K$, we have $H_i(\widetilde{S^3 \setminus K}) = 0$ for $i \geq 3$. Thus $\pi_i(\widetilde{S^3 \setminus K}) = 0$ $(i \geq 3)$ by a theorem of Whitehead. So the result follows. $\square$

Corollary 3.5.2. $S^3 \setminus K_1$ *is homotopy-equivalent to* $S^3 \setminus K_2$ *if and only if* $\pi_1(S^3 \setminus K_1)$ *is isomorphic to* $\pi_1(S^3 \setminus K_2)$.

Proof: Two simplicial complexes are Eilenberg-MacLane ($K(\pi, n)$-space) spaces of the same type, then they are homotopy-equivalent. Hence the result follows by Theorem 3.5.1. □

Proposition 3.5.3. *Let $E(K) = \overline{S^3 \setminus N(K)}$ be the knot exterior, where $N(K)$ is a tubular neighborhood of the knot K. Then we have*

$$H_n(E(K); \mathbb{Z}) = \begin{cases} \mathbb{Z} & n = 0, 1 \\ 0 & n \geq 2 \end{cases}$$

Proof: We have $N(K)$ is a solid torus which is a $K(\mathbb{Z}, 1)$-space. Denote $T^2 = N(K) \cap E(K) = \partial N(K) = \partial E(K)$ a torus. Since $E(K)$ is connected, $H_0(E(K), \mathbb{Z}) = 0$. We apply the Mayer-Vietoris sequence to $S^3 = E(K) \cup N(K)$,

$$0 = H_2(S^3) \to H_1(T^2) \to H_1(N(K)) \oplus H_1(E(K)) \to H_1(S^3) = 0.$$

Thus $\mathbb{Z} \oplus H_1(E(K)) = \mathbb{Z} \oplus \mathbb{Z} = H_1(T^2)$ implies $H_1(E(K)) = \mathbb{Z}$. Continuing the Mayer-Vietoris sequence, we obtain

$$0 = H_3(T^2) \to H_3(N(K)) \oplus H_3(E(K)) \to H_3(S^3) \to$$

$$H_2(T^2) \to H_2(N(K)) \oplus H_2(E(K)) \to H_2(S^3) = 0.$$

Since $T^2 = \partial E(K)$, the group $H_2(T^2)$ is mapped by the inclusion $T^2 \hookrightarrow E(K)$ to $0 \in H_2(E(K))$. So $H_2(E(K)) = 0$. Note that $H_3(S^3) \to H_2(T^2)$ is surjective and $H_3(N(K)) = 0$. So $H_3(E(K)) = 0$. Since $E(K)$ is a 3-manifold, $H_n(E(K)) = 0$ for $n > 3$. □

Consider the isomorphism $\mathbb{Z} \oplus \mathbb{Z} = H_1(T^2) \to H_1(N(K)) \oplus H_1(E(K))$ in the Mayer-Vietoris sequence. The generators of $H_1(N(K)) = \mathbb{Z}$ and $H_1(E(K)) = \mathbb{Z}$ are determined up to their inverses. Let l be a simple closed curve in $N(K)$ which represents the generator of $H_1(N(K))$. So l is also a nontrivial class in $\partial N(K)$ and trivial homology class in $E(K)$. Let m be a curve in $E(K)$ which represents the generator in $H_1(E(K))$. Thus the pair (l, m) determines the isomorphism of $H_1(T^2) \cong \mathbb{Z} \oplus \mathbb{Z}$. By the well-known result, we may choose l, m as simple closed curves on $\partial N(K) = \partial E(K) = T^2$ which intersect at one point.

The curve m is homotopically trivial in $N(K)$ and is bounded by a disk. It is a meridian of the solid torus $N(K)$ with $lk(l, m) = \pm 1$. One may orient m so that $lk(l, m) = 1$ which determines m up to an isotopy of $T^2 = \partial N(K)$. The pair (l, m) is called a longitude and a meridian pair of the knot K. The knot K and the longitude l bound an annulus in $N(K)$. Since l is a simple closed curve on $\partial E(K)$ which is homologous to 0 in

$E(K)$ and not homologous to zero in $\partial E(K)$, the longitude l is determined up to isotopy and orientation by $E(K)$. The meridian m is a simple closed curve on $\partial E(K)$ intersect l at one point. The meridian is not determined by $E(K)$ because simple closed curves on $\partial E(K)$ which are homologous to $m^{\pm 1}l^r, r \in \mathbb{Z}$, have the same properties.

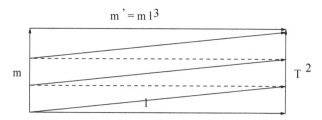

Fig. 3.16   Meridian is not determined by $\partial E(K)$

The Wirtinger presentation is a presentation of the group of a knot (or link) in S^3 (or $\mathbb{R}^3$), provided the regular projection "picture" of the knot. Let the base point be ∞. Each arc α_i is assumed connected to α_{i-1} and α_{i+1} (mod n) where n is the number of arcs in the presentation of the knot. Orient α_i by assigning a direction on the knot K, and draw a short arrow x_i under the arc α_i in the right-left direction. This x_i represents a loop in $S^3 \setminus K$ based at ∞. Now at each crossing, there is a certain relation among the x_i's.

For n crossings, there are n-relations.

Theorem 3.5.4. *(Wirtinger Presentation) The group $\pi_1(S^3 \setminus K)(= \pi_1(\mathbb{R}^3 \setminus K))$ is generated by homotopy classes x_i and has presentation*

$$\langle x_1, x_2, \cdots, x_n | r_1, r_2, \cdots, r_n \rangle.$$

Proof: See [Rolfsen, 1976, pp. 58–60] and [Burde and Zieschang, 1985, p. 33].   □

Example 3.5.5. Trefoil knot 3_1.

There are 3 arcs α_1, α_2 and α_3 with underlying three generators x_1, x_2 and x_3. The relation $r_1 : x_1 x_2 = x_2 x_3$ (or $r_1 : x_1 x_2 x_3^{-1} x_2^{-1}$). Similarly we work out $r_2 : x_2 x_3 x_1^{-1} x_3^{-1}$ and $r_3 : x_3 x_1 x_2^{-1} x_1^{-1}$. Now we have a Wirtinger presentation for trefoil knot

$$\langle x_1, x_2, x_3 | r_1, r_2, r_3 \rangle = \langle x_1, x_2, x_3 | r_1, r_3 \rangle.$$

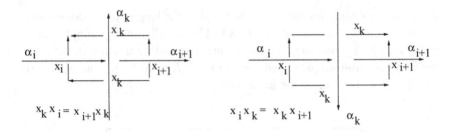

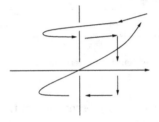

Fig. 3.17   Wirtinger presentation of knot group

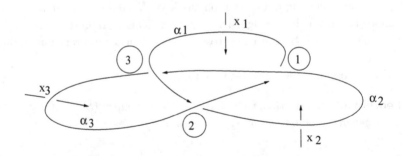

Fig. 3.18   Wirtinger presentation of trefoil knot

Note that $r_3 \cdot r_1 = x_3 x_1 x_3^{-1} x_2^{-1} = r_2^{-1}$. r_2 is a consequence of r_1 and r_3. In fact each relator is a consequence of the other relators. Hence the group can be presented by, $x_3 = x_1 x_2 x_1^{-1}$,

$$\langle x_1, x_2, x_3 | x_1 x_2 x_3^{-1} x_2^{-1}, x - 3 x_1 x_2^{-1} x_1^{-1} \rangle = \langle x_1, x_2 | x_1 x_2 x_1 x_2^{-1} x_1^{-1} x_2^{-1} \rangle.$$

Let $a = x_1 x_2$ and $b = x_2^{-1} x_1^{-1} x_2^{-1}$. So $x_1 = a x_2^{-1}$ and $b = x_2^{-1} x_2 a^{-1} x_2^{-1} = a^{-1} x_2^{-1}$. We obtain $x_2 = b^{-1} a^{-1}$ and $x_1 = a^2 b$. Thus the relation reduces to

$$r = x_1 x_2 x_1 (x_2^{-1} x_1^{-1} x_2^{-1})$$
$$= a^2 b \cdot b^{-1} a^{-1} \cdot a^2 b \cdot b$$
$$= a^3 b^2.$$

Therefore the group has a presentation $\langle a, b | a^3 b^2 \rangle$.

Exercise 3.5.6. Show that $\langle x, y | xyx = yxy \rangle$ is isomorphic to $\langle a, b | a^3 b^2 \rangle$.

The group is a free product of $\mathbb{Z}\langle a \rangle$ and $\mathbb{Z}\langle b \rangle$ with amalgamated subgroup $\langle a^3 \rangle = \langle b^{-2} \rangle$. Hence it is not commutative. Thus the trefoil knot is not the unknot.

Example 3.5.7. Figure eight knot 4_1:

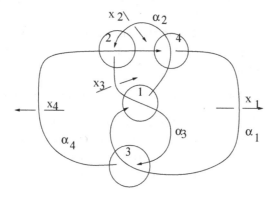

Fig. 3.19   Wirtinger presentation of figure eight knot

$$r_1 : x_1 x_3 = x_3 x_2; \quad r_2 : x_4 x_2 = x_3 x_4;$$

$$r_3 : x_3 x_1 = x_1 x_4; \quad r_4 : x_2 x_4 = x_1 x_2.$$

So the figure eight knot group has a presentation

$$\pi_1(4_1) = \langle x_1, x_2, x_3, x_4 | r_1, r_2, r_3, r_4 \rangle.$$

Note that the relation r_4 is a consequence of others. Hence $x_4 = x_2^{-1} x_1 x_2$ plugging into the other relation and reducing the relation r_4, we have

$$\langle x_1, x_2, x_3 | x_1 x_3 = x_3 x_2, x_2^{-1} x_1 x_2^2 = x_3 x_2^{-1} x_1 x_2, x_3 x_1 = x_1 x_2^{-1} x_1 x_2 \rangle.$$

Eliminate $x_2 = x_3^{-1}x_1x_3$ from r_1 further,

$$\langle x_1, x_3 | r_2 = x_1^{-1}x_3x_1x_3^{-1}x_1x_3 = x_3x_1^{-1}x_3x_1 \rangle.$$

(See [Burde and Zieschang, 1985] for the nontriviality of this group.)

We have for square knots $3_1 \# 3_1, \overline{3}_1 \# \overline{3}_1$, by Seifert-Van Kampen theorem,

$$\pi_1(3_1 \# 3_1) = \langle x, y, u, v | xyx = yxy, uvu = vuv, x = u \rangle.$$

We can eliminate u to have a presentation

$$\langle x, y, v | xyx = yxy, xvx = vxv \rangle.$$

For granny knot $3_1 \# \overline{3}_1$, we have the same presentation

$$\langle x, y, v | xyx = yxy, xvx = vxv \rangle.$$

By a method of signature of knots, we can distinguish the square knots and granny knot. But their fundamental groups are the same. Hence the knot group $f : \mathcal{L}(S^3) \to \mathcal{G}$ by $f(K) = \pi_1(S^3 \setminus K)$ invariant is not one-to-one.

The complement of $3_1 \# 3_1$ and the complement of $3_1 \# \overline{3}_1$ are not homeomorphic by using peripheral structure of π_1 in [Fox, 1952], although they are homotopic to each other.

Exercise 3.5.8. Compute the groups of Hopf link and Borromean rings.

Definition 3.5.9. (Tietze transformations) T1: Adding and removing a superfluous relation $\langle X | R \rangle$ to $\langle X | R \cup \{r\} \rangle$, where r is a consequence of R.

T2: Adding and removing a superfluous generator $\langle X | R \rangle$ to $\langle X, g | R \cup \{wg^{-1}\} \rangle$, where w is a word in other generators.

Note that any presented group is invariant under the Tietze transformation in Definition 3.5.9. Any two different presentations of the same group can be interchanged from each other by a finite sequence of Tietze transformations.

Example 3.5.10. Let $G = \langle a, b, c, d | ab = c, bc = d, cd = a, da = b \rangle$

$$r_1 : ab = c \to b = a^{-1}c$$
$$r_3 : cd = a \to d = c^{-1}a = (a^{-1}c)^{-1} = b^{-1}$$

Hence r_1 and r_3 shows that d is a superfluous generator and r_3 is a superfluous relator. By Tietze transformations, we have

$$G = \langle a, b, c | ab = c, bc = b^{-1}, b^{-1}a = b \rangle.$$

Now $c = b^{-2}$ and $a = b^2$. From r_1, $ab = c$ is identical to $b^2b = b^{-2}$, and $G = \langle b | b^5 = 1 \rangle \cong \mathbb{Z}_5$.

If G has a presentation $\langle X|R \rangle$, then its abelianization G_{ab} has a presentation $\langle X|R \cup \{[x,y]:x,y \in X\} \rangle$.

Let D_n be a dihedral group of order $2n$ with a presentation

$$D_n = \langle x,y|x^n = y^2 = (xy)^2 = 1 \rangle.$$

Its abelianization $(D_n)_{ab}$ has a presentation

$$(D_n)_{ab} = \langle x,y|x^n = y^2 = (xy)^2 = [x,y] = 1 \rangle.$$

In $(D_n)_{ab}$, $xy = yx$, hence

$$(xy)^2 = xy \cdot xy$$
$$= x^2 \cdot y^2$$
$$= x^2.$$

Thus from relations, we obtain $x^n = x^2 = 1$ and $y^2 = 1$. If n is odd, then $x = 1$ and $(D_n)_{ab} = \langle y|y^2 = 1 \rangle \cong \mathbb{Z}_2$; if n is even, then $x^2 = 1$ and $(D_n)_{ab} = \langle x,y|x^2 = y^2 = 1 \rangle \cong \mathbb{Z}_2 \oplus \mathbb{Z}_2$.

Proposition 3.5.11. *All Wirtinger presentations of a link L can be interchanged from each other by a finite sequence of Tietze transformations.*

Proof: The idea is based on the basic result in §2.4 on the Reidemeister moves for oriented links. A regular projection of a link L can be transformed to another regular projection by a finite sequence of Reidemeister moves.

For the Reidemeister move of type I, let

$$\{x_1, \cdots, x_{n-1}, x_n, y|r_1, \cdots, r_{n-1}, yx_n^{-1}\} \tag{3.2}$$

be the Wirtinger presentation of a link L associated to a regular projection containing a part of the left of Figure 1.3. By replacing the part of the projection in the middle of Figure 1.3, the Wirtinger presentation changes to

$$\{x_1, \cdots, x_{n-1}, x_n|r_1', \cdots, r_{n-1}'\}, \tag{3.3}$$

where r_j' $(1 \leq j \leq n-1)$ is obtained from r_j by replacing all y by x_n. The presentation (3.2) can be interchanged to (3.3) by Tietze transformations. If r_j contains the letter y and is written as $r_j = uy^{\pm 1}v$ for $u,v \in \langle x_1, \cdots, x_{n-2}, x_n, y \rangle$. By replacing r_j by r_j^{-1} if necessary (call it T1a), one has $r_j = uy^{-1}v$. Set T1b as replacing one of relators r_j by its conjugate wr_jw^{-1}; and T1c as replacing one of relators r_j by r_jr_i for $i \neq j$. Note that T1a, T1b and T1c are just specific methods in T1. By applying T1b and T1c, we can replace r_j by

$$v^{-1}\{(vr_jv^{-1})yx_n^{-1}\}v = ux_n^{-1}v.$$

It changes the letter y by x_n in the relators $r_1, \cdots, r_{n-2}$ by T1. So we have transformed (3.2) to

$$\{x_1, \cdots, x_n, y | r_1', \cdots, r_{n-1}', yx_n^{-1}\}.$$

By applying the inverse transformation of T2, we get the presentation (3.3).

Similarly, one can get the Tietze transformations for presentations under the Reidemeister moves of type II and III. We leave it as an exercise. □

3.6   Free differential calculus

In covering space, there is a one-to-one correspondence between finitely presented groups and fundamental groups of 2-complexes, between normal subgroups and regular coverings of such complexes [Massey, 1977, Chapters 5-7].

Let $p : \tilde{X} \to X$ be a regular covering of a connected 2-complex. Assume that X is a finite CW-complex with one 0-cell P. The fundamental group $\pi_1(X, P)$ is obtained by assigning a generator x_i to each oriented 1-cell (also denoted by x_i), and a defining relation to each 2-cell e_j of X. So $\pi_1(X, P)$ has a presentation

$$\langle x_1, x_2, \cdots, x_n | \; r_1, r_2, \cdots, r_n \rangle.$$

Choose a base point $\tilde{P} \in \tilde{X}$ over P. Then $p_*(\pi_1(\tilde{X}, \tilde{P})) \lhd \pi_1(X, P)$ is a normal subgroup and its quotient group as the group of covering transformations. Let $\phi : \pi_1(X, P) \to G = \pi_1(X, P)/p_*(\pi_1(\tilde{X}, \tilde{P}))$ be the projection of the quotient group. It can be extended to a group ring homomorphism $\phi : \mathbb{Z}\pi_1(X, P) \to \mathbb{Z}G$.

The edge x_i lifts to an edge $\tilde{x}_i$ with initial point $\tilde{P}$. For a word w in $\pi_1(X, P)$, we denote a closed path in the 1-skeleton X^1 of X, and the element it represents in the free group $F_n \cong \pi_1(X^1, P) = \langle x_1, \cdots, x_n \rangle$. There is a unique lift $\tilde{w}$ of w starting at $\tilde{P}$. Hence $\tilde{w}$ is a homotopy 1–chain in $Z_1(\tilde{X}, \tilde{P})$. Every 1-chain can be written in the form $\sum_{i=1}^{n} a_i \tilde{x}_i$, where $a_i \in \mathbb{Z}G$.

For $w_1 w_2$, first lift w_1 to $\tilde{w}_1$ with end point $w_1^\phi \cdot \tilde{P}$. The covering transformation w_1^ϕ maps $\tilde{w}_2$ (started at $\tilde{P}$ and ended at $w_2^\phi \cdot \tilde{P}$) onto a chain $w_1^\phi \cdot \tilde{w}_2$ (started at $w_1^\phi \cdot \tilde{P}$ and ended at $w_1^\phi \cdot w_2^\phi \cdot \tilde{P}$). Thus the lift $\widetilde{w_1 w_2}$ is given by $\tilde{w}_1 + w_1^\phi \cdot \tilde{w}_2$ by the uniqueness.

Let $\tilde{w}_k = \sum_{i=1}^n a_{ki} \tilde{x}_i$ with $a_{ki} \in \mathbb{Z}G$. Then we get

$$w_1\tilde{w}_2 = \sum_{i=1}^n A_i \tilde{x}_i$$
$$= \tilde{w}_1 + w_1^\phi \cdot \tilde{w}_2$$
$$= \sum_{i=1}^n a_{1i}\tilde{x}_i + w_1^\phi \cdot \sum_{i=1}^n a_{2i}\tilde{x}_i$$
$$= \sum_{i=1}^n (a_{1i} + w_1^\phi \cdot a_{2i}) \cdot \tilde{x}_i$$

Hence $A_i = a_{1i} + w_1^\phi \cdot a_{2i}$ for each $1 \le i \le n$. Now we have the partial derivative

$$(\frac{\partial}{\partial x_i})^\phi : \pi_1(X, P) \to \mathbb{Z}G; \quad w \mapsto \tilde{w} = \sum_{i=1}^n a_i \tilde{x}_i \mapsto a_i,$$

satisfying the linearity and product rule:

$$(\frac{\partial}{\partial x_i})^\phi(w_1 + w_2) = (\frac{\partial}{\partial x_i})^\phi(w_1) + (\frac{\partial}{\partial x_i})^\phi(w_2),$$

$$(\frac{\partial}{\partial x_i})^\phi(w_1 w_2) = (\frac{\partial}{\partial x_i})^\phi(w_1) + w_1^\phi \cdot (\frac{\partial}{\partial x_i})^\phi(w_2).$$

Example 3.6.1. Let $X = \mathbb{R}^3 \setminus \{(1,0,0), (0,1,0), (0,0,1)\}$ and $P = (0,0,0)$. Then we have the fundamental group $\pi_1(X, P) = \langle x_1, x_2 \rangle$ the free group of 2-generators. Its universal covering is given by the tree as in Figure 3.20.

Example 3.6.2. Figure eight knot 4_1:

Now x_1 has a lift x and x_2 has a lift y. Note that $\tilde{1} = 0$ and

$$0 = x_1\tilde{x}_1^{-1} = \tilde{x}_1 + x_1 \cdot x_1^{-1}.$$

So $\tilde{x}_1^{-1} = -x_1^{-1} \cdot x$. Let $w = x_1 y_1^2 x_1^{-1}$. Then we have

$$\tilde{w} = \tilde{x}_1 + x_1 \cdot y_1 \cdot (\widetilde{y_1 x_1^{-1}})$$
$$= x + x_1(y + y_1 \cdot \widetilde{y_1 x_1^{-1}})$$
$$= x + x_1 y + x_1 y_1 (y + y_1 \cdot \tilde{x}_1^{-1})$$
$$= x + x_1 y + x_1 y_1 y - x_1 y_1^2 x_1^{-1} x$$
$$= (1 - x_1 y_1^2 x_1^{-1})x + (x_1 + x_1 y_1)y$$

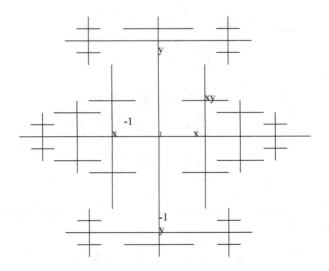

Fig. 3.20   Universal cover of the 8

So

$$\left(\frac{\partial}{\partial x_1}\right)^P w = 1 - x_1 y_1^2 x_1^{-1},$$

$$\left(\frac{\partial}{\partial x_2}\right)^P w = x_1 + x_1 y_1.$$

Let F_n be a free group of rank n, and $\phi \in AutF_n$. Let F_n^ϕ be the image of F_n under the automorphism ϕ, $\mathbb{Z}F_n^\phi$ be the group ring with integral coefficients. An element of $\mathbb{Z}F_n^\phi$ is a sum $\sum a_g \cdot g$ with $a_g \in \mathbb{Z}$ and $g \in F_n^\phi$. It is the free Abelian group on the set of elements of F_n^ϕ as basis, the product of two basis elements is given by the product in F_n^ϕ:

$$\left(\sum \phi(x_i)x_i\right)\left(\sum \phi(y_i)y_i\right) = \sum_{x_i, y_j \in F_n} \phi(x_i)\phi(y_j)x_i y_j$$

$$= \sum_{x_i, y_j \in F_n} \phi(x_i \cdot y_j)x_i \cdot y_j$$

This group ring is characterized by the following universal property. Let $i : F_n^\phi \to \mathbb{Z}F_n^\phi$ be the obvious embedding.

Proposition 3.6.3. *For any function* $f : F_n^\phi \to R$ *with* $f(g_1 g_2) = f(g_1)f(g_2)$ *and* $f(1) = 1_R$ *and a ring* R, *there exists a unique ring homo-*

morphism $F : \mathbb{Z}F_n^\phi \to R$ such that $F \circ i = f$ (the diagram is commutative)

$$F_n^\phi \xrightarrow{\quad i \quad} \mathbb{Z}F_n^\phi$$
$$\downarrow{\scriptstyle f} \qquad\qquad \downarrow{\scriptstyle F}$$
$$R \xrightarrow{\quad = \quad} R$$

Proof: Define $F(\sum_g a_g \cdot g) = \sum_g a_g f(g)$ which obviously is the desired ring homomorphism. $\qquad\square$

Definition 3.6.4. There is a group homomorphism $\varepsilon : \mathbb{Z}F_n^\phi \to \mathbb{Z}$ defined by

$$\varepsilon(\sum_g a_g \cdot g) = \sum_g a_g \in \mathbb{Z}.$$

The map ε is called the *augmentation* of $\mathbb{Z}F_n^\phi$. The kernel $\ker \varepsilon = IF_n^\phi$ is called the *augmentation ideal* of $\mathbb{Z}F_n^\phi$.

Lemma 3.6.5. *(1) As an Abelian group IF_n^ϕ is free on the set $W = \{x^\phi - 1 | 1 \neq x^\phi \in F_n^\phi\}$.*
(2) As F_n^ϕ-module, IF_n^ϕ is generated by $\{x_i^\phi - 1 : 1 \leq i \leq n\}$.

Proof: (1) It is clear that W is linear independent. It suffices to show that W generates IF_n^ϕ. If $w = \sum a(x^\phi) \cdot x^\phi \in \ker \varepsilon$, then $\sum_{x^\phi \in F_n^\phi} a(x^\phi) = 0$ and

$$w = \sum_{x^\phi \in F_n^\phi} a(x^\phi) \cdot x^\phi - \sum_{x^\phi \in F_n^\phi} a(x^\phi) = \sum_{x^\phi \in F_n^\phi} a(x^\phi)(x^\phi - 1).$$

(2) F_n is generated by x_i and F_n^ϕ is generated by $\phi(x_i) = x_i^\phi$. We have to show that if $g \in F_n^\phi$, then $g - 1$ belongs to the module generated by $\{x_i^\phi - 1 : 1 \leq i \leq n\}$. Since $gh - 1 = g(h - 1) + (g - 1)$ and $g^{-1} - 1 = -g^{-1}(g - 1)$, this follows from the representation of g as $\prod \phi(x_i)^{\varepsilon_i}$. $\qquad\square$

Definition 3.6.6. A function $d : G \to A$ from the group G into the G-module A with the property (product rule):

$$d(g_1 \cdot g_2) = d(g_1) + g_1 \cdot d(g_2),$$

is called a *derivation* from G to A.

The set of derivations from G to A has an Abelian group structure.

Lemma 3.6.7. *For any derivation d, we have, for $n \in \mathbb{Z}, x \in G$,*

$$d(x^n) = \frac{x^n - 1}{x - 1} \cdot dx.$$

Proof: For $n = 0$, $d(1) = d(1 \cdot 1) = d(1) + 1 \cdot d(1)$. Thus $d(1) = 0$. For $n \in \mathbb{N}$,

$$d(x^n) = d(x^{n-1} \cdot x) = d(x^{n-1}) + x^{n-1}dx.$$

Thus we obtain $d(x^n) - d(x^{n-1}) = x^{n-1}dx$, and

$$\sum_{i=1}^{n}(d(x^i) - d(x^{i-1})) = \sum_{i=1}^{n} x^{i-1}dx.$$

Hence $d(x^n) - d(1) = (\sum_{i=1}^{n} x^{i-1})dx$ which is equivalent to

$$d(x^n) = \frac{x^n - 1}{x - 1} \cdot dx.$$

If n is negative, and $n = -m$, we have $d(x^n x^m) = 0$. Thus

$$\begin{aligned}
d(x^n) &= -x^n \cdot dx^m \\
&= -x^n \cdot \frac{x^m - 1}{x - 1} \cdot dx \\
&= \frac{-x^{n+m} + x^n}{x - 1} \cdot dx \\
&= \frac{x^n - 1}{x - 1} \cdot dx.
\end{aligned}$$

$\square$

Example 3.6.8. (1) Let A be a real algebra of differentiable functions of $\mathbb{R}$ into $\mathbb{R}$ and $x_0 \in \mathbb{R}$; $\mathbb{R}$ can be considered as an A-module with $(f, a) \mapsto f(x_0)a$. Thus the mapping $f \mapsto Df(x_0)$ from A to $\mathbb{R}$ is a derivation since $D(fg)(x_0) = Df(x_0)g(x_0) + f(x_0)Dg(x_0)$.

(2) Let $d^\varepsilon : \mathbb{Z}F_n^\phi \to \mathbb{Z}F_n^\phi$ be a mapping defined by $d^\varepsilon(\sum a_g g) = \sum a_g(g - 1)$. Note that $\varepsilon \circ d^\varepsilon = 0$. The map d^ε is a derivation.

Definition 3.6.9. (Partial Derivation) The derivations

$$\frac{\partial}{\partial x_i} : \mathbb{Z}F_n \to \mathbb{Z}F_n, \quad x_j \mapsto \begin{cases} 1 \ i = j \\ 0 \ i \neq j \end{cases}$$

of the group ring of a free group F_n are called *partial derivations*.

We also have a total derivation $d : \mathbb{Z}F_n \to \mathbb{Z}F_n$ given by $d = \sum_{i=1}^{n} \frac{\partial}{\partial x_i}$.

Lemma 3.6.10. *There is a uniquely determined derivation $d : \mathbb{Z}F_n \to \mathbb{Z}F_n$ with $dx_i = w_i$ for arbitrary element $w_i \in \mathbb{Z}F_n$.*

Proof: $d(x_i^{-1}) = -x_i^{-1}dx_i = -x_i^{-1} \cdot w_i$. The product rule implies the uniqueness on

$$d(x_{i_1}^{\varepsilon_1} \cdots x_{i_k}^{\varepsilon_k}) = dx_{i_1}^{\varepsilon_1} + x_{i_1}^{\varepsilon_1} dx_{i_2}^{\varepsilon_2} + \cdots + x_{i_1}^{\varepsilon_1} \cdots x_{i_{k-1}}^{\varepsilon_{k-1}} dx_{i_k}^{\varepsilon_k}.$$

Then the product rule follows from $w = w_1 w_2$ to $dw = dw_1 + w_1 dw_2$. The equations

$$d(w_1 x_i x_i^{-1} w_2) = dw_1 + w_1 dx_i + w_1 x_i dx_i^{-1} + w_1 dw_2 = dw_1 + w_1 dw_2,$$

and $d(w_1 x_i^{-1} x_i w_2) = dw_1 + w_1 dw_2$ verify the well-definiteness of d. $\qquad \square$

Corollary 3.6.11. *(1) For any derivation* $d: \mathbb{Z}F_n \to \mathbb{Z}F_n$, *we have*

$$d = \sum_{i=1}^{n} \frac{\partial}{\partial x_i} \cdot d(x_i).$$

(2) $\sum_{i=1}^{n} \frac{\partial}{\partial x_i} \cdot w_i = 0$ *if and only if* $w_i = 0$ *for all* i.

Proof: (1) Let $d(x_i) = w_i \in \mathbb{Z}F_n$. Then the derivation (total derivation) $\sum_{i=1}^{n} \frac{\partial}{\partial x_i} w_i$ with the property

$$(\sum_{i=1}^{n} \frac{\partial}{\partial x_i} w_i)(x_i) = (\sum_{i \neq j} \frac{\partial}{\partial x_j} w_j)(x_i) + w_i \frac{\partial x_i}{\partial x_i} = w_i.$$

Hence $d = \sum_{i=1}^{n} \frac{\partial}{\partial x_i} w_i$ by the uniqueness of Lemma 3.6.10.

(2) Note that $d = 0$ is a trivial derivation with the same property. Then the result follows from $d(x_i) = w_i = 0$ for all i. $\qquad \square$

Theorem 3.6.12. *(Fundamental Theorem of Free Calculus) (1) Let* $d^{\varepsilon}: \mathbb{Z}F_n \to \mathbb{Z}F_n$ *be the map defined by* $w = \sum a_i x_i \mapsto w - \varepsilon(w) = \sum a_i(x_i - 1)$. *Then* d^{ε} *is a derivation with*

$$d^{\varepsilon} w = \sum_{i=1}^{n} \frac{\partial w}{\partial x_i}(x_i - 1).$$

(2) $d^{\varepsilon} w = \sum_{i=1}^{n} v_i(x_i - 1)$ *if and only if* $v_i = \frac{\partial w}{\partial x_i}$ *for all* i.

Proof: (1) By Corollary 3.6.11, all we need to check is that $d^{\varepsilon}(x_i) = x_i - 1 = w_i$. Thus the result follows from Corollary 3.6.11.

(2) $\frac{\partial}{\partial x_j}(w - \varepsilon(w)) = \frac{\partial w}{\partial x_j} - \varepsilon(w) \cdot \frac{\partial 1}{\partial x_j} = \frac{\partial w}{\partial x_j}$. Also we have

$$\frac{\partial}{\partial x_j}(\sum_{i=1}^{n} v_i(x_i - 1)) = v_j.$$

So both sides define derivations with $x_i \mapsto x_i - 1$. Hence the result follows from the uniqueness in Lemma 3.6.10. $\qquad \square$

The abstract version of Theorem 3.6.12 is that the augmentation ideal IF_n represents the functor $D(F_n, \cdot)$. Denote $D(G, A)$ to be the set of all derivations from G to A. Let $\mathcal{M}_G$ be the category of G-module and $\mathcal{A}$ be the category of Abelian groups. For a G-module morphism $\alpha : A_1 \to A_2$, $D(G, \alpha) : D(G, A_1) \to D(G, A_2)$ by $D(G, \alpha)(d) = \alpha \circ d : G \to A_2$ (again a derivation). Thus $D(G, \cdot) : \mathcal{M}_G \to \mathcal{A}$ is a functor.

Theorem 3.6.13. *The functor $D(G, \cdot)$ can be represented by the functor $Hom_G(IG, \cdot)$. That is there is a natural isomorphism $f : D(G, A) \to Hom_G(IG, A)$ defined by $f(d)(x - 1) = dx$ for $x \in G$ and any G-module A.*

Proof: $\Longrightarrow$ Let us show that $f(d)(x-1) = dx$ is a G-module homomorphism.

$$\begin{aligned}
f(d)(y(x-1)) &= f(d)(yx - 1 - (y - 1)) \\
&= f(d)(yx - 1) - f(d)(y - 1) \\
&= d(yx) - dy = ydx \\
&= yf(d)(x - 1)
\end{aligned}$$

Hence it gives a G-module homomorphism.

$\Longleftarrow$ Let $\phi \in Hom_G(IG, A)$, define $d_\phi \in D(G, A)$ by $d_\phi(x) = \phi(x - 1)$. We need to show that d_ϕ is a derivation.

$$\begin{aligned}
d_\phi(xy) &= \phi(xy - 1) \\
&= \phi(x(y - 1) + (x - 1)) \\
&= \phi(x(y - 1)) + \phi(x - 1) \\
&= x\phi(y - 1) + d_\phi(x) \\
&= d_\phi(x) + xd_\phi(y)
\end{aligned}$$

Thus f is well-defined. Verify that f is a homomorphism:

$$f(d_1 + d_2)(x - 1) = (d_1 + d_2)(x) = d_1(x) + d_2(x) = f(d_1)(x - 1) + f(d_2)(x - 1).$$

There exists an inverse homomorphism $f^{-1} : \phi \to d_\phi$ which shows that f is an isomorphism.

Note that $f \circ \alpha(d) = f(\alpha \circ d)$ and $f(\alpha \circ d)(x - 1) = \alpha \circ d(x)$.

$$Hom_G(IG, \alpha)(f(d)(x - 1)) = \alpha \circ f(d)(x - 1) = \alpha \circ d(x).$$

The following diagram is commutative:

$$\begin{array}{ccc}
D(G, A_1) & \xrightarrow{\ f\ } & Hom_G(IG, A_1) \\
\downarrow{\alpha} & & \downarrow{Hom_G(IG, \alpha)} \\
D(G, A_2) & \xrightarrow{\ f\ } & Hom_G(IG, A_2).
\end{array}$$

Hence f is a natural isomorphism. □

Exercise 3.6.14. (1) Compute the derivations of $\frac{\partial}{\partial x}$ and $\frac{\partial}{\partial y}$ for the following.

a. $xyxx^{-1}y^{-1}$.

b. $(xy)^n x(xy)^{-n}$.

(2) Prove the Chain rule for $w, v_1, \cdots, v_m$ words in F_n:

$$\frac{\partial}{\partial x_i} w(v_1, v_2, \cdots, v_m) = \sum_{j=1}^{m} \frac{\partial w}{\partial v_j} \frac{\partial v_j}{\partial x_i}.$$

(3) Derive the quotient rule for derivation $d(uv^{-1})$ and $d(v^{-1}u)$.

3.7 Magnus representations

For $w \in F_n$, define a map from the free group F_n to the multiplicative group of 2×2 matrices over the polynomial group ring,

$$M : F_n \to M_{2\times 2}(P(\mathbb{Z}F_n; z)),$$

where $P(\mathbb{Z}F_n; z)$ is the set of polynomials with coefficients in the group ring $\mathbb{Z}F_n$ and with degree in $z = (z_1, \cdots, z_n)$ non-negative.

$$M(w) = \begin{pmatrix} w & \sum_{i=1}^{n} \frac{\partial w}{\partial x_i} \cdot z_i \\ 0 & 1 \end{pmatrix} = \begin{pmatrix} w & d^z w \\ 0 & 1 \end{pmatrix}.$$

Note that d^z is a derivation defined on $x_i \mapsto z_i$. In particular, we have $M(1) \begin{pmatrix} 1 & 0 \\ 0 & 1 \end{pmatrix}$, and

$$M(w_1 w_2) = \begin{pmatrix} w_1 w_2 & d^z(w_1 w_2) \\ 0 & 1 \end{pmatrix}$$

$$= \begin{pmatrix} w_1 w_2 & d^z w_1 + w_1 d^z w_2 \\ 0 & 1 \end{pmatrix}$$

$$= \begin{pmatrix} w_1 & d^z w_1 \\ 0 & 1 \end{pmatrix} \begin{pmatrix} w_2 & d^z w_2 \\ 0 & 1 \end{pmatrix}$$

$$= M(w_1) M(w_2).$$

Definition 3.7.1. The map $M : F_n \to M_{2\times 2}(P(\mathbb{Z}F_n; z))$ is called the Magnus representation.

Lemma 3.7.2. For $\phi \in \mathrm{Aut}F_n$, we have $M^\phi = M \circ \phi$ the Magnus ϕ-representation of F_n in $M_{2\times 2}(P(\mathbb{Z}F_n; z))$.

Proof: By definition, we have

$$M^\phi(w) = \begin{pmatrix} w^\phi & \sum_{i=1}^n (\frac{\partial w}{\partial x_i})^\phi z_i \\ 0 & 1 \end{pmatrix}.$$

Using the Magnus representation M,

$$M^\phi(1) = M \circ \phi(1) = M(1) = Id_{2\times 2},$$

$$\begin{aligned} M^\phi(w_1 w_2) &= M \circ \phi(w_1 w_2) \\ &= M(\phi(w_1)\phi(w_2)) \\ &= M(\phi(w_1))M(\phi(w_2)) \\ &= M^\phi(w_1)M^\phi(w_2). \end{aligned}$$

$\square$

Theorem 3.7.3. *(1)* $\ker M^\phi = [\ker \phi, \ker \phi]$

(2) $\begin{pmatrix} m & \sum_{i=1}^n m_i z_i \\ 0 & 1 \end{pmatrix}$ *is a Magnus ϕ-representation if and only if* $m \in F_n^\phi, m_i \in \mathbb{Z}F_n^\phi$ *satisfy*

$$\sum_{i=1}^n m_i(x_i^\phi - 1) = m - 1.$$

Proof: (1) Elements in $\ker M^\phi$ if and only if $w^\phi = 1$ and $(\frac{\partial w}{\partial x_i})^\phi = 0$ $(1 \le i \le n)$. By a theorem of Blachfield [Birman, 1974], $(\frac{\partial w}{\partial x_i})^\phi = 0$ if and only if $w \in [\ker \phi, \ker \phi]$.

(2) The matrix is a Magnus ϕ-representation if and only if there exists a derivation d such that $m = w^\phi$ and $d(x_i)^\phi = m_i$. By the fundamental theorem of free calculus $(\sum_{i=1}^n d(x_i)(x_i - 1) = w - 1)$, we have

$$\sum_{i=1}^n d(x_i)^\phi(x_i^\phi - 1^\phi) = w^\phi - 1,$$

which is equivalent to the result desired. $\square$

Now we can extend the Magnus representation to $k \times k$ matrix.

Proposition 3.7.4. *Let* $M : F_n \to M_{k\times k}(P(\mathbb{Z}F_n, z))$ *be a mapping defined by*

$$M(w) = \begin{pmatrix} w & d^z w & (d^z)^2 w & \cdots & (d^z)^{k-1}w \\ 0 & 1 & d_\varepsilon^z w & \cdots & (d^z)_\varepsilon^{k-2}w \\ \cdot & \cdot & \cdot & \cdots & \cdot \\ 0 & 0 & 0 & \cdots & 1 \end{pmatrix},$$

where $(d^z)^i$ *is a composite derivation i-times of d^z and $(d^z)_\varepsilon^i w = (d^z)^i w - \varepsilon \circ (d^z)^i w$. Then M is a representation in $M_{k\times k}(P(\mathbb{Z}F_n, z))$.*

Proof: It is clear that $M(1) = Id_{k \times k}$. It suffices to check $M(w_1 w_2) = M(w_1)M(w_2)$. But this follows from

$$(d^z)^{i+1}(w) = d^z((d^z)^i w) = \sum_{1 \leq i_j \leq n} \frac{\partial^q w}{\partial x_{i_1} \partial x_{i_2} \cdots \partial x_{i_q}} z_{i_1} z_{i_2} \cdots z_{i_q},$$

and Leibniz rule $d^k(xy) = \sum_{i+j=k} d^i x \cdot d^j y$ for any derivation d with $d^0 x = x$ and $d^0 y = 1$. $\qquad \square$

Similarly we have $M^\phi : F_n \to M_{k \times k}(P(\mathbb{Z}F_n^\phi; z))$ for any $\phi \in AutF_n$ by $M^\phi = M \circ \phi$.

For example, define $d^z(x_i) = z_i$. So $\frac{\partial}{\partial x_j}(x_i)\delta_{ij}$. Hence $(d^z)^i = 0$ for $i \geq 2$. The representation defined by the matrix

$$M(x_i) = \begin{pmatrix} 1 & z_i & \cdots & 0 \\ 0 & 1 & \cdots & 0 \\ \cdot & \cdot & \cdots & z_i \\ 0 & 0 & \cdots & 1 \end{pmatrix},$$

is a faithful representation of F_n modulo the k-th group of the lower central series of F_n over the ring $P(\mathbb{Z}; z)$ for $k \geq 2$ [Birman, 1974, Corollary 3.8.1].

Let $\phi \in AutF_n$ and $A_\phi = \{a \in AutF_n | \phi(x) = a \circ \phi(x), x \in F_n\}$. If $\phi : F_n \to (F_n)_{ab} = F_n/[F_n, F_n]$, then $A_\phi = Aut(F_n)_{ab}$. If $a \in A_\phi$, define

$$\|a\| = \det((\frac{\partial a(x_i)}{\partial x_j})^\phi).$$

Note that $(\frac{\partial a(x_i)}{\partial x_j})^\phi \in \mathbb{Z}F_n^\phi$. Thus we have $m : A_\phi \to \mathbb{Z}F_n^\phi$ given by $m(a) = \|a\|$ or $m(a) = (\frac{\partial a(x_i)}{\partial x_j})^\phi$. For the matrix, we have the chain rule to obtain the representation,

$$\frac{\partial a_1 \circ a_2(x_i)}{\partial x_j} = \frac{\partial a_1(a_2(x_i))}{\partial x_j} \tag{3.4}$$

$$= \sum_{k=1}^n \frac{\partial a_1(a_2(x_k))}{\partial v_k} \cdot \frac{\partial v_k}{\partial x_j}, \tag{3.5}$$

where $a_1(v_1, v_2, \cdots, v_k) = a_1(a_2(x_i))$. So the matrix $m(a_1 \circ a_2) = m(a_1)m(a_2)$. For the determinant, $m(Id) = \|Id\| = \det(\frac{\partial x_i}{\partial x_j})^\phi = 1$ and

$$m(a_1 \circ a_2) = \|a_1 \circ a_2\| = \det((\frac{\partial a_1 \circ a_2(x_i)}{\partial x_j})^\phi)$$

$$= \det((\frac{\partial a_1(v_k)}{\partial v_k})^\phi \cdot (\frac{\partial v_k}{\partial x_j})^\phi)$$

$$= \det((\frac{\partial a_1(v_k)}{\partial v_k})^\phi) \cdot \det((\frac{\partial v_k}{\partial x_j})^\phi)$$

$$= \|a_1\| \cdot \|a_2\| = m(a_1) \cdot m(a_2),$$

where the second equality follows from (3.4). Hence we have representations $m : A_\phi \to \mathbb{Z}F_n^\phi$ as the determinant and $m : A_\phi \to M(P(\mathbb{Z}F_n^\phi; z))$.

Example 3.7.5. Let $\phi : F_n \to 1$ be a trivial homomorphism. Then

$$A_\phi = \{a \in AutF_n | \phi(x) = a \circ \phi(x)\} = AutF_n.$$

If $a \in A_\phi = AutF_n$ with $a(x_i) = w_i(x_1, \cdots, x_n)$ and

$$(\frac{\partial a(x_i)}{\partial x_j})^\phi = (\frac{\partial w_i}{\partial x_j})^\phi = \varepsilon_j,$$

where ε_j is the exponent sum of x_j appearing in w_i. Thus $m : A_\phi = AutF_n \to \mathbb{Z}F_n^\phi \to \mathbb{Z}$ is a integral valued representation. Since $a \in AutF_n$ has $a^{-1} \in A_\phi = AutF_n$, $m(a) \cdot m(a^{-1}) = Id$ as matrix. So the matrix $m(a) = (\frac{\partial a(x_i)}{\partial x_j})^\phi$ with integer entries is invertible and determinant is ± 1. So the image of F_n under $m(a)$ is a subgroup of the unimodular group. Each $n \times n$ matrix with integral entries and determinant ± 1 can be associated with an automorphism of $(F_n)_{ab}$. Jakob Nielsen proved that every automorphism of $(F_n)_{ab}$ is induced by an element in $AutF_n$. Hence the Magnus representation of $AutF_n \to M_{n \times n}(P(\mathbb{Z}F_n^\phi; z))$ is a faithful representation of the full group $Aut(F_n)_{ab}$ by the full group of $n \times n$ unimodular matrices.

Example 3.7.6. (Burau representation) There is a faithful representation $\rho_B : B_n \to AutF_n$ which allows us to think the braid group B_n as a subgroup of $AutF_n$. Let $\mathbb{Z}\langle t \rangle$ be an infinite cyclic group and let $\phi : F_n \to \mathbb{Z}\langle t \rangle$ be defined by $\phi(x_i) = t(1 \le i \le n)$. Thus $A_\phi = \{a \in B_n \subset AutF_n | \phi(x_i) = \sigma_j \circ \phi(x_i)\}$ since B_n is generated by $\sigma_1, \cdots, \sigma_{n-1}$ and so is a under the faithful representation ρ_B. We have $m_{B_n} : B_n \subset A_\phi \to \mathbb{Z}F_n^\phi$. Recall that $\sigma_i(x_i) = x_i x_{i+1} x_i^{-1}$, $\sigma_i(x_{i+1}) = x_i$ and $\sigma_i(x_j) = x_j$ for $j \ne i, i+1$. Thus

$$\frac{\partial \sigma_i(x_i)}{\partial x_j} = 0, \quad j \ne i, i+1,$$

$$\frac{\partial \sigma_i(x_i)}{\partial x_i} = \frac{\partial(x_i x_{i+1} x_i^{-1})}{\partial x_i}$$

$$= 1 + x_i \frac{\partial}{\partial x_i}(x_{i+1} x_i^{-1})$$

$$= 1 + x_i(-x_{i+1} x_i^{-1}).$$

Hence we have $(\frac{\partial \sigma_i(x_i)}{\partial x_i})^\phi = 1 - \phi(x_i x_{i+1} x_i^{-1}) = 1 - t \cdot t \cdot t^{-1} = 1 - t.$

$$\frac{\partial \sigma_i(x_i)}{\partial x_{i+1}} = \frac{\partial}{\partial x_{i+1}}(x_i x_{i+1} x_i^{-1})$$

$$= 0 + x_i \frac{\partial}{\partial x_{i+1}}(x_{i+1} x_i^{-1})$$

$$= x_i(1 + x_{i+1} \frac{\partial}{\partial x_{i+1}} x_i^{-1})$$

$$= x_i.$$

So $(\frac{\partial \sigma_i(x_i)}{\partial x_{i+1}})^\phi = \phi(x_i) = t$, and $\frac{\partial \sigma_i(x_{i+1})}{\partial x_i} = 1, \frac{\partial \sigma_i(x_{i+1})}{\partial x_j} = 0.$ Thus we have the matrix

$$m_{B_n}(\sigma_i) = \begin{pmatrix} Id_{(i-1)\times(i-1)} & 0 & 0 & 0 \\ 0 & 1-t & t & 0 \\ 0 & 1 & 0 & 0 \\ 0 & 0 & 0 & Id_{(n-i-1)\times(n-i-1)} \end{pmatrix}.$$

The representation $m_{B_n} : B_n \to M_{n\times n}(\mathbb{Z}F_n^\phi; t)$ is called the *Burau representation* of B_n.

Mechanical interpretation of the Burau representation: For a positive braid (no negative exponents), lay the braid out flat as n bowling lanes, the lanes going over each other according to the braid. If a ball traveling along a lane has probability t of falling off the top lane (and continuing the lane below) at every crossing, then the (i,j) entry of the Burau representation m_{B_n} is the probability that a ball bowled in the ith lane will end up in the jth lane. So the polynomial entry of the Burau matrix m_{B_n} of a positive braid will be non-negative for $0 \le t \le 1$.

Example 3.7.7. (Gassner representation) Recall $P_n = \ker(B_n \to \Sigma_n) \subset B_n$ the pure braid group has a faithful representation as a subgroup of $AutF_n$ with the action of the pure braid generators $\{A_{rs}\}$. Let $\mathbb{Z}\langle t_1, \cdots, t_n \rangle$ be a free Abelian group with basis $t_1, \cdots, t_n$. Let $a : F_n \to \mathbb{Z}\langle t_1 \cdots, t_n \rangle$ be defined by $a(x_i) = t_i \ (1 \le i \le n)$. Since the pure braid generators map each generator $x_i \in F_n$ into a conjugate of itself, thus $a(x_i) = A_{rs} \circ a(x_i)$ for all $1 \le i \le n, 1 \le r < s \le n$. Thus

$$A_a = \{\phi \in P_n \subset AutF_n | a(x_i) = \phi \circ a(x_i)\} = P_n,$$

since P_n is generated by A_{rs}. The map

$$m_{P_n} : P_n \to M(\mathbb{Z}F_n^a; t_1, \cdots, t_n), \quad (A_{rs}) \mapsto ((\frac{\partial A_{rs}(x_i)}{\partial x_j})^a),$$

is a representation of P_n.

Exercise 3.7.8. Verify that (see Lemma 2.3.1)

$$\left(\frac{\partial A_{rs}(x_i)}{\partial x_j}\right)^a = \begin{cases} \delta_{ij} & \text{if } s < i \text{ or } i < r \\ (1-t_i)\delta_{ir} + t_r\delta_{ij} & \text{if } s = i \\ (1-t_i)(\delta_{ij} + t_i\delta_{sj}) + t_i t_s\delta_{ij} & \text{if } r = i \\ (1-t_i)(1-t_s)\delta_{rj} - (1-t_r)\delta_{sj} + \delta_{ij} & \text{if } r < i < s \end{cases}$$

This representation was first discovered by B. J. Gassner in 1961.

Conjecture 3.7.9. *The Gassner representation is faithful.*

Lemma 3.7.10. *If $\beta \in B_n$ with $\overline{\beta}$ a knot, then $m_{B_n}(\beta) - Id$ is a Alexander matrix for the knot $\overline{\beta}$; if $\beta \in P_n$, then $m_{P_n}(\beta) - Id$ is an Alexander matrix for the pure link $\overline{\beta}$.*

Proof: The braid acts on F_n given by Proposition 2.5.1 through ρ_F, and the knot group $\pi_1(S^3 \setminus \overline{\beta})$ admits the presentation given by Theorem 2.6.2 and Theorem 2.7.4. Thus $\left(\frac{\partial \beta(x_i)}{\partial x_j}\right)^\phi$ is a Jacobian matrix for the knot $\overline{\beta}$. Note that $\beta(x_i) : A_i x_{\mu_i} A_i^{-1} = x_i$. The relation $\beta(x_i) = A_i x_{\mu_i} A_i^{-1} x_i^{-1}$ and

$$\frac{\partial}{\partial x_j}\beta(x_i) = \frac{\partial A_i}{\partial x_j} + A_i\frac{\partial x_{\mu_i}}{\partial x_j} - A_i x_{\mu_i} A_i^{-1}\frac{\partial A_i}{\partial x_j} + A_i x_{\mu_i} A_i^{-1}\frac{\partial x_i^{-1}}{\partial x_j}.$$

Thus the result follows by evaluating $\left(\frac{\partial\beta(x_i)}{\partial x_j}\right)^\phi$ and comparing with $m_{B_n}(\beta) - Id$. $\qquad\square$

Lemma 3.7.11. *m_{B_n} and m_{P_n} are reducible to $(n-1) \times (n-1)$ representations.*

Proof: For F_n with different generators $\langle g_1, \cdots, g_n \rangle$ defined by $g_i = \prod_{j=1}^i x_j$. Let $\phi : F_n \to \mathbb{Z}\langle t \rangle$ be $\phi(x_i) = t$. So $\phi(g_i) = \prod_{j=1}^i \phi(x_j) = t^i$. The generator $\sigma_1, \cdots, \sigma_{n-1}$ of B_n acts on $g_1, \cdots, g_n$ as

$$\sigma_i : \begin{cases} g_k \mapsto g_k & k \neq i \\ g_i \mapsto g_{i+1}g_i^{-1}g_{i+1} & i \neq 1 \\ g_1 \mapsto g_2 g_1^{-1} & i = 1. \end{cases}$$

The corresponding representation m_{B_n} with respect to the new basis $g_1, \cdots, g_n$ is $\left(\frac{\partial\beta(g_i)}{\partial g_j}\right)^\phi$.

$$\frac{\partial\sigma_1(g_1)}{\partial g_j} = \frac{\partial g_2 g_1^{-1}}{\partial g_j} = \begin{cases} -g_2 g_1^{-1} & j = 1 \\ 1 & j = 2 \\ 0 & j \neq 1, 2. \end{cases}$$

Also $\frac{\partial \sigma_1(g_k)}{\partial g_j} = \delta_{kj}$, and

$$m_{B_n}(\sigma_1) = (\frac{\partial \sigma_1(g_i)}{\partial g_j})^\phi = \begin{pmatrix} -\phi(g_2 g_1^{-1}) & 1 & 0 \\ 0 & 1 & 0 \\ 0 & 0 & Id_{(n-2)\times(n-2)} \end{pmatrix}$$

$$= \begin{pmatrix} -t^2 \cdot t^{-1} & 1 & 0 \\ 0 & 1 & 0 \\ 0 & 0 & Id_{(n-2)\times(n-2)} \end{pmatrix}.$$

Similarly, we obtain

$$m_{B_n}(\sigma_k) = \begin{pmatrix} Id_{(k-2)\times(k-2)} & 0 & 0 & 0 & 0 \\ 0 & 1 & 0 & 0 & 0 \\ 0 & t & -t & 1 & 0 \\ 0 & 0 & 0 & 1 & 0 \\ 0 & 0 & 0 & 0 & Id_{(n-k-1)\times(n-k-1)} \end{pmatrix}.$$

The last row in $m_{B_n}(\sigma_k)$ $(1 \le k \le n-1)$ is always $(0, 0, \cdots, 0, 1)$ which fixes a 1-dimensional subspace. Hence m_{B_n} can be reduced to a $(n-1) \times (n-1)$ representation $m_{B_n}^r$ of B_n by deleting the last row and column. Similarly the method works for P_n. □

It is clear that the Burau representation m_{B_n} fixes the subspace spanned by the vector $(1, 1, \cdots, 1)$. The reduced Burau representation $m_{B_n}^r$ is obtained from the Burau representation on the quotient space of $\mathbb{C}^n$ by the vector $(1, 1, \cdots, 1)$. The reduced Burau representation $m_{B_n}^r$ is

$$m_{B_n}^r(\sigma_1) = \begin{pmatrix} -t & 0 & 0 \\ -1 & 1 & 0 \\ 0 & 0 & Id_{n-3} \end{pmatrix}, \quad m_{B_n}^r(\sigma_i) = \begin{pmatrix} Id_{i-2} & 0 & 0 & 0 \\ & -1 & -t & 0 \\ & 0 & -t & 0 \\ & 0 & -1 & 1 \\ 0 & 0 & 0 & Id_{n-i-2} \end{pmatrix},$$

$$m_{B_n}^r(\sigma_{n-1}) = \begin{pmatrix} Id_{n-3} & 0 & 0 \\ 0 & 1 & -t \\ 0 & 0 & -t \end{pmatrix},$$

where $2 \le i \le n-2$ and the diagonal $-t$ in $m_{B_n}^r$ is in the (i, i) entry. This is the simplest nontrivial representation of B_n. Its importance is also given in the following theorem.

Theorem 3.7.12. *[Burau, 1936] If $\tilde{\Delta}_\beta(t)$ is the reduced Alexander polynomial of $\overline{\beta}$, then*

$$\frac{t^n - 1}{t - 1} \cdot \tilde{\Delta}_\beta(t) = \det(m_{B_n}^r(\beta) - Id).$$

If $\beta \in P_n$ and $\Delta_\beta(t_1, \cdots, t_n)$ is the Alexander polynomial of $\overline{\beta}$, then

$$(1 + t_1 + t_1 t_2 + \cdots + t_1 t_2 \cdots t_{n-1}) \cdot \Delta_\beta(t_1, \cdots, t_n) = \det(m_{P_n}(\beta) - Id).$$

Proof: See [Birman, 1974, Theorem 3.11] for details. $\qquad\square$

For $n = 3$, Magnus and Peluso [1969] showed that both the Burau representation m_{B_n} and Gassner representation m_{P_n} are faithful. In general $n \geq 4$ people expect m_{P_n} faithful. Moody [1993] showed that m_{B_n} is not faithful for $n \geq 9$; and Long and Paton [1993] improved Moody's result to show that m_{B_n} is not faithful for $n \geq 6$. Bigelow [1999] showed that m_{B_n} is not faithful for $n \geq 5$. It is not known for $n = 4$ case yet.

3.8   Alexander ideals and Alexander polynomials

We give another way to obtain the Alexander polynomial via the presentation of the link group. Let L be a link with $\mu(L)$-components and exterior $X = S^3 \setminus L$. Let P be a base point of X and $p : \tilde{X} \to X$ be the maximal Abelian covering space with $\pi_1(X) \xrightarrow{\phi} H_1(X) \cong \mathbb{Z}^{\mu(L)}$. Let ψ be the canonical homomorphism of the link group $\psi : F_n \to \pi_1(X) = \langle x_1, \cdots, x_n | r_1, \cdots, r_m \rangle$. Thus we have the natural extension to group rings

$$\mathbb{Z}F_n \xrightarrow{\psi} \mathbb{Z}\pi_1(X) \xrightarrow{\phi} \mathbb{Z}G,$$

where $G = \pi_1(X)/p_*(\pi_1(\tilde{X}, \tilde{P}))$ and $\mathbb{Z}G = \mathbb{Z}[t_1^{\pm 1}, \cdots, t_{\mu(L)}^{\pm 1}]$ the ring of integral Laurent polynomial in $\mu(L)$-variables.

Lemma 3.8.1. *The presented matrix* $((\frac{\partial r_j}{\partial x_i})^{\phi\psi})$ *($1 \leq i \leq n, 1 \leq j \leq m$) is a presentation matrix of* $H_1(\tilde{X}, \tilde{P})$ *as a G-module.*

Proof: For $w \in \mathbb{Z}\pi_1(X)$, $\tilde{w}$ its lift, by the fundamental theorem of free calculus,

$$\partial \tilde{w} = (w^{\phi\psi} - 1)\tilde{P} = \sum (\frac{\partial w}{\partial x_i})^{\phi\psi}(x_i^{\phi\psi} - 1)\tilde{P} = \sum (\frac{\partial w}{\partial x_i})^{\phi\psi} \partial \tilde{x}_i.$$

We have the exact homology sequence

$$0 = H_1(\tilde{P}) \to H_1(\tilde{X}) \to H_1(\tilde{X}, \tilde{P}) \xrightarrow{\partial} H_0(\tilde{P}) = \mathbb{Z}G \xrightarrow{i_*} H_0(\tilde{X}, \tilde{P}) \to 0.$$

Note that $\ker i_* = Im\partial = (I\pi_1(X))^{\phi\psi}$ by Lemma 3.6.5 and Theorem 3.6.12. The fundamental theorem shows that $\ker i_*$ is generated by $\{(x_i^{\phi\psi} - 1)\tilde{P} | 1 \leq i \leq n\}$ as a G-module. Then there is a short exact sequence

$$0 \to H_1(\tilde{X}) \to H_1(\tilde{X}, \tilde{P}) \xrightarrow{\partial} \ker i_* \to 0. \qquad (3.6)$$

Note that $G = \mathbb{Z}^{\mu(L)} \cong \mathbb{Z}\langle t_1, \cdots, t_{\mu(L)} \rangle$ is the Abelian group of rank $\mu(L)$ and $\ker i_*$ is a free $\mathbb{Z}^{\mu(L)}$-module generated by $(t_i - 1)\tilde{P}$. The sequence (3.6) splits with right inverse f $(\partial \circ f = Id)$ and

$$H_1(\tilde{X}, \tilde{P}) \cong H_1(\tilde{X}) \oplus f(\mathbb{Z}G \cdot (t_i - 1)\tilde{P}, 1 \leq i \leq n).$$

Hence the result follows. $\qquad\qquad\qquad\qquad\qquad\qquad\qquad\qquad\quad\Box$

Let M be a finitely generated R-module. Let R_0 be the fraction field of R and $M_0 = M \otimes_R R_0$ as R_0-vector space. Thus the rank $rk(M) = \dim_{R_0} M_0$ and $Tor(M) = \{m \in M | \exists r \in R \ni r \cdot m = 0\}$. The annihilator ideal of M is $Ann(M) = \{r \in R | r \cdot m = 0, \forall m \in M\}$. Let $R^p \overset{m}{\to} R^q \overset{\phi}{\to} M \to 0$ be a finite presentation for M ($p - q$ is called the *deficiency*, a *short free resolution* of M if m is injective). For each $k \geq 0$, the k-th elementary ideal of M is the ideal $E_k(M)$ generated by $(q-k) \times (q-k)$ subdeterminants of the matrix representing m if $0 < q - k \leq p$; and R if $q - k \leq 0$; 0 if $q - k > p$ or $k < 0$.

Definition 3.8.2. The Alexander module of L is the $\mathbb{Z}G$-module $A(L) = H_1(\tilde{X}, \tilde{P})$. The k-th Alexander ideal of L is $E_k(L) = E_k(A(L))$ and the k-th Alexander polynomial of L is $\Delta_k(L) = \Delta_k(A(L))$ (greatest common divisor of elements of $E_k(L)$)

Hence every $(n-1) \times (n-1)$ minor Δ_{ij} of the Jacobian $(\frac{\partial r_i}{\partial x_j})^{\phi\psi}_{n \times n}$ of a Wirtinger presentation of a knot group is a presentation matrix of $H_1(\tilde{X})$. $\Delta(t) = \det \Delta_{ij}$. The elementary ideals of the Jacobian are the elementary ideals of the knot.

Example 3.8.3. A Wirtinger presentation of the group of the trefoil is

$$\langle x_1, x_2, x_3 | r_1 = x_1 x_2 x_3^{-1} x_2^{-1}, r_2 = x_2 x_3 x_1^{-1} x_3^{-1}, r_3 = x_3 x_1 x_2^{-1} x_1^{-1} \rangle.$$

Now we use free calculus to compute.

$$\frac{\partial r_1}{\partial x_1} = 1, \frac{\partial r_1}{\partial x_2} = x_1 - x_1 x_2 x_3^{-1} x_2^{-1}, \frac{\partial r_1}{\partial x_3} = -x_1 x_2 x_3^{-1}$$

$$(\frac{\partial r_1}{\partial x_1})^{\phi\psi} = 1, (\frac{\partial r_1}{\partial x_2})^{\phi\psi} = t - t \cdot t \cdot t^{-1} \cdot t^{-1}, (\frac{\partial r_1}{\partial x_3})^{\phi\psi} = -t$$

$$\frac{\partial r_2}{\partial x_1} = -x_2 x_3 x_1^{-1}, \frac{\partial r_2}{\partial x_2} = 1, \frac{\partial r_2}{\partial x_3} = x_2 - x_2 x_3 x_1^{-1} x_3^{-1}$$

$$(\frac{\partial r_2}{\partial x_1})^{\phi\psi} = -t, (\frac{\partial r_2}{\partial x_2})^{\phi\psi} = 1, (\frac{\partial r_2}{\partial x_3})^{\phi\psi} = t - 1$$

$$\frac{\partial r_3}{\partial x_1} = x_3 - x_3 x_1 x_2^{-1} x_1^{-1}, \frac{\partial r_3}{\partial x_2} = -x_3 x_1 x_2^{-1}, \frac{\partial r_3}{\partial x_3} = 1$$

$$(\frac{\partial r_3}{\partial x_1})^{\phi\psi} = t - 1, (\frac{\partial r_3}{\partial x_2})^{\phi\psi} = -t, (\frac{\partial r_3}{\partial x_3})^{\phi\psi} = 1.$$

Hence the Jacobian of the presentation matrix $(\frac{\partial r_j}{\partial x_i})^{\phi\psi}$ is

$$\begin{pmatrix} 1 & t-1 & -t \\ -t & 1 & t-1 \\ t-1 & -t & 1 \end{pmatrix}.$$

The Alexander ideal $E_k = 0$ if $k < 0$; $E_k = R = \mathbb{Z}\langle t \rangle$ if $k \geq 3$. When $k = 1$, look at 2×2 minors. $\det \Delta_{ij}$ are $-t^2 + t - 1$, $t^2 - t + 1$. Hence $E_1(t) = (1 - t + t^2)$. When $k = 2$, look at 1×1 minors, $E_2(t) = (1) = \mathbb{Z}\langle t \rangle$.

Note that $(r+1) \times (r+1)$ minors can be expanded as a linear combination over $\mathbb{Z}G$ of $r \times r$ minors of the presentation matrix. Therefore

$$0 \subset E_0(L) \subset E_1(L) \subset \cdots \subset E_n(L) = \mathbb{Z}G.$$

Example 3.8.4. Compute all the elementary ideals of the matrix

$$M = \begin{pmatrix} 3 & 0 \\ 0 & 1+t \\ 1+t & 1+t \end{pmatrix}.$$

$E_0(M)$ is generated by 2×2 minors

$$\begin{vmatrix} 3 & 0 \\ 0 & 1+t \end{vmatrix}, \begin{vmatrix} 0 & 1+t \\ 1+t & 1+t \end{vmatrix}, \begin{vmatrix} 3 & 0 \\ 1+t & 1+t \end{vmatrix},$$

by $3(1+t)$, $-(1+t)^2$, $3(1+t)$ over $\mathbb{Z}$. Thus $E_0(M) = \langle 3(1+t), (1+t)^2 \rangle$. Over $\mathbb{Z}(\mathbb{Z}_2)$ $(t^2 \mapsto 1)$, $E_0(M) = \langle 3(1+t), 2(1+t) \rangle = \langle 1+t \rangle$ since $(1+t)^2 = 1 + 2t + t^2 = 2(1+t)$ and $(1+t) = 3(1+t) - 2(1+t)$.

$E_1(M)$ is generated by 1×1 minors (entries of the matrix M). Hence $E_1(M) = \langle 3, (1+t) \rangle$, and $E_k(M) = \mathbb{Z}(\mathbb{Z}_2)$ for $k \geq 2$.

$$0 \subset E_0(M) = \langle (1+t) \rangle \subset E_1(M) = \langle 3, (1+t) \rangle \subset E_2(M) = \mathbb{Z}(\mathbb{Z}_2).$$

Example 3.8.5. Torus knot $T_{p,q}$ has presentation

$$\pi_1 = \langle x, y | r = x^p y^{-q} \rangle, \quad p, q > 0, \quad g.c.d(p,q) = 1.$$

The projection homomorphism $\phi : \pi_1 \to H_1 = \mathbb{Z}\langle t \rangle$ is defined by

$$\phi(x) = t^q, \quad \phi(y) = t^p.$$

Now we compute the Jacobian of the presentation.

$$\frac{\partial r}{\partial x} = \frac{x^p - 1}{x - 1}, \quad \frac{\partial r}{\partial y} = x^p \frac{y^{-q} - 1}{y - 1}.$$

So the Jacobian is the 1×2 matrix

$$(\frac{\partial r}{\partial x}, \frac{\partial r}{\partial y})^{\phi\psi} = (\frac{t^{pq} - 1}{t^q - 1}, t^{pq} \frac{t^{-pq} - 1}{t^p - 1}).$$

Since there exists $\alpha(t), \beta(t) \in \mathbb{Z}(t)$ such that

$$\alpha(t)\frac{t^p - 1}{t - 1} + \beta(t)\frac{t^q - 1}{t - 1} = 1,$$

$$\alpha(t) \cdot \frac{t^{pq} - 1}{t^q - 1} + \beta(t) \cdot \frac{t^{pq} - 1}{t^p - 1} = \Delta_{p,q}(t),$$

where $\Delta_{p,q}(t) = g.c.d(\frac{t^{pq}-1}{t^q-1}, \frac{t^{pq}-1}{t^p-1}))$ is the Alexander polynomial of the torus knot $T_{p,q}$. By multiplying $\frac{(t^p-1)(t^q-1)}{(t^{pq}-1)(t-1)}$ on the second identity, we obtain

$$\alpha(t)\frac{t^p - 1}{t - 1} + \beta(t)\frac{t^q - 1}{t - 1} = \Delta_{p,q}(t) \cdot \frac{(t^p - 1)(t^q - 1)}{(t^{pq} - 1)(t - 1)} = 1.$$

So $\Delta_{p,q}(t) = \frac{(t^{pq}-1)(t-1)}{(t^p-1)(t^q-1)}$ and the Alexander module of $T_{p,q}$ is $A(T_{p,q}) = \mathbb{Z}(t)/\Delta_{p,q}(t)$.

Example 3.8.6. Borromean link has the Wirtinger presentation

$$\langle x_1, x_2, x_3, y_1, y_2, y_3 | r_1 = y_1^{-1}x_2^{-1}y_1y_2, r_2 = y_2^{-1}x_3y_2y_3,$$

$$r_3 = x_2^{-1}x_3x_2y_3^{-1}, r_4 = x_1^{-1}x_3y_1y_3^{-1}, r_5 = x_1^{-1}x_2x_1y_2^{-1}\rangle.$$

Note that $y_1 = x_3^{-1}x_1x_3, y_2 = x_1^{-1}x_2x_1, y_3 = x_2^{-1}x_3x_2$ are superfluous generators. By the Tietze transformations, the presentation of the Borromean link is reduced to

$$\langle x_1, x_2, x_3 | R_1 = x_3^{-1}x_1^{-1}x_3x_2^{-1}x_3^{-1}x_1x_3x_1^{-1}x_2x_1,$$

$$R_2 = x_1^{-1}x_2^{-1}x_1x_3x_1^{-1}x_2x_1x_2^{-1}x_3^{-1}x_2\rangle.$$

Now we compute the Jacobian of the presentation.

$$\frac{\partial R_1}{\partial x_1} = -x_3^{-1}x_1^{-1} + x_3^{-1}x_1^{-1}x_3x_2^{-1}x_3^{-1}$$

$$-x_3^{-1}x_1^{-1}x_3x_2^{-1}x_3^{-1}x_1x_3x_1^{-1} + x_3^{-1}x_1^{-1}x_3x_2^{-1}x_3^{-1}x_1x_3x_1^{-1}x_2.$$

With $\phi \circ \psi(x_i) = t_i$ $(1 \leq i \leq 3)$, we get

$$(\frac{\partial R_1}{\partial x_1})^{\phi\psi} = -t_1^{-1}t_3^{-1} + t_1^{-1}t_2^{-1}t_3^{-1} - t_1^{-1}t_2^{-1} + t_1^{-1}$$
$$= t_1^{-1}t_2^{-1}t_3^{-1}(t_2 - 1)(t_3 - 1),$$

$$\frac{\partial R_1}{\partial x_2} = -x_3^{-1}x_1^{-1}x_3x_2^{-1} + x_3^{-1}x_1^{-1}x_3x_2^{-1}x_3^{-1}x_1x_3x_1^{-1},$$

$$(\frac{\partial R_1}{\partial x_2})^{\phi\psi} = -t_1^{-1}t_2^{-1} + t_1^{-1}t_2^{-1} = 0,$$

$$\frac{\partial R_1}{\partial x_3} = -x_3^{-1} + x_3^{-1}x_1^{-1} - x_3^{-1}x_1^{-1}x_3x_2^{-1}x_3^{-1} + x_3^{-1}x_1^{-1}x_3x_2^{-1}x_3^{-1}x_1,$$

$$(\frac{\partial R_1}{\partial x_3})^{\phi\psi} = -t_3^{-1} + t_3^{-1}t_1^{-1} - t_1^{-1}t_2^{-1}t_3^{-1} + t_2^{-1}t_3^{-1}$$
$$= -t_1^{-1}t_2^{-1}t_3^{-1}(t_1 - 1)(t_2 - 1).$$

Similarly, we have $\frac{\partial R_2}{\partial x_1}$, $\frac{\partial R_2}{\partial x_2}$, $\frac{\partial R_2}{\partial x_3}$ with the 2×3 Jacobian $(\frac{\partial R_j}{\partial x_i})^{\phi\psi}$ given by

$$\begin{pmatrix} t_1^{-1}t_2^{-1}t_3^{-1}(t_2 - 1)(t_3 - 1) & 0 & -t_1^{-1}t_2^{-1}t_3^{-1}(t_1 - 1)(t_2 - 1) \\ t_1^{-1}t_2^{-1}(t_2 - 1)(t_3 - 1) & -t_1^{-1}t_2^{-1}(t_1 - 1)(t_3 - 1) & 0 \end{pmatrix}.$$

Thus $E_0(L)$ is generated by 2×2 minors $-t_1^{-2}t_2^{-2}t_3^{-1}(t_1 - 1)(t_2 - 1)(t_3 - 1)^2$ and

$$t_1^{-2}t_2^{-2}t_3^{-1}(t_2 - 1)(t_3 - 1), t_1^{-2}t_2^{-2}t_3^{-1}(t_1 - 1)(t_2 - 1)^2(t_3 - 1).$$

Over $\mathbb{Z}\langle t_1^{\pm 1}, t_2^{\pm 1}, t_3^{\pm 1}\rangle$, we have

$$E_0(L) = \langle \Delta \cdot (t_1 - 1), \Delta \cdot (t_2 - 1), \Delta \cdot (t_3 - 1)\rangle = \langle\Delta\rangle,$$

where $\Delta = (t_1 - 1)(t_2 - 1)(t_3 - 1)$ is the Alexander polynomial of the Borromean link. Now the $E_1(L)$ is generated by 1×1 minors

$$E_1(L) = \langle (t_2 - 1)(t_3 - 1), (t_1 - 1)(t_2 - 1), (t_1 - 1)(t_3 - 1)\rangle.$$

$E_k(L) = \mathbb{Z}\langle t_1^{\pm 1}, t_2^{\pm 1}, t_3^{\pm 1}\rangle$ for $k \geq 2$.

Exercise 3.8.7. (1) Compute the element ideals of the matrix over $\mathbb{Z}(t)$:

$$\begin{pmatrix} 2 & 1+t & -t \\ -t & 1 & 1-t \\ -1+t & 0 & 1 \end{pmatrix}.$$

(2) Compute the Alexander ideals and polynomials of the figure eight knot and Hopf link.

3.9 Twisted Alexander polynomials

To each representation of the knot group, Lin [2001] assigned a twisted Alexander polynomial for the pair of the knot and the representation. The definition in [Lin, 2001] used regular Seifert surfaces. Wada [1994, §4.5] extended this in terms of free differential calculus.

Let $\pi_1(X)$ be the group of a link L. By Theorem 2.8.1 and Theorem 3.5.4, it is a finitely presented group, $\pi_1(X) = \{x_1, \cdots, x_n | r_1 \cdots r_m\}$. As in §4.7, there are group ring homomorphisms

$$\mathbb{Z}F_n \xrightarrow{\psi} \mathbb{Z}\pi_1(X) \xrightarrow{\phi} \mathbb{Z}G,$$

where $G = H_1(X)$ and $\mathbb{Z}G = \mathbb{Z}[t_1^{\pm 1}, \cdots, t_{\mu(L)}^{\pm 1}]$ is the ring of integral Laurent polynomial in $\mu(L)$-variables.

Let ρ be a representation of $\pi_1(X)$ on a finitely generated free module V over some unique factorization domain R. Choosing a basis for V with $\dim_R V = N$, ρ can be realized as a homomorphism

$$\rho : \pi_1(X) \to AutV = GL_N(R).$$

The corresponding ring homomorphism is

$$\overline{\rho} : \mathbb{Z}\pi_1(X) \to \mathbb{Z}GL_N(R) = M_N(R),$$

where $M_N(R)$ is a matrix algebra.

The twisted version of Alexander polynomials defined in §4.7 is by working on the following group ring homomorphism

$$\mathbb{Z}F_n \xrightarrow{\psi} \mathbb{Z}\pi_1(X) \xrightarrow{\overline{\rho}\otimes\phi} M_N(R) \otimes \mathbb{Z}G \cong M_N(R[t_1^{\pm 1}, \cdots, t_{\mu(L)}^{\pm 1}]). \qquad (3.7)$$

Thus the presented matrix $((\frac{\partial r_j}{\partial x_i})^{(\overline{\rho}\otimes\phi)\circ\psi})(1 \leq i \leq n, 1 \leq j \leq m)$ is called the Alexander matrix of $\pi_1(X)$ associated to the representation ρ. Denote $\Phi = (\overline{\rho} \otimes \phi) \circ \psi$ and $R[G] = R[t_1^{\pm 1}, \cdots, t_{\mu(L)}^{\pm 1}]$.

Lemma 3.9.1. *(1)* $((\frac{\partial r_j}{\partial x_i})^{\Phi})(1 \leq i \leq n, 1 \leq j \leq m)$ *can be viewed as a linear transformation from* $R[G]^{Nn}$ *to* $R[G]^{Nm}$. *Then the kernel of this linear transformation is one-to-one corresponding to the set of derivations of* $\pi_1(X)$ *with values in* $R[G]^n$.

(2) The matrix $((\frac{\partial r_j}{\partial x_i})^{\Phi})$ *is a presentation matrix of* $H_1(\tilde{X}, \tilde{P})$ *as* $M_N(R[G])$-module.

Proof: For $w \in \mathbb{Z}\pi_1(X)$, $\tilde{w}$ its lift with respect to $p : \tilde{X} \to X$, by Theorem 3.6.12,

$$\partial \tilde{w} = (w - 1)^\Phi \tilde{P}$$
$$= \sum (\frac{\partial w}{\partial x_i})^\Phi (x_i - 1)^\Phi \tilde{P}$$
$$= \sum (\frac{\partial w}{\partial x_i})^\Phi \partial \tilde{x}_i,$$

where $\partial \tilde{x}_i \in R[G]^N$. By the exact sequence, $0 = H_1(\tilde{P})$,

$$0 \to H_1(\tilde{X}; R[G]^N) \to H_1(\tilde{X}, \tilde{P}; R[G]^N) \overset{\partial}{\to} H_0(\tilde{P}; R[G]^N)$$

$$\overset{i_*}{\to} H_0(\tilde{X}, \tilde{P}; R[G]^N) \to 0.$$

The kernel $\ker i_* = \operatorname{Im} \partial = (\mathcal{I}\pi_1(X))^\Phi$ by Lemma 3.6.5 and Theorem 3.6.12. The fundamental theorem shows that $\ker i_*$ is generated by $\{\partial \tilde{x}_i : 1 \leq i \leq n\}$ as a $R[G]$-module. Then the short exact sequence

$$0 \to H_1(\tilde{X}; R[G]^N) \to H_1(\tilde{X}, \tilde{P}; R[G]^N) \overset{\partial}{\to} \ker i_* \to 0,$$

splits since $R[G]$ is also a unique factorization domain. Note that (1) and (2) are equivalent with different statements [Wada, 1994]. $\square$

Definition 3.9.2. The twisted Alexander module of L associated to ρ is the $R[G]$-module $A(L, \rho) = H_1(\tilde{X}, \tilde{P}; R[G]^N)$.

For $1 \leq j \leq n$, let M_j be the $m \times (n-1)$ matrix obtained from ∂ by removing the j-th column. M_j is a $Nm \times N(n-1)$ matrix with entries in $R[G]$. Let $I = (i_1, \cdots, i_{N(n-1)})$ for $1 \leq i_1 < \cdots < i_{N(n-1)} \leq Nm$, and M_j^I be a $N(n-1) \times N(n-1)$ matrix consisting of the i_k-th row of the matrix M_j where $k = 1, 2, \cdots, N(n-1)$. The matrix M_j^I is a $(n-1) \times (n-1)$ minor of the Jacobian $(\frac{\partial r_j}{\partial x_i})^\Phi_{n \times n}$ of a Wirtinger presentation of a knot group.

Lemma 3.9.3. *(1)* $\det(x_i^\Phi - Id) \neq 0$ *for some i;*
(2) $\det M_j^I \cdot \det(x_k^\Phi - Id) = \pm \det M_k^I \cdot \det(x_j^\Phi - Id)$ *for $1 \leq j < k \leq n$ and I.*

Proof: (1) Since $\phi(x_i) = t_1^{e_1} t_2^{e_2} \cdots t_{\mu(L)}^{e_{\mu(L)}} \neq 1$ from the surjectivity of ϕ,

$$(x_k - 1)^\Phi = t_1^{e_1} t_2^{e_2} \cdots t_{\mu(L)}^{e_{\mu(L)}} \rho(\phi(x_i)) - Id,$$

hence (1) follows.

(2) Assume that $j = 1$ and $k = 2$ without loss of generality. Since we have relation $r_j = 1$ in $\mathbb{Z}\pi_1(X)$,

$$0 = \partial r_j = \sum_i \frac{\partial r_j}{\partial x_i}(x_i - 1).$$

Applying the homomorphism $\Phi = (\bar{\rho} \otimes \phi) \circ \psi$ to the above equality, we get

$$\sum_i (\frac{\partial r_j}{\partial x_i})^\Phi (x_i - 1)^\Phi = 0. \tag{3.8}$$

Let M_2 be the matrix obtained from M_2^I by replacing the first column

$$((\frac{\partial r_1}{\partial x_1})^\Phi, (\frac{\partial r_2}{\partial x_1})^\Phi, \cdots, (\frac{\partial r_n}{\partial x_1})^\Phi)^t$$

with the following column

$$((\frac{\partial r_1}{\partial x_1})^\Phi (x_1 - 1)^\Phi, (\frac{\partial r_2}{\partial x_1})^\Phi (x_1 - 1)^\Phi, \cdots, (\frac{\partial r_n}{\partial x_1})^\Phi (x_1 - 1)^\Phi)^t.$$

Then the difference between M_2 and M_2^I is the multiplication $(x_1 - 1)$ in the first column, and

$$\det M_2 = \det M_2^I \cdot \det(x_1^\Phi - Id). \tag{3.9}$$

By (3.8), we have

$$(\frac{\partial r_j}{\partial x_1})^\Phi (x_1 - 1)^\Phi = -\sum_{i \geq 2} (\frac{\partial r_j}{\partial x_i})^\Phi (x_i - 1)^\Phi$$

$$= -(\frac{\partial r_j}{\partial x_2})^\Phi (x_2 - 1)^\Phi - \sum_{i \geq 3} (\frac{\partial r_j}{\partial x_i})^\Phi (x_i - 1)^\Phi,$$

for each $1 \leq j \leq m$. Hence the first column of M_2 can be added a linear combination of other columns $\sum_{i \geq 3} (\frac{\partial r_j}{\partial x_i})^\Phi (x_i - 1)^\Phi$, and the resulting matrix $\tilde{M}_1$ has its first column

$$(-(\frac{\partial r_1}{\partial x_2})^\Phi (x_2 - 1)^\Phi, -(\frac{\partial r_2}{\partial x_2})^\Phi (x_2 - 1)^\Phi, \cdots, -(\frac{\partial r_n}{\partial x_2})^\Phi (x_2 - 1)^\Phi)^t.$$

Note that M_1^I also has the property similar to (3.9) as

$$\det M_1 = \det M_1^I \cdot \det(x_2 - 1)^\Phi. \tag{3.10}$$

The matrix $\tilde{M}_1$ has the following identity:

$$\det \tilde{M}_1 = \det M_1^I \cdot \det(1 - x_2)^\Phi = \det M_2.$$

Therefore, by (3.9) and (3.10), one has

$$\det M_2^I \cdot \det(x_1^\Phi - Id) = \det M_2$$
$$= \det \tilde{M}_1$$
$$= \frac{\det(1 - x_2)^\Phi}{\det(x_2 - 1)^\Phi} \cdot \det M_1$$
$$= \pm \det M_1^I \cdot \det(x_2 - 1)^\Phi.$$

Note that the sign is given by $\det(-1)^\Phi$.   □

Definition 3.9.4. The twisted Alexander polynomial of $\pi_1(X)$ associated to the representation ρ is defined to be

$$\Delta_{L,\rho}(t_1, \cdots, t_{\mu(L)}) = \frac{Q_j(t_1, \cdots, t_{\mu(L)})}{\det(x_j^\Phi - Id)},$$

where $\det(x_j^\Phi - Id) \neq 0$ as in Lemma 3.9.3 (1), $\pi_1(X)$ is the link group of L and $Q_j \in R[G]$ is the greatest common divisor of $\det M_j^I$ for all the choice of the indices I.

Remarks: (1) If ρ is the trivial 1-dimensional representation and L is a knot, then $\Delta_{L,\rho} = \Delta_L/(t-1)$. In general, the twisted Alexander polynomial is a rational function. Our definition works for any finitely presentable group.

(2) By Lemma 3.9.3, $\Delta_{L,\rho}$ is an invariant of $\pi_1(X)$ and the associated homomorphism ϕ and ρ, up to $\varepsilon t_1^{e_1} \cdots t_{\mu(L)}^{e_{\mu(L)}}$ with ε a unit element of R and $e_k \in \mathbb{Z}$ $(1 \leq k \leq \mu(L))$.

(3) For a Wirtinger presentation of $\pi_1(X)$ of the link complement in S^3, one has $\pi_1(X) = \{x_1, \cdots, x_n | r_1, \cdots, r_{n-1}\}$ and hence each matrix M_j is a square matrix. So for $1 \leq j \leq n$, $Q_j(t_1, \cdots, t_{\mu(L)}) = \det M_j$ and

$$\Delta_{L,\rho}(t_1, \cdots, t_{\mu(L)}) = \frac{\det M_j}{\det(x_j^\Phi - Id)}.$$

Proposition 3.9.5. *The twisted Alexander polynomial $\Delta_{L,\rho}$ is independent of the choice of the presentation of $\pi_1(X)$.*

Proof: Let $P = \{x_1, \cdots, x_n | r_1, \cdots, r_m, r\}$ be a presentation obtained from the presentation of $\pi_1(X)$ by the Tietze transformation T1 in Definition 3.5.9. Thus

$$r = \prod_{k=1}^p w_k r_{i_k}^{\varepsilon_k} w_k^{-1},$$

where $w_k \in F_n, \varepsilon_k = \pm 1$ and $1 \le i_k \le m$ for $1 \le k \le p$. From §4.5,

$$\partial r = \sum_{k=1}^{p} (\prod_{l=1}^{k-1} w_l r_{i_l}^{\varepsilon_l} w_l^{-1})(z_k \partial r_{i_k} + (1 - w_k r_{i_k}^{\varepsilon_k} w_k^{-1})\partial w_k),$$

where $z_k = w_k$ if $\varepsilon_k = 1$ and $z_k = -w_k r_{i_k}^{-1}$ if $\varepsilon_k = -1$. So the differential with respect to x_i is given by

$$\frac{\partial r}{\partial x_i} = \sum_{k=1}^{p} (\prod_{l=1}^{k-1} w_l r_{i_l}^{\varepsilon_l} w_l^{-1})(z_k \frac{\partial r_{i_k}}{\partial x_i} + (1 - w_k r_{i_k}^{\varepsilon_k} w_k^{-1})\frac{\partial w_k}{\partial x_i}).$$

Note that $\Phi(r_j) = 1$ for all j. One gets

$$(\frac{\partial r}{\partial x_i})^{\Phi} = \sum_{k=1}^{p} \varepsilon_k w_k^{\Phi} (\frac{\partial r_{i_k}}{\partial x_i})^{\Phi}.$$

Let $M(P)$ be the Alexander matrix obtained from the presentation of P. Then the first Nn rows of $M(P)$ are exactly the same as the matrix of M, and the above equality shows that the last N rows of $M(P)$ are linear combinations of the first Nm rows of M. Hence $\det M_j^I = \det M(P)_j^I$ and the twisted Alexander polynomials unchanged from the calculation of the matrix $M(P)$.

Let $P' = \{x_1, \cdots, x_n, g | r_1, \cdots r_m, gw^{-1}\}$ be a presentation obtained from the presentation of $\pi_1(X)$ by the Tietze transformation T2. Now the Alexander matrix $M(P')$ obtained from the presentation P' is of the form

$$M(P') = \begin{pmatrix} M & 0 \\ * & 1 \end{pmatrix}.$$

For $\det(x_j - 1)^{\Phi} \ne 0$, the determinant of the matrix $M(P')_j^J$ consisting of the rows of $M(P')$ indicated by the Nn-tuples: $J = (i_1, \cdots, i_{Nn})$ $(1 \le i_1 < \cdots < i_{Nn} \le N(m+1))$ can be non-zero only if J is of the form

$$J = (i_1, \cdots, i_{N(n-1)}, Nm + 1, \cdots, N(m+1)),$$

and $\det M(P')_j^J = \det M_j^I$ for $I = (i_1, \cdots, i_{N(n-1)})$. Thus the twisted Alexander polynomial unchanged from the calculation of the matrix $M(P')$. $\square$

Note that a relation $u = v$ corresponds to the relator uv^{-1}. From $\partial(uv^{-1}) = \partial u - (uv^{-1})\partial v$, one has

$$(\frac{\partial uv^{-1}}{\partial x_i})^{\Phi} = (\frac{\partial u}{\partial x_i})^{\Phi} - (uv^{-1})^{\Phi}(\frac{\partial v}{\partial x_i})^{\Phi} = (\frac{\partial u}{\partial x_i})^{\Phi} - (\frac{\partial v}{\partial x_i})^{\Phi},$$

for $1 \leq i \leq n$. This shows that one may use $u - v$ instead of uv^{-1} for computing the twisted Alexander polynomials.

Example 3.9.6. Let $L = 3_1$ and $\pi_1(X) = \{x_1, x_2 | x_1 x_2 x_1 = x_2 x_1 x_2\}$. Note that $\phi(x_1) = \phi(x_2) = t$. Let $\rho : \pi_1(X) \to GL_2(\mathbb{Z}[s^{\pm 1}])$ be a 2-dimensional representation over the Laurent polynomial ring $\mathbb{Z}[s^{\pm 1}]$. Given the representation ρ as follows:

$$\rho(x_1) = \begin{pmatrix} -s & 1 \\ 0 & 1 \end{pmatrix}, \quad \rho(x_2) = \begin{pmatrix} 1 & 0 \\ s & -s \end{pmatrix}.$$

$$\left(\frac{\partial r}{\partial x_1}\right)^\Phi = \left(\frac{\partial}{\partial x_1}(x_1 x_2 x_1)\right)^\Phi - \left(\frac{\partial}{\partial x_1}(x_2 x_1 x_2)\right)^\Phi$$

$$= (1 + x_1 x_2)^\Phi - x_2^\Phi$$

$$= Id_{2\times 2} + \begin{pmatrix} -s & 1 \\ 0 & 1 \end{pmatrix} t \begin{pmatrix} 1 & 0 \\ s & -s \end{pmatrix} t - \begin{pmatrix} 1 & 0 \\ s & -s \end{pmatrix} t$$

$$= \begin{pmatrix} 1 - t & -st^2 \\ -st + st^2 & 1 + st - st^2 \end{pmatrix}.$$

Now we have $\det(x_1 - 1)^\Phi = \det \begin{pmatrix} -st - 1 & t \\ 0 & t - 1 \end{pmatrix} = (1 - t)(1 + st)$. Thus the corresponding Alexander polynomial is

$$\Delta_{3_1, \rho}(t) = \frac{\det(\frac{\partial r}{\partial x_1})^\Phi}{\det(x_1 - 1)^\Phi} = 1 - st^2.$$

Exercise 3.9.7. (1) Compute $\left(\frac{\partial r}{\partial x_2}\right)^\Phi$ for the trefoil knot 3_1 and the representation ρ given in the Example 3.9.6.

(2) Find the twisted Alexander matrix and $\frac{\det(\frac{\partial r}{\partial x_2})^\Phi}{\det(x_2 - 1)^\Phi}$.

Note that the twisted Alexander polynomial does not depend on the choice of the basis for the representation space V. If there is an automorphism $f : V \to V$ such that $\rho'(\gamma) = f \circ \rho(\gamma) \circ f^{-1}$ for all $\gamma \in \pi_1(X)$, then representations ρ and ρ' are equivalent up to conjugacy. It defines an equivalent relation $\rho \sim \rho'$.

Exercise 3.9.8. If $\rho_1 \sim \rho_2$, then $\Delta_{L,\rho_1} = \Delta_{L,\rho_2}$.

We give the following theorem from [Li and Xu, 2003].

Theorem 3.9.9. *There exist two inequivalent representations ρ_1 and ρ_2 such that*

$$\Delta_{T_{n,m}, \rho_1}(t) = \Delta_{T_{n,m}, \rho_2}(t).$$

Proof: For the torus knot group $T_{n,m} = \{a, b \,|\, a^n = b^m\}$ with two generators and one relation $r = a^n - b^m$, we have

$$\frac{\partial r}{\partial a} = \frac{a^n - 1}{a - 1} = a^{n-1} + a^{n-2} + \cdots + a + 1,$$

$$\frac{\partial r}{\partial b} = -\frac{b^m - 1}{b - 1} = -(b^{m-1} + b^{m-2} + \cdots + b + 1).$$

Let $\phi : T_{n,m} \to \mathbb{Z}\langle t \rangle$ be a homomorphism $\phi(a) = \phi(b) = t$ (therefore $t^n = t^m$ from the relation).

Let $\rho_1 : T_{n,m} \to SL_2(F_{2^s})$ be the trivial 2-dimensional representation as $\rho_1(a) = Id_{2\times 2}$ and $\rho_1(b) = Id_{2\times 2}$. The corresponding Alexander matrix is

$$\left(\left(\frac{\partial r}{\partial a}\right)^\Phi, \left(\frac{\partial r}{\partial b}\right)^\Phi \right) = \left(\frac{t^n - 1}{t - 1} Id_{2\times 2},\ -\frac{t^m - 1}{t - 1} Id_{2\times 2} \right).$$

Since $\det(1 - a)^\Phi = \det(1 - b)^\Phi = (1 - t)^2 \neq 0$, the twisted Alexander polynomial $\Delta_{T_{n,m},\rho_1}(t)$ for the torus knot group associated to the trivial $SL_2(F_{2^s})$ representation ρ_1 is

$$\Delta_{T_{n,m},\rho_1}(t) = \frac{(t^n - 1)^2}{(t - 1)^4}.$$

Similarly, one can get $\Delta_{T_{n,m},\rho_1}(t) = \frac{(t^m - 1)^2}{(t-1)^4}$ with noticing the relation $t^n = t^m$.

For both n and m odd, there is an irreducible (nontrivial) representation ρ_2 [Li and Xu, 2003, §3.2] such that

$$\rho_2(a) = \begin{pmatrix} 1 & 1 \\ 0 & 1 \end{pmatrix}, \quad \rho_2(b) = \begin{pmatrix} 1 & 1 \\ 0 & 1 \end{pmatrix}.$$

Then we have the following

$$\left(\frac{\partial r}{\partial a}\right)^\Phi = (a^{n-1} + a^{n-2} + \cdots + a + 1)^\Phi$$

$$= \rho_2(a)^{n-1} t^{n-1} + \rho_2(a)^{n-2} t^{n-2} + \cdots + \rho_2(a)t + Id_{2\times 2}$$

$$= \begin{pmatrix} t^{n-1} & 0 \\ 0 & t^{n-1} \end{pmatrix} + \begin{pmatrix} t^{n-2} & t^{n-2} \\ 0 & t^{n-2} \end{pmatrix} + \cdots + \begin{pmatrix} t & t \\ 0 & t \end{pmatrix} + Id_{2\times 2}$$

$$= \begin{pmatrix} \frac{t^n - 1}{t - 1} & t\frac{t^{n-1}-1}{t^2-1} \\ 0 & \frac{t^n - 1}{t-1} \end{pmatrix} \in M_2(F_{2^s}[t^{\pm 1}]).$$

Similarly we have $\left(\frac{\partial r}{\partial b}\right)^\Phi = \begin{pmatrix} \frac{t^m - 1}{t - 1} & t\frac{t^{m-1}-1}{t^2-1} \\ 0 & \frac{t^m - 1}{t-1} \end{pmatrix} \in M_2(F_{2^s}[t^{\pm 1}])$. For $1 \leq j \leq g = 2$, $(M_1, M_2) = \left(\left(\frac{\partial r}{\partial b}\right)^\Phi, \left(\frac{\partial r}{\partial a}\right)^\Phi \right)$ is the Alexander matrix of $T_{n,m}$

associated to the irreducible representation ρ_2. Since $r = g - 1$, $M_1^I = M_1$ and $M_2^I = M_2$, and $\det(1 - a)^\Phi = \det(1 - b)^\Phi = (1 - t)^2$ [Waldhausen, 1970, §4 and §5]. We have

$$\Delta_{T_{n,m},\rho_2} = \frac{(t^n - 1)^2}{(t - 1)^4}.$$

The result follows from the fact that ρ_1 is the 2-dimensional trivial representation and ρ_2 is an irreducible representation (see [Li and Xu, 2003] for more details). $\square$

Chapter 4

Jones polynomials

By studying representations of the braid group satisfying a certain quadratic relation, Jones [1987] found a polynomial invariant in two variables for oriented links. The Jones polynomial is expressed in terms of Ocneanu trace over the Hecke algebras of type A. Kauffman also found a very elementary way to obtain the one-variable polynomial. Another approach is the skein invariant studied by many mathematicians: P. Freyd, D. Yetter, J. Hoste, W. Lickorish, K. Millett, J. Prztycki and P. Traczyk. Finding a topological or geometric interpretation of the Jones polynomial remains one of the most important problems in the knot theory and low dimensional topology. We will follow Jones' original approach to study the invariant and its applications in [Harpe, Kervaire and Weber, 1986; Jones, 1987; Murasugi, 1987].

4.1 Hecke algebra

Generators $\sigma_i \in B_n$ are all conjugate, so that all 1-dimensional representations of a braid group B_n are classified by non-zero scalars. If one enforces to have all 2-dimensional representations for each generator in B_n, then there is much richer structure for each σ_i having exact 2-dimensional representation. Denote $-g_i$ for the image of σ_i under the 2-dimensional representation. Thus $-g_i$ must satisfy the equation $g_i^2 + ag_i + b = 0$ by the Cayley-Hamilton Theorem for a and b scalars. In terms of eigenvalues of $-g_i$, we have

$$(-g_i - \lambda_1)(-g_i - \lambda_2) = (-g_i)^2 + ag_i + b,$$

$$a = -Tr(-g_i) = \lambda_1 + \lambda_2, \quad b = \det(g_i) = \lambda_1 \cdot \lambda_2.$$

If we normalize $\lambda_1 = 1$, we have $\lambda_2 = b = a - 1$, i.e., $a - b = 1$ is the relation for g_i with an eigenvalue one. Pick $b = -q$ (q is a scalar). Then $a = -q + 1$.

The quadratic representation of σ_i can be expressed into, after recall $-g_i$ to be g_i,

$$g_i^2 - (q-1)g_i - q = 0 \Leftrightarrow g_i^2 = (q-1)g_i + q.$$

Let K be a field and $q \in K$ be some element of K.

Definition 4.1.1. The Hecke algebra $H_n(q)$ over K is the associated K-algebra with unit 1 and generated by $g_1, \cdots, g_{n-1}$ subject to the following relations:

$$\begin{cases} g_i g_j = g_j g_i & |i - j| \geq 2 \\ g_i g_{i+1} g_i = g_{i+1} g_i g_{i+1} & i = 1, 2, \cdots, n-2 \\ g_i^2 = (q-1)g_i + q & i = 1, 2, \cdots, n-1. \end{cases} \quad (4.1)$$

Remark 4.1.2.

(1) As we explained before, the information of B_n with 2-dimensional representation for each generator is completely equivalent to the information of the Hecke algebra $H_n(q)$. Also we observe that $B_n \subset H_n(q)$ as embedded into the K-algebra.

(2) For $q = 1$, we have $H_n(1) = \mathbb{C}\Sigma_n$. This corresponds to g_i with eigenvalues ± 1. As group algebra, $H_n(1) = \Sigma_n$ is semisimple. The semisimplicity is an open condition, hence $H_n(q)$ is also semisimple if $\|q - 1\| < \varepsilon$ for sufficiently small $\varepsilon > 0$ [Jones, 1987, Theorem 4.5].

(3) There is a natural embedding as K-algebra $H_n(q) \hookrightarrow H_{n+1}(q)$. In fact, we will show that the natural embedding makes H_{n+1} a $(H_n(q), H_n(q))$-bimodule.

(4) Wenzl [1985] has written explicitly and demonstrably on irreducible representations of $H_n(q)$ for each Young diagram. By the formulae of [Wenzl, 1985], one can make the precise meaning of "sufficiently close to 1" to invert q and $1 - q^p$ for the appropriate p. With $q \neq 0$ or q not a root of unity, the simple $H_n(q)$ modules (or quadratic irreducible representations of B_n) are in one-to-one correspondence with Young diagrams. Their decomposition rule and hence their dimensions are the same as for the symmetric group S_n [Jones, 1987, Theorem 4.5].

The Hecke algebra $H_n(q)$ is of type A_{n-1} since the defining relation fits into the Coxeter-Dynkin picture, as arises from the representations of $GL(n, q)$, where they define the centralizer of the natural representation on the set of flags. Other types of Hecke algebras exist, and it would be interesting to see if the similar Jones polynomial can be constructed from other types of Hecke algebras.

Note that $H_n(q) \oplus (H_n(q) \otimes_{H_{n-1}(q)} H_n(q))$ is a $(H_n(q), H_n(q))$-bimodule. We can define a natural map

$$\phi : H_n(q) \oplus (H_n(q) \otimes_{H_{n-1}(q)} H_n(q)) \to H_{n+1}(q)$$

by $a + \sum_i b_i \otimes c_i \mapsto a + \sum_i b_i g_n c_i$.

Lemma 4.1.3. ϕ *is well-defined K-algebra homomorphism.*

Proof: If $u \in H_{n-1}(q)$, then we have

$$\phi(bu \otimes c) = bug_n c, \quad \phi(b \otimes uc) = bg_n uc.$$

By Definition 4.1.1, u is a K-linear combination of monomials in $g_1, \cdots, g_{n-2}$ which commute with g_n in $H_{n+1}(q)$ ($ug_n = g_n u$). Hence we have $\phi(bu \otimes c) = \phi(b \otimes uc)$. Thus ϕ is well-defined. $\qquad \square$

Lemma 4.1.4. ϕ *is a K-algebra epimorphism.*

Proof: By induction on n, the result is true for $n = 1$ with $H_0(q)$ the trivial K-algebra. Assume that the result is true for $n - 1$. Let g be a monomial in $g_1, g_2, \cdots, g_n$ in $H_{n+1}(q)$. If g does not contain g_n, then g is in the image of ϕ by the induction. If g contains only one g_n as $g = bg_n c$, then $g = \phi(b \otimes c)$. If g contains g_n at least twice in its expression $g = h_1 g_n h_2 g_n h_3$, then $h_i (i = 1, 2, 3)$ is monomial in $g_1, \cdots, g_{n-1}$. If h_2 does not contain g_{n-1}, then we have

$$\begin{aligned}
g &= h_1 g_n h_2 g_n h_3 \\
&= h_1 h_2 g_n^2 h_3 \\
&= h_1 h_2 ((q - 1) g_n + q) h_3 \\
&= (q - 1) h_1 h_2 g_n h_3 + q h_1 h_2 h_3.
\end{aligned}$$

So $g = \phi(q h_1 h_2 h_3 + (q - 1) h_1 h_2 \otimes h_3)$. If h_2 contains g_{n-1} exactly once, then $h_2 = h_4 g_{n-1} h_5$, and

$$\begin{aligned}
g &= h_1 g_n (h_4 g_{n-1} h_5) g_n h_3 \\
&= h_1 h_4 (g_n g_{n-1} g_n) h_5 h_3 \\
&= h_1 h_4 g_{n-1} g_n g_{n-1} h_5 h_3.
\end{aligned}$$

Again we have $g = \phi(h_1 h_4 g_{n-1} \otimes g_{n-1} h_5 h_3)$. For g with more than two g_n, one can inductively apply the previous argument. Hence ϕ is surjective.

Every element in $H_{n+1}(q)$ is a sum of $a + \sum_i b_i g_n c_i$ as an image of ϕ. Combining Lemma 4.1.3, we have ϕ is a K-algebra epimorphism. $\square$

Lemma 4.1.4 shows that $H_{n+1}(q)$ is generated as a vector space over K by the monomials with at most one occurrence of g_n.

Lemma 4.1.5. *As a K-algebra, $\dim_K H_{n+1}(q) \leq (n+1)!$.*

Proof: Define a system of linearly independent monomials
$$S_i = \{1, g_i, g_i g_{i-1}, \cdots, g_i g_{i-1} \cdots g_1\}.$$
Note that $v \in S_i$ implies $g_{i+1} v \in S_{i+1}$. $S_i (i = 1, 2, \cdots, n)$ is a basis for $H_{n+1}(q)$. Again we prove the result by induction on n. Assume the result is true for $H_n(q)$.

As $H_{n+1}(q)$ is generated over K by monomials $a \in H_n(q)$, $bg_n c$ with $b, c \in H_n(q)$. For $a \in H_n(q)$, the basis for $H_n(q)$ is also linearly independent over K for $H_{n+1}(q)$. By induction hypothesis, c is a K-linear combination of monomials of the form $v_i \in S_i (i = 1, \cdots, n-1)$. So we get
$$bg_n c = bg_n v_1 v_2 \cdots v_{n-1} = b v_1 v_2 \cdots v_{n-2} g_n v_{n-1}.$$
But $g_n v_{n-1} = u_n \in S_n$. By induction again, $b v_1 v_2 \cdots v_{n-2}$ is a K-linear combination of monomials of the form $u_1, u_2, \cdots, u_{n-1}$ with $u_i \in S_i$. Thus $bg_n c$ is a K-linear combination of monomials $u_1, \cdots, u_n$. Thus
$$\dim_K H_{n+1}(q) \leq |S_n| \cdot \dim_K H_n(q) \leq (n+1) \cdot n!,$$
where $|S_n|$ is the cardinality of the set S_n. $\square$

Let Σ_{n+1} be the symmetric group on $\{1, \cdots, n+1\}$ generated by $s_1, \cdots, s_n$. Let $l : \Sigma_{n+1} \to \mathbb{N}$ be the word length in Σ_{n+1} with respect to the generators $\{s_1, s_2, \cdots, s_n\}$. Define $L_i \in End_K(K\Sigma_{n+1})$ by
$$L_i(\pi) = \begin{cases} s_i \pi & l(s_i \pi) > l(\pi) \\ q s_i \pi + (q-1)\pi & l(s_i \pi) < l(\pi) \end{cases}$$
for every $\pi \in \Sigma_{n+1}$.

Lemma 4.1.6. *The following identity holds: $L_i^2 = (q-1)L_i + q$ for $1 \leq i \leq n$.*

Proof: For $\pi \in \Sigma_{n+1}$, (1) if $l(s_i \pi) > l(\pi)$, then from $s_i^2 \pi = \pi$ and $l(s_i(s_i \pi)) < l(s_i \pi)$, we have
$$\begin{aligned} L_i^2(\pi) &= L_i(L_i(\pi)) \\ &= L_i(s_i \pi) \\ &= q s_i(s_i \pi) + (q-1)(s_i \pi) \\ &= (q-1)L_i(\pi) + q\pi \\ &= ((q-1)L_i + q)(\pi). \end{aligned}$$

(2) if $l(s_i\pi) < l(\pi)$, then from $l(s_i^2\pi) > l(s_i\pi)$, we have

$$\begin{aligned}
L_i^2(\pi) &= L_i(L_i(\pi)) \\
&= L_i(qs_i\pi + (q-1)\pi) \\
&= qs_i^2\pi + (q-1)L_i(\pi) \\
&= ((q-1)L_i + q)(\pi).
\end{aligned}$$

□

Lemma 4.1.7. *Define $R_j \in End_K(K\Sigma_{n+1})$ by*

$$R_j(\pi) = \begin{cases} \pi s_j & l(\pi s_j) > l(\pi) \\ q\pi s_j + (q-1)\pi & l(\pi s_j) < l(\pi) \end{cases}$$

Then we have $L_iR_j = R_jL_i$ for all $i,j \in \{1,\cdots,n\}$.

Proof: Let $i,j \in \{1,\cdots,n\}$ and $\pi \in \Sigma_{n+1}$. The verification of the identity $L_iR_j(\pi) = R_jL_i(\pi)$ reduces into six cases.

$$l(s_i\pi s_j) = l(\pi) + 2, \tag{4.2}$$

$$l(s_i\pi s_j) = l(\pi) - 2, \tag{4.3}$$

$$l(s_i\pi s_j) = l(\pi), \tag{4.4}$$

$$l(s_i\pi) = l(\pi) + (\pm 1) = l(\pi s_j), \tag{4.5}$$

$$l(s_i\pi) = l(\pi) + 1, l(\pi s_j) = l(\pi) - 1, \tag{4.6}$$

$$l(s_i\pi) = l(\pi) - 1, l(\pi s_j) = l(\pi) + 1. \tag{4.7}$$

In (4.5), when $l(s_i\pi) = l(\pi s_j) = l(\pi) + \pm 1$, then identities applied to Σ_{n+1} viewed as a Coxeter group imply that in these cases we have $s_i\pi = \pi s_j$. This is the desired property to complete the identity. We leave the rest of the cases as Exercises for the reader. □

Lemma 4.1.8. *The identities hold:*

$$\begin{cases} L_iL_j = L_jL_i & |i-j| \geq 2, \\ L_iL_{i+1}L_i = L_{i+1}L_iL_{i+1} & 1 \leq i \leq n-2. \end{cases}$$

Proof: Let $\pi \in \Sigma_{n+1}$ and write it as $s_{i_1} s_{i_2} \cdots s_{i_r}$ with $r = l(\pi)$ in a reduced form. Thus $\pi = R_{i_r} \cdots R_{i_2} R_{i_1}(1)$.

$$
\begin{aligned}
L_i L_j(\pi) &= L_i L_j (R_{i_r} \cdots R_{i_2} R_{i_1}(1)) \\
&= R_{i_r} \cdots R_{i_2} R_{i_1} L_i L_j(1) \\
&= R_{i_r} \cdots R_{i_2} R_{i_1}(s_i s_j) \\
&= R_{i_r} \cdots R_{i_2} R_{i_1}(s_j s_i),
\end{aligned}
$$

where the second identity follows from Lemma 4.1.6 and the third from $|i - j| \geq 2$. Then we obtain

$$
\begin{aligned}
L_j L_i(\pi) &= L_j L_i (R_{i_r} \cdots R_{i_2} R_{i_1}(1)) \\
&= R_{i_r} \cdots R_{i_2} R_{i_1} (L_j L_i(1)) \\
&= R_{i_r} \cdots R_{i_2} R_{i_1}(s_j s_i) \\
&= L_i L_j(\pi).
\end{aligned}
$$

$$
\begin{aligned}
L_i L_{i+1} L_i(\pi) &= L_i L_{i+1} L_i (R_{i_r} \cdots R_{i_2} R_{i_1}(1)) \\
&= R_{i_r} \cdots R_{i_2} R_{i_1} (L_i L_{i+1} L_i(1)) \\
&= R_{i_r} \cdots R_{i_2} R_{i_1}(s_i s_{i+1} s_i) \\
&= R_{i_r} \cdots R_{i_2} R_{i_1}(s_{i+1} s_i s_{i+1}) \\
&= R_{i_r} \cdots R_{i_2} R_{i_1} (L_{i+1} L_i L_{i+1}(1)) \\
&= L_{i+1} L_i L_{i+1} (R_{i_r} \cdots R_{i_2} R_{i_1}(1)) \\
&= L_{i+1} L_i L_{i+1}(\pi).
\end{aligned}
$$

$\square$

Thus there is an algebra map $L : H_{n+1}(q) \to End_K(K\Sigma_{n+1})$ such that $L(g_i) = L_i$ is well-defined and preserving the algebraic relations from Lemma 4.1.6, Lemma 4.1.7 and Lemma 4.1.8.

Proposition 4.1.9. *The K-algebra epimorphism ϕ is an isomorphism, and* $\dim_K H_{n+1}(q) = (n+1)!$.

Proof: Lemma 4.1.5 shows that $H_n(q) \otimes_{H_{n-1}(q)} H_n(q)$ is spanned over K by the subspaces $H_n(q) \otimes u_{n-1}$ with $u_{n-1} \in S_{n-1}$. The monomials $u_1, \cdots, u_{n-1}, u_n$ with $u_i \in S_i$ are also K-linear independent. If $U_i = g_i g_{i-1} \cdots g_{i-j}$, then $L(U_i) = w_i = s_i s_{i-1} \cdots s_{i-j}$. Any of the $(n+1)!$ elements of Σ_{n+1} is of the form $w_1, w_2, \cdots, w_n$ so that they are K-linearly independent in $K\Sigma_{n+1}$. Now the map L defined $x \mapsto L(x)(1)$ is K-linear, this shows that the set $\{U_1, U_2, \cdots, U_n\}$ is K-linearly independent. Hence $\dim_K H_{n+1}(q) = (n+1)!$, so is $\dim_K \{H_n(q) \oplus (H_n(q) \otimes_{H_{n-1}(q)} H_n(q))\} = (n+1)!$. This is equivalent to the isomorphism property of ϕ by the dimension counting. $\square$

4.2   Ocneanu trace on $H_n(q)$

The fundamental ideal of Jones polynomials is the construction of the Ocneanu trace on $H_n(q)$ with invariance under the Markov moves.

Theorem 4.2.1. *There exists a trace $Tr : H_n(q) \to K$ compatible with the embedding:*

$$
\begin{array}{ccc}
H_n(q) & \longrightarrow & H_{n+1}(q) \\
\Big\downarrow {\scriptstyle Tr} & & \Big\downarrow {\scriptstyle Tr} \\
K & \xrightarrow{\ =\ } & K,
\end{array}
$$

such that

  (1) $Tr(1) = 1$,
  (2) Tr is K-linear and $Tr(ab) = (ba)$,
  (3) If $a, b \in H_n(q)$, then $Tr(ag_nb) = zTr(ab)$ for $z \in K$.

Proof: By Proposition 4.1.9, we identify $H_{n+1}(q)$ by the isomorphism

$$\phi : H_n(q) \oplus (H_n(q) \otimes_{H_{n-1}(q)} H_n(q)) \to H_{n+1}(q).$$

The K-linear map $Tr : H_{n+1}(q) \to K$ is defined by induction on n.

$$Tr(x) = Tr(\phi(a + \sum_i b_i \otimes c_i)) = Tr(a) + \sum_i zTr(b_ic_i).$$

If $a, b \in H_n(q)$, $\phi(a \otimes b) = ag_nb$, then

$$Tr(ag_nb) = Tr(\phi(a \otimes b)) = zTr(ab).$$

(3) follows. By induction, the commutative diagram follows by the definition of Tr. (1) is trivial. Hence all we need to check is the property (2).

By applying this to $S = H_n(q) \cup \{g_n\}$, the only case which (2) does not follow trivially from the definition of Tr is $Tr(g_nxg_ny) = Tr(xg_nyg_n)$ for $x, y \in H_n(q)$.

  (i) If $x \in H_{n-1}(q), y \in H_{n-1}(q)$, then

$$g_nxg_ny = xg_ng_ny = xg_nyg_n.$$

  (ii) If $x = ag_{n-1}b$ and a, b and $y \in H_{n-1}(q)$, then

$$
\begin{aligned}
Tr(g_nxg_ny) &= Tr(g_nag_{n-1}bg_ny) \\
&= Tr(a(g_ng_{n-1}g_n)by) \\
&= Tr(ag_{n-1}g_ng_{n-1}by) \\
&= zTr(ag_{n-1}^2by) \\
&= z(q-1)Tr(ag_{n-1}by) + qzTr(aby)
\end{aligned}
$$

$$Tr(xg_nyg_n) = Tr(ag_{n-1}bg_nyg_n)$$
$$= Tr(ag_{n-1}bg_n^2y)$$
$$= Tr(ag_{n-1}b(q-1)g_ny + ag_{n-1}bqy)$$
$$= (q-1)zTr(ag_{n-1}by) + qTr(ag_{n-1}by)$$
$$= z(q-1)Tr(ag_{n-1}by) + qzTr(aby)$$

(iii) If $x \in H_{n-1}(q)$ and $y = cg_{n-1}d$ for $c, d \in H_{n-1}(q)$, then we have $Tr(g_nxg_ny) = Tr(xg_nyg_n)$ via the same method in (ii).

(iv) If $x = ag_{n-1}b$ and $y = cg_{n-1}d$ with $a, b, c, d \in H_{n-1}(q)$, then we have

$$Tr(g_nxg_ny) = Tr((g_nag_{n-1}b)g_n(cg_{n-1}d))$$
$$= Tr((ag_ng_{n-1}g_n)(bcg_{n-1}d))$$
$$= Tr(ag_{n-1}g_n(g_{n-1}bcg_{n-1}d))$$
$$= zTr(ag_{n-1}(g_{n-1}bcg_{n-1}d))$$
$$= zTr(a(q-1)g_{n-1}bcg_{n-1}d) + qzTr(abcg_{n-1}d)$$
$$= (q-1)zTr(ag_{n-1}bcg_{n-1}d) + qz^2Tr(abcd).$$

$$Tr(xg_nyg_n) = Tr(ag_{n-1}bg_ncg_{n-1}dg_n)$$
$$= Tr(ag_{n-1}bcg_ng_{n-1}g_nd)$$
$$= Tr((ag_{n-1}bcg_{n-1})g_n(g_{n-1}d))$$
$$= zTr((ag_{n-1}bcg_{n-1})(g_{n-1}d))$$
$$= zTr(ag_{n-1}bc(q-1)g_{n-1}d) + zTr(ag_{n-1}bcqd)$$
$$= (q-1)zTr(ag_{n-1}bcg_{n-1}d) + qz^2Tr(abcd).$$

□

This is a result of Ocneanu [Jones, 1987, Theorem 5.1]. Hence for every $z \in \mathbb{C}$ there is a linear trace functional on $\cup_{n=1}^{\infty} H_n(q)$ uniquely defined by Theorem 4.2.1.

Example 4.2.2. We compute the trace for g_1^3.

$$Tr(g_1^3) = Tr(g_1((q-1)g_1 + q))$$
$$= (q-1)Tr(g_1^2) + qTr(g_1)$$
$$= (q-1)Tr((q-1)g_1 + q) + qz$$
$$= (q-1)^2z + (q-1)q + qz$$
$$= (q^2 - q + 1)z + (q-1)q$$

Another method to evaluate the Ocneanu trace is to restrict the trace Tr on the Hecke algebra through the direct product of matrix algebras indexed by the Young diagram. The Hecke algebra is a direct product of matrix algebras, and the restrictions to these matrix algebras and a trace on a matrix algebra is a multiple of the usual trace. If the scaling factors (weights) associated with each diagram can be determined, then the calculation of the trace of any element of the Hecke algebra is by decomposing it into the matrix algebras and taking the weighted sum of the usual matrix traces. The weights have been determined by Ocneanu.

4.3   Two-variable Jones polynomial

Using Theorem 4.2.1, we define a link polynomial for braids which is invariant under the Markov moves.

Definition 4.3.1. For $\alpha \in B_n$ and $\rho : B_n \to H_n(q)$ $(\sigma_i \mapsto g_i)$, the Jones polynomial of two variables (q, z) is defined by

$$V_\alpha(q, z) = (\frac{1}{z})^{(n(\alpha)+e(\alpha)-1)/2} \cdot (\frac{q}{w})^{(n(\alpha)-e(\alpha)-1)/2} \cdot Tr(\rho(\alpha)),$$

where $w = 1 - q + z$ and $e(\alpha)$ is the exponent sum of α.

Theorem 4.3.2. *The two-variable Jones polynomial $V_\alpha(q, z)$ is an invariant of the link $\overline{\alpha}$ given by the closure of the braid α.*

Proof: We have to verify that $V_\alpha(q, z)$ is invariant under the Markov moves by Theorem 2.6.4.

Markov move I: $\alpha \in B_n$, $\gamma \in B_n$, $\beta = \gamma\alpha\gamma^{-1}$. We have $n(\alpha) = n(\beta) = n$ both in B_n and $e(\beta) = e(\gamma) + e(\alpha) + e(\gamma^{-1}) = e(\alpha)$.

$$Tr(\rho(\beta)) = Tr(\rho(\gamma)\rho(\alpha)\rho(\gamma)^{-1}) = Tr(\rho(\alpha)).$$

Hence we have $V_\alpha(q, z) = V_\beta(q, z)$

Markov move II: $\beta = \alpha \cdot \sigma_n \in B_{n+1}$. Then we have $n(\beta) = n(\alpha) + 1$ and $e(\beta) = e(\alpha) + 1$. So

$$V_\beta(q, z) = (\frac{1}{z})^{(n(\beta)+e(\beta)-1)/2} \cdot (\frac{q}{w})^{(n(\beta)-e(\beta)-1)/2} \cdot Tr(\rho(\beta))$$

$$= (\frac{1}{z})^{(n(a)+1+e(\alpha)+1-1)/2} \cdot (\frac{q}{w})^{(n(a)+1-e(\alpha)-1-1)/2} \cdot Tr(\rho(\alpha)g_n)$$

$$= (\frac{1}{z})^{(n(\alpha)+e(\alpha)-1)/2} \cdot \frac{1}{z} \cdot (\frac{q}{w})^{(n(a)-e(\alpha)-1)/2} \cdot zTr(\rho(\alpha))$$

$$= V_\alpha(q, z).$$

Markov move II^{-1}: $\beta = \alpha \cdot \sigma_n^{-1} \in B_{n+1}$. Then we have $n(\beta) = n(\alpha) + 1$ and $e(\beta) = e(\alpha) - 1$. So

$$V_\beta(q, z) = (\frac{1}{z})^{(n(\beta)+e(\beta)-1)/2} \cdot (\frac{q}{w})^{(n(\beta)-e(\beta)-1)/2} \cdot Tr(\rho(\beta))$$

$$= (\frac{1}{z})^{(n(\alpha)+e(\alpha)-1)/2} \cdot (\frac{q}{w})^{(n(a)-e(\alpha)-1)/2} \cdot \frac{q}{w} Tr(\rho(\alpha)g_n^{-1})$$

$$= V_\alpha(q, z).$$

The last equality follows from the following calculation. $g_n^{-1} = \frac{1}{q}(1 - q + g_n)$. Thus

$$Tr(\rho(\alpha)g_n^{-1}) = Tr(\rho(\alpha)\frac{1}{q}(1 - q + g_n))$$

$$= \frac{1}{q}((1 - q)Tr(\rho(\alpha)) + zTr(\rho(\alpha)))$$

$$= \frac{1 - q + z}{q}Tr(\rho(\alpha))$$

$$= \frac{w}{q}Tr(\rho(\alpha)).$$

Hence the result follows. $\qquad\square$

Using changing variables $z = \frac{(q-1)}{1-\lambda q}$, one can see that

$$V_\alpha(q, z) = X_\alpha(q, \lambda) = (-\frac{1 - \lambda q}{\sqrt{\lambda}(1 - q)})^{n(\alpha)-1}(\sqrt{\lambda})^{e(\alpha)} Tr(\rho(\alpha)), \qquad (4.8)$$

as defined in [Jones, 1987, Definition 6.1].

Example 4.3.3. The right-handed trefoil knot $K = 3_1 = \overline{\sigma_1^3}$. That is $\alpha = \sigma_1^3 \in B_2$. So $n(\alpha) = 2$, $e(\alpha) = 3$ and $\rho(\alpha) = g_1^3$. By Example 4.2.2, we have

$$V_{\sigma_1^3}(q, z) = X_{3_1}(q, z)$$

$$= (-\frac{1 - \lambda q}{\sqrt{\lambda}(1 - q)})^{2-1}(\sqrt{\lambda})^3 Tr(g_1^3)$$

$$= -\frac{1 - \lambda q}{(1 - q)}\lambda((q^2 - q + 1)z + (q - 1)q)$$

$$= -\lambda\frac{1 - \lambda q}{(1 - q)}((q^2 - q + 1)\frac{q - 1}{1 - \lambda q} + (q - 1)q)$$

$$= -\lambda(-q^2 + q - 1 + (\lambda q - 1)q)$$

$$= \lambda q(q + q^{-1} - \lambda q)$$

$$= (\lambda q)((q^{1/2} - q^{-1/2})^2 + 2 - (\lambda q)).$$

If we set $t = \sqrt{\lambda q}$ and $x = q^{1/2} - q^{-1/2}$, then

$$X_{3_1}(q, z) = V_{\sigma_1^3}(t, x) = t^2(x^2 + 2 - t^2).$$

Example 4.3.4. The figure eight knot $K = 4_1 = \overline{\sigma_1\sigma_2^{-1}\sigma_1\sigma_2^{-1}}$. That is $\alpha = \sigma_1\sigma_2^{-1}\sigma_1\sigma_2^{-1} \in B_3$. So $n(\alpha) = 3$, $e(\alpha) = 0$ and $\rho(\alpha) = g_1 g_2^{-1} g_1 g_2^{-1}$. By using $g_2^{-1} = \frac{1}{q}(1 - q + g_2)$, we have,

$$g_1 g_2^{-1} g_1 g_2^{-1} = \frac{1}{q^2} g_1(1 - q + g_2) g_1(1 - q + g_2)$$

$$= \frac{1}{q^2}((1-q)^2 g_1^2 + (1-q)g_1^2 g_2 + (1-q)g_1 g_2 g_1 + g_1 g_2 g_1 g_2),$$

and $Tr(g_1^2 g_2) = zTr(g_1^2)$, $Tr(g_1 g_2 g_1) = Tr(g_1^2 g_2) = zTr(g_1^2)$ and

$$Tr(g_1 g_2 g_1 g_2) = Tr(g_1 g_1 g_2 g_1) = zTr(g_1^3).$$

$$Tr(g_1 g_2^{-1} g_1 g_2^{-1}) = \frac{1}{q^2}((1-q)^2 Tr(g_1^2) + 2(1-q)zTr(g_1^2) + zTr(g_1^3))$$

$$= \frac{1}{q^2}((1-q)(1 - q + 2z)Tr(g_1^2) + zTr(g_1^3))$$

$$= \frac{1}{q^2}((1-q)(1 - q + 2z)(q + (q - 1)z)$$

$$+ z((q^2 - q + 1)z + (q - 1)q)),$$

where $Tr(g_1^2) = q + (q-1)z$ from $g_1^2 = q + (q-1)g_1$ and $Tr(g_1^3)$ follows from Example 4.2.2. Substitute $z = (q - 1)/(1 - \lambda q)$, we obtain the following

$$(\frac{1 - \lambda q}{1 - q})^2 (1 - q)(1 - q + 2z)Tr(g_1^2)$$

$$= \frac{(1 - \lambda q)^2}{1 - q}(1 - q + 2z)(q + (q - 1)z)$$

$$= \frac{(1 - \lambda q)^2}{1 - q} \frac{-(1 + \lambda q)}{1 - \lambda q} \frac{q - \lambda q^2 + (q - 1)^2}{1 - \lambda q}$$

$$= -(1 + \lambda q)(q^2 - q + 1 - \lambda q^2).$$

$$(\frac{1 - \lambda q}{1 - q})^2 zTr(g_1^3) = -\frac{1 - \lambda q}{1 - q}((q^2 - q + 1)z + (q - 1)q)$$

$$= -\frac{1 - \lambda q}{1 - q} \frac{(q - 1)(q^2 + 1 - \lambda q^2)}{1 - \lambda q}$$

$$= q^2 + 1 - \lambda q^2.$$

Now we have our two-variable Jones polynomial of the figure eight knot.

$$
\begin{aligned}
X_{4_1}(q, \lambda) &= (-\frac{1 - \lambda q}{\sqrt{\lambda}(1 - q)})^{3-1}(\sqrt{\lambda})^0 Tr(\rho(\alpha)) \\
&= \frac{(1 - \lambda q)^2}{\lambda(1 - q)^2} Tr(g_1 g_2^{-1} g_1 g_2^{-1}) \\
&= \frac{1}{\lambda q^2} \cdot (\frac{1 - \lambda q}{1 - q})^2 ((1 - q)(1 - q + 2z) Tr(g_1^2) + z Tr(g_1^3)) \\
&= \frac{1}{\lambda q^2} \cdot (-(1 + \lambda q)(q^2 - q + 1 - \lambda q^2) + q^2 + 1 - \lambda q^2) \\
&= \frac{1}{\lambda q^2}(\lambda^2 q^3 - \lambda(q^3 - q^2 + q) + q) \\
&= \frac{1}{\lambda q}(\lambda^2 q^2 - \lambda q(q - 1 + q^{-1}) + 1) \\
&= \frac{1}{t^2}(t^4 - t^2(x^2 + 1) + 1) \\
&= (t^2 - 1 + t^{-2}) - x^2.
\end{aligned}
$$

Note that $q - 1 + q^{-1} = (q^{1/2} - q^{-1/2})^2 + 1 = x^2 + 1$.

Exercise 4.3.5. (1) Compute $X_{5_1}(q, z)$, where $K = 5_1 = \overline{\sigma_1^5}$.
(2) Compute $X_{6_1}(q, z)$, where $6_1 = \overline{\sigma_1^2 \sigma_2^2 \sigma_1^{-1} \sigma_2}$.

The n-component unlink is given by the closure of the trivial braid $1 \in B_n$. Hence $\alpha = 1$, $n(\alpha) = n$, $e(\alpha) = 0$ and $Tr(\rho(\alpha)) = 1$.

$$
\begin{aligned}
X_{\overline{1}}(q, \lambda) &= (-\frac{1 - \lambda q}{\sqrt{\lambda}(1 - q)})^{n-1}(\sqrt{\lambda})^0 Tr(\rho(\alpha)) \\
&= (-\frac{1 - \lambda q}{\sqrt{\lambda q}(q^{-1/2} - q^{1/2})})^{n-1} \\
&= (\frac{1 - t^2}{tx})^{n-1} = (\frac{t^{-1} - t}{x})^{n-1}.
\end{aligned}
$$

Example 4.3.6. The Hopf link $L = 2_1 = \overline{\alpha}$ with $\alpha = \sigma_1^2$. So $n(\alpha) = 2$, $e(\alpha) = 2$ and $Tr(\rho(\alpha)) = Tr(g_1^2)$. Thus the two-variable Jones polynomial of the Hopf link is

$$
\begin{aligned}
X_L(q,\lambda) &= (\frac{1-\lambda q}{\sqrt{\lambda}(1-q)})^{2-1}(\sqrt{\lambda})^2 Tr(g_1^2) \\
&= -\frac{(1-\lambda q)\sqrt{\lambda}}{1-q}(q+(q-1)z) \\
&= -\frac{\sqrt{\lambda}}{1-q}(q-\lambda q^2+(q-1)^2) \\
&= \frac{\sqrt{\lambda q}}{(q^{1/2}-q^{-1/2})}(-\lambda q+q-1+q^{-1}) \\
&= \frac{t}{x}(x^2-t^2+1).
\end{aligned}
$$

Remark 4.3.7. Let $f : \cup_{n=1}^{\infty}B_n \to \mathbb{C}$ be a composition of the quotient map $\cup_{n=1}^{\infty}B_n \to \cup_{n=1}^{\infty}B_n/\sim$ and $\phi : \cup_{n=1}^{\infty}B_n/\sim \to \mathbb{C}$, where $\cup_{n=1}^{\infty}B_n/\sim$ is the set of equivalence classes under Markov moves. Let $\rho_i : \mathbb{C}B_n \to M_{n_i \times n_i}(\mathbb{C})$ be a group ring representation such that

$$
f_n(\beta) = \sum_{i=1}^{r_n} a_{n,i} Tr(\rho_i(\beta)),
$$

for each element $\beta \in B_n$ and n. Let $\mathcal{C}_n(\phi)$ be the algebra $\oplus_{i=1}^{r_n}\rho_i(\mathbb{C}B_n)$ and

$$
v_n = \sum_{i=1}^{r_n} a_{n,i}\rho_i : B_n \to \mathcal{C}_n(\phi).
$$

The link invariant ϕ is of trace type if there exists an algebra homomorphism $j_n : \mathcal{C}_n(\phi) \to \mathcal{C}_{n+1}(\phi)$ such that

$$
\begin{array}{ccc}
B_n & \xrightarrow{\;v_n\;} & \mathcal{C}_n(\phi) \\
\downarrow{\scriptstyle i_n} & & \downarrow{\scriptstyle j_n} \\
B_{n+1} & \xrightarrow{\;v_{n+1}\;} & \mathcal{C}_{n+1}(\phi).
\end{array}
$$

Thus Jones polynomial is of trace type with algebra $\mathcal{C}_n(\phi) = H_n(q)$; the Kauffman polynomial is of trace type with algebra $\mathcal{C}_n(\phi)$ as a q-analogue of Brauer algebra.

If L_1 and L_2 are oriented links, then $X_{L_1 \# L_2}(q,\lambda) = X_{L_1}(q,\lambda) \cdot X_{L_2}(q,\lambda)$, where $L_1 \# L_2$ is the connected sum of the two links. The result follows from the braid expression of the connected sum and the independence of trace. If $L_1 = \overline{\alpha}_1, \alpha_1 \in B_n$ and $L_2 = \overline{\alpha}_2, \alpha_2 \in B_m$, then $L_1 \# L_2 = \alpha_1 \sum^{n-1}(\alpha_2), \alpha_1 \sum^{n-1}(\alpha_2) \in B_{n+m}$ with $\sum(\sigma_i) = \sigma_{i+1}$. If w_1

is a word on $1, g_1, \cdots, g_n$ and w_2 is a word on $1, g_{n+1}, \cdots, g_{n+m}$, then $tr(w_1 w_2) = tr(w_1) \cdot tr(w_2)$. This property of connected sum will be proved from skein relations in the next section.

If L is an oriented link, let $\tilde{L}$ be the oriented link obtained by viewing L in a mirror (mirror image of L by reversing all the crossings in the same projection of L). If $\alpha \in B_n$, then the mirror image of $\overline{\alpha}$ is $\theta(\alpha^{-1}) \in B_n$.

Proposition 4.3.8.

$$X_{\tilde{L}}(q, \lambda) = (-\frac{1 - \lambda q}{\sqrt{\lambda}(1 - q)})^{n(\alpha)-1}(\sqrt{\lambda})^{-e(\alpha)} Tr(\rho(\alpha^{-1})) = X_L(1/q, 1/\lambda),$$

where $n(\alpha) = n(\alpha^{-1})$ and $e(\alpha^{-1}) = -e(\alpha)$.

Proof: Let $f(q, \lambda) = Tr(\rho(\alpha))$ and express $\rho(\alpha)$ as a product of each g_i written $(q + 1)e_i - 1$, where $e_i = \frac{1+g_i}{1+q}$. By the definition of the Hecke algebra $H_n(q)$, we have

$$
\begin{aligned}
e_i^2 - e_i &= e_i(e_i - 1) \\
&= \frac{1 + g_i}{1 + q} \cdot \frac{g_i - q}{1 + q} \\
&= \frac{g_i^2 - (q - 1)g_i - q}{(1 + q)^2} = 0;
\end{aligned}
$$

$$Tr(e_i) = \frac{q(1 - \lambda)}{(1 + q)(1 - \lambda q)};$$

$$e_i e_j = e_j e_i, \quad |i - j| \geq 2;$$

$$e_i e_{i+1} e_i - e_{i+1} e_i e_{i+1} = \frac{q}{(1 + q)^2}(e_i - e_{i+1});$$

where the above identities follow from straightforward calculations by the definition of the Hecke algebra $H_n(q)$. Then $\rho(\alpha^{-1})$ has the same expression on the e_i's with only q replaced by q^{-1}. Both expressions $\frac{q(1-\lambda)}{(1+q)(1-\lambda q)}$ and $\frac{q}{(1+q)^2}$ are invariant under the change of variables $q \mapsto \frac{1}{q}, \lambda \mapsto \frac{1}{\lambda}$. Furthermore, those relations with e_i's are equivalent to relations to define the Hecke algebra $H_n(q)$ as long as $q \neq -1$. Hence, it is sufficient to compute the trace of words on the e_i's, which will be the sum of powers of q times powers of $Tr(e_i)$. Therefore $Tr(\rho(\alpha^{-1})) = f(1/q, \lambda) = f(1/q, 1/\lambda)$. The expression $-\frac{1 - \lambda q}{\sqrt{\lambda}(1 - q)}$ is also invariant under the change variables $q \mapsto \frac{1}{q}, \lambda \mapsto \frac{1}{\lambda}$. The result follows from the clear relation $e(\alpha^{-1}) = -e(\alpha)$. $\square$

Corollary 4.3.9. *If $|e(\alpha)| > n(\alpha) - 1$, then the link $\overline{\alpha}$ is not amphicheiral ($\overline{\alpha} \neq \overline{\alpha^{-1}}$, $\alpha \in B_n$).*

Proof: If $e(\alpha) > n - 1$ ($n(\alpha) = n$ for $\alpha \in B_n$), then

$$\lim_{\lambda \to 0} X_L(q, \lambda) = \lim_{\lambda \to 0} (-\frac{1 - \lambda q}{\sqrt{\lambda}(1 - q)})^{n-1}(\sqrt{\lambda})^{e(\alpha)} Tr(\rho(\alpha))$$

$$= \lim_{\lambda \to 0} (-\frac{1 - \lambda q}{\sqrt{\lambda}(1 - q)}\sqrt{\lambda})^{n-1}(\sqrt{\lambda})^{e(\alpha} - (n - 1))Tr(\rho(\alpha))$$

$$= 0,$$

where $Tr(\rho(\alpha^{\pm}))$ as a function of q and λ has a finite limit for fixed $q \neq 0$ as $\lambda \to 0$. Hence the Jones polynomial as a Laurent polynomial of $\sqrt{\lambda}$ has the only strictly positive powers of λ. If $\overline{\alpha}$ is amphicheiral, then $X_{\overline{L}}(q, \lambda) = X_L(q, \lambda)$ has the only strictly positive powers of λ. By Proposition 4.3.8, we also have $X_{\overline{L}}(q, \lambda) = X_L(1/q, 1/\lambda)$ which leads to the strictly negative powers of λ. The contradiction shows that $X_{\overline{L}}(q, \lambda) \neq X_L(q, \lambda)$. Similarly for $e(\alpha) < -(n-1)$, we use the limit of $X_{\overline{L}}(q, \lambda)$ to conclude the result. □

This gives a different proof of Bennequin's result that $\overline{\alpha}$ is nontrivial if $|e(\alpha)| > n - 1$.

4.4 Skein relation

Definition 4.4.1. A link invariant $P : \mathcal{L}(Y) \to A$ (A is a commutative ring) is a linear skein invariant if

(1) $P(O) = 1$ for the unknot O,

(2) There exist three invertible elements $a_{\pm}, a_0$ such that for the skein related $L_{\pm}, L_0$ (see Figure 3.15)

$$a_+ P(L_+) + a_- P(L_-) + a_0 P(L_0) = 0.$$

Let $l = i\sqrt{z/w}$ ($-l^2 = z/w = z/(1 - q + z)$) and $im = \sqrt{q} - 1/\sqrt{q}$. Define

$$P_\alpha(l, m) = V_\alpha(q, z),$$

as a Laurent polynomial in the commutative ring $\mathbb{Z}[l, l^{-1}, m, m^{-1}]$.

The skein rule can be described in terms of braids. The link L_+ is the closure of $\alpha_+ = \beta\sigma_k\gamma$, and L_- is the closure of $\alpha_- = \beta\sigma_k^{-1}\gamma$ and $L_0 = \overline{\beta\gamma}$ for $\beta, \gamma \in B_n$. We define an element in the Hecke algebra $H_n(q)$ over the field extension $K(\sqrt{q/zw})$ by

$$W_\alpha(q, z) = (\frac{1}{z})^{(n(\alpha)+e(\alpha)-1)/2} \cdot (\frac{q}{w})^{(n(\alpha)-e(\alpha)-1)/2} \cdot (\rho(\alpha)).$$

Hence $V_\alpha(q, z) = Tr(W_\alpha(q, z))$.

Lemma 4.4.2. *(Skein Invariance)*

$$lW_{\alpha_+}(q, z) + l^{-1}W_{\alpha_-}(q, z) + mW_{\alpha_0}(q, z) = 0,$$

in the Hecke algebra over the field $K(\sqrt{q/zw})$.

Proof: Set $e = e(\alpha)$ and $n = n(\alpha)$. Thus

$$e(\alpha_+) = e + 1 \qquad n(\alpha_+) = n \tag{4.9}$$
$$e(\alpha_-) = e - 1 \qquad n(\alpha_-) = n \tag{4.10}$$

$$
\begin{aligned}
W_{\alpha_+}(q, z) &= (\tfrac{1}{z})^{(n+e-1)/2} \cdot \sqrt{\tfrac{1}{z}} \cdot (\tfrac{q}{w})^{(n-e-1)/2} \cdot (\tfrac{q}{w})^{-1/2} \cdot \rho(\beta \sigma_k \gamma) \\
&= (\tfrac{w}{qz})^{1/2}(\tfrac{1}{z})^{(n+e-1)/2} \cdot (\tfrac{q}{w})^{(n-e-1)/2} \cdot \rho(\beta) g_k \rho(\gamma) \\
W_{\alpha_-}(q, z) &= (\tfrac{qz}{w})^{1/2}(\tfrac{1}{z})^{(n+e-1)/2} \cdot (\tfrac{q}{w})^{(n-e-1)/2} \cdot \rho(\beta) g_k^{-1} \rho(\gamma)
\end{aligned}
$$

$$l W_{\alpha_+}(q, z) + l^{-1} W_{\alpha_-}(q, z) + m W_{\alpha_0}(q, z)$$

$$
\begin{aligned}
&= i(\tfrac{z}{w})^{1/2}(\tfrac{w}{qz})^{1/2}(\tfrac{1}{z})^{(n+e-1)/2} \cdot (\tfrac{q}{w})^{(n-e-1)/2} \cdot \rho(\beta) g_k \rho(\gamma) \\
&\quad - i(\tfrac{z}{w})^{-1/2}(\tfrac{qz}{w})^{1/2}(\tfrac{1}{z})^{(n+e-1)/2} \cdot (\tfrac{q}{w})^{(n-e-1)/2} \cdot \rho(\beta) g_k^{-1} \rho(\gamma) \\
&\quad + i(1/\sqrt{q} - \sqrt{q})(\tfrac{1}{z})^{(n+e-1)/2} \cdot (\tfrac{q}{w})^{(n-e-1)/2} \cdot \rho(\beta) \rho(\gamma) \\
&= i(\tfrac{1}{z})^{(n+e-1)/2} \cdot (\tfrac{q}{w})^{(n-e-1)/2} \rho(\beta) \cdot H \cdot \rho(\gamma),
\end{aligned}
$$

where $H = q^{-1/2} g_k - q^{1/2} g_k^{-1} + q^{-1/2} - q^{1/2}$. Note that $g_k^{-1} = \frac{1}{q}(1 - q + g_k)$.

$$
\begin{aligned}
H &= q^{-1/2} g_k - q^{1/2} g_k^{-1} + q^{-1/2} - q^{1/2} \\
&= q^{-1/2} g_k - q^{1/2}(\tfrac{1}{q}(1 - q + g_k)) + q^{-1/2} - q^{1/2} \\
&= q^{-1/2} g_k - q^{-1/2} + q^{1/2} - q^{-1/2} g_k + q^{-1/2} - q^{1/2} \\
&= 0.
\end{aligned}
$$

Thus we obtain the skein relation for $W_\alpha(q, z)$. $\qquad\square$

Corollary 4.4.3. *The two-variable Jones polynomial* $V_\alpha(q, z) = P_\alpha(l, m)$ *is a skein invariant in the commutative ring* $\mathbb{Z}[l^{\pm 1}, m^{\pm 1}]$.

Proof: All we need to check is the condition (1) in Definition 4.4.1 since (2) in Definition 4.4.1 follows from applying the trace function. The unknot

$O = \overline{\sigma_1}$ is the closure of $\sigma_1 \in B_2$. Thus we have $n(\sigma_1) = 2$ and $e(\sigma_1) = 1$, and

$$
\begin{aligned}
V_{\sigma_1}(q, z) &= (\frac{1}{z})^{(2+1-1)/2}(\frac{q}{w})^{(2-1-1)/2}Tr(\rho(\sigma_1)) \\
&= \frac{1}{z}Tr(g_1) \\
&= 1.
\end{aligned}
$$

$\square$

Remark 4.4.4.

(1) The Alexander polynomial of a link is related to $P_\alpha(q, z)$

$$
\Delta_K(t) = P_\alpha(i, i(t^{1/2} - t^{-1/2})),
$$

with $K = \overline{\alpha}$.

(2) The one-variable Jones polynomial $V_\alpha(t)$ is defined by

$$
V_\alpha(t) = P_\alpha(it^{-1}, i^{-1}(t^{1/2} - t^{-1/2})).
$$

The skein invariance gives the relation

$$
t^{-1}V_{\alpha_+}(t) - tV_{\alpha_-}(t) = (t^{1/2} - t^{-1/2})V_{\alpha_0}(t),
$$

which is the formula (11.1) in [Jones, 1987].

4.5 Uniqueness and universal property of skein invariant

Every link in S^3 or $\mathbb{R}^3$ has a regular projection. Hence there is a finite set of crossing points.

Lemma 4.5.1. *There is a choice of overcrossing and undercrossing at each crossing point which produces the unlink O^r.*

Proof: Order the unlink O^r arbitrarily. Running along the image of the circles, one after the other, denote that each new crossing is an underpass. The link corresponding to this choice of crossings is the unlink O^r with r components. It is clear that the various components are heritaged one above the other in their chosen order. $\square$

Theorem 4.5.2. *(Uniqueness) If P is a linear skein invariant in Definition 4.4.1, then P is uniquely determined by $a_\pm$ and a_0 of the skein invariance relation.*

L_+ L_- L_0

Fig. 4.1   Skein the unknot

Proof: See Figure 4.1. we have, $L_+ = L_- = O, L_0 = O^2$,

$$a_+ P(L_+) + a_- P(L_-) + a_0 P(L_0) = 0.$$

$$a_+ P(O) + a_- P(O) + a_0 P(O^2) = 0.$$

From $P(O) = 1$, $P(O^2) = -\frac{a_+ + a_-}{a_0}$. Inductively we show that $P(O^r) = (-\frac{a_+ + a_-}{a_0})^{r-1}$. Assume that it is true for r.

$$a_+ P(L_+ \coprod O^{r-1}) + a_- P(L_- \coprod O^{r-1}) + a_0 P(L_0 \coprod O^{r-1}) = 0.$$

$$a_+ P(O^r) + a_- P(O^r) + a_0 P(O^{r+1}) = 0.$$

Thus we get

$$P(O^{r+1}) = (-\frac{a_+ + a_-}{a_0})P(O^r)$$

$$= (-\frac{a_+ + a_-}{a_0})(-\frac{a_+ + a_-}{a_0})^{r-1}$$

$$= (-\frac{a_+ + a_-}{a_0})^r.$$

For each over and under the change $L_\pm$, we get the first two terms of a skein relation (in a definite order). The third term L_0 of the skein related diagrams $L_\pm$ and L_0 has one less crossing than $L_=$ and L_-. Define the complexity of a link diagram to be the pair (N, S), where N is the number of crossings and S is the number of crossing changing into the unlink. Define an order in the complexity by

$$(N, S) < (N', S'), \quad \text{if } N < N' \text{ or } N = N' \text{ with } S < S'.$$

Then by the induction on the alphabetical order of the complexity we have the linear skein invariant P completely determined by $P(O) = 1$ and (2) in Definition 4.4.1. $\qquad\square$

Example 4.5.3. Let us compute the skein linear invariant for the right handed trefoil. From Figure 4.2, we have, $L_- = O$ the unknot,

$$a_+ P(L_+) + a_- P(O) + a_0 P(L_0) = 0.$$

Fig. 4.2   Skein crossing the trefoil

For the Hopf link L_0, further changing in Figure 4.3, we have, $L_0 = L_0^+, L_0^- = O^2, L_0^0 = O$,

$$a_+ P(L_0^+) + a_- P(L_0^-) + a_0 P(O) = 0$$

Thus we get

$$P(L_0) = -\frac{a_0 + a_-(-\frac{a_+ + a_-}{a_0})}{a_+}.$$

Plugging into the first skein relation, we obtain

$$P(L_+) = \frac{a_-}{a_+} + \frac{a_0}{a_+^2}(a_0 + a_-(-\frac{a_+ + a_-}{a_0}))$$

$$= -a_- a_+^{-1} + a_0^2 a_+^{-2} - \frac{a_-}{a_+^2}(a_+ + a_-)$$

$$= -2a_- a_+^{-1} + a_0^2 a_+^{-2} - a_-^2 a_+^{-2}.$$

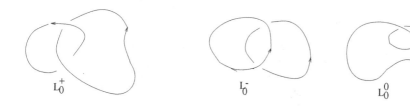

Fig. 4.3   Skein crossing the Hopf link

Lemma 4.5.4. *Let* $P : \mathcal{L} \to \mathbb{Z}[l^{\pm 1}, m^{\pm 1}]$ *be a skein invariant as in Definition 4.4.1. Then each monomial* $l^a m^b$ *occurring in* P *satisfies* $a \equiv b$ (mod 2).

Proof: It is true for the unknot. For the unlink O^r with r components,

$$P(O^r) = (-\frac{l+l^{-1}}{m})^{r-1} = (-lm^{-1} - (lm)^{-1})^{r-1}.$$

Thus it is also true for the unlink. By Definition 4.4.1, we have

$$P(L_+) = -l^{-2}P(L_-) - l^{-1}mP(L_0).$$

Hence $a \equiv b \pmod 2$ follows by induction on the complexity (N, S) of the link diagram with $r(L_+) = r(L_-) = r(L_0) \pm 1$. $\qquad\square$

Theorem 4.5.5. *(Universal Property) (1) The skein invariant P determines a unique skein invariant*

$$T : \mathcal{L} \to \mathbb{Z}[x^{\pm 1}, y^{\pm 1}, z^{\pm 1}]$$

such that

$$xT(L_+) + yT(L_-) + zT(L_0) = 0$$

for every triple of skein related link diagrams.

(2) If $P_A : \mathcal{L} \to A$ is any skein invariant with respect to three invertible elements $a_\pm$ and a_0 in A, then there exists a unique $s : \mathbb{Z}[x^{\pm 1}, y^{\pm 1}, z^{\pm 1}] \to A$ such that

$$
\begin{array}{ccc}
\mathcal{L} & \xrightarrow{\ T\ } & \mathbb{Z}[x^{\pm 1}, y^{\pm 1}, z^{\pm 1}] \\
\downarrow{\scriptstyle P_A} & & \downarrow{\scriptstyle s} \\
A & \xrightarrow{\ =\ } & A
\end{array}
$$

and $s(x) = a_+, s(y) = a_-$ and $s(z) = a_0$.

Proof: (1) Define $T(l^a m^b) = x^{\frac{1}{2(a-b)}} y^{-\frac{1}{2(a-b)}} x^b$ as a map from $\mathcal{L} \to \mathbb{Z}[l^{\pm 1}, m^{\pm 1}]$ to $\mathbb{Z}[x^{\pm 1}, y^{\pm 1}, z^{\pm 1}]$. By Lemma 4.5.4, T is a Laurent polynomial in x, y, z. So

$$T(x, y, z) = P((\frac{x}{y})^{1/2}, z \cdot (\frac{y}{x})^{1/2}),$$

is a homogeneous function of degree zero. It is clear that $T(O) = 1$. With $l = (\frac{x}{y})^{1/2}$ and $m = z \cdot (yx)^{-1/2}$), we have

$$lP(L_+) + l^{-1}P(L_-) + mP(L_0) = 0.$$

$$(\frac{x}{y})^{1/2}P(L_+) + (\frac{x}{y})^{-1/2}P(L_-) + z \cdot (yx)^{-1/2}P(L_0) = 0.$$

Multiplying $(\frac{x}{y})^{1/2}$ on both sides, we get

$$(\frac{x}{y})P(L_+) + P(L_-) + \frac{z}{y}P(L_0) = 0$$

which is equivalent to the result.

(2) Let us define $s(x) = a_+, s(y) = a_-$ and $s(z) = a_0$ and a natural extension $s : \mathbb{Z}[x^{\pm 1}, y^{\pm 1}, z^{\pm 1}] \to A$. Then we have both $s \circ T$ and P_A satisfy the same skein invariance with respect to $a_\pm$ and a_0, by Theorem 4.5.2, for all link diagram

$$s \circ T(L) = P_A(L).$$

$\square$

We have seen in the previous sections that there are skein invariants (Existence result).

4.6 Properties of Jones polynomial

In this section, we prove some basic properties of the polynomials $P_\alpha(l, m)$ and $V_\alpha(q, z)$ as we defined in previous sections.

Lemma 4.6.1. *Let L' be the oriented link obtained from L by reversing the orientations of all the components. Then $P_{L'}(l, m) = P_L(l, m)$.*

Proof: Let L_+, L_-, L_0 be three skein related links. By reversing the orientation of all components of L, we obtain the corresponding three skein related links L'_+, L'_-, L'_0. Hence we have

$$lP_{L'_+}(l, m) + l^{-1}P_{L'_-}(l, m) + mP_{L'_0}(l, m) = 0,$$

and $P_{O'}(l, m) = P_O(l, m) = 1$. Thus we get $P_{L'}(l, m) = P_L(l, m)$ by the Uniqueness Theorem 4.5.2. $\square$

Remark 4.6.2.

(1) Lemma 4.6.1 can be derived from the braid calculation as in [Jones, 1987, Example 6.8]. This property can be used to detect the difference between two orientations of a link.

(2) If $L = \bar{\alpha}$ with $\alpha = \sigma_{i_1}^{\varepsilon_1} \cdots \sigma_{i_r}^{\varepsilon_r}$, then

$$L' = \bar{\alpha}', \quad \alpha' = \sigma_{i_r}^{\varepsilon_r} \cdots \sigma_{i_1}^{\varepsilon_1}.$$

Note that the mapping $\alpha \mapsto \alpha'$ is a well-defined anti-automorphism of B_n. Applying the map $\rho : B_n \to H_n(q)$, we have an extended anti-automorphism sending $g_{i_1}^{\varepsilon_1} \cdots g_{i_r}^{\varepsilon_r}$ to $g_{i_r}^{\varepsilon_r} \cdots g_{i_1}^{\varepsilon_1}$. Hence for $x \in H_n(q)$, $Tr(x) = Tr(x')$ by Theorem 4.2.1.

Lemma 4.6.3. *Let $\tilde{L}$ be the mirror image of L. Then we have $P_{\tilde{L}}(l, m) = P_L(l^{-1}, m)$.*

Proof: Again we correspond skein links in the order $\tilde{L}_-, \tilde{L}_+, \tilde{L}_0$. Hence

$$lP_{\tilde{L}_-}(l,m) + l^{-1}P_{\tilde{L}_+}(l,m) + mP_{\tilde{L}_0}(l,m) = 0.$$

Comparing with $lP_{L_+}(l,m) + l^{-1}P_{L_-}(l,m) + mP_{L_0}(l,m) = 0$ and $P_{\tilde{O}} = P_O = 1$,

$$P_{\tilde{L}}(l,m) = P_L(l^{-1},m)$$

follows from Uniqueness Theorem 4.5.2.   □

See [Jones, 1987, Example 6.9 and Proposition 6.10].

Lemma 4.6.4. *Let $L_1 \coprod L_2$ be the disjoint unlinked link from L_1 and L_2. Then*

$$P_{L_1 \coprod L_2}(l,m) = -\frac{l+l^{-1}}{m}P_{L_1}(l,m) \cdot P_{L_2}(l,m).$$

Proof: Let $L_i = \overline{\alpha}_i$ with $\alpha_1 \in B_n$ and $\alpha_2 \in B_m$. We define a shift map $s : B_n \to B_{n+m}$ by $s(\sigma_i) = \sigma_{m+i}$ for $i = 1, 2, \cdots, n-1$. Thus $L_1 \coprod L_2$ is the closure of the new braid $\alpha_1 \cdot s(\alpha_2)$. It follows that α_1 and $s(\alpha_2)$ commute in B_{n+m} and

$$Tr(\rho(\alpha_1 \cdot s(\alpha_2))) = Tr(\rho(\alpha_1)) \cdot Tr(\rho(\alpha_2)).$$

By definition, we have

$$V_{\alpha_1 \cdot s(\alpha_2)}(q,z) = (\frac{q}{zw})^{1/2}V_{\alpha_1}(q,z) \cdot V_{\alpha_2}(q,z).$$

With the changing variable $l = i(z/w)^{1/2}, m = i(q^{-1/2} - q^{1/2})$,

$$-\frac{l+l^{-1}}{m} = -\frac{(z/w)^{1/2} - (z/w)^{-1/2}}{q^{-1/2} - q^{1/2}}$$

$$= -(\frac{zq}{w})^{1/2} \cdot \frac{1 - w/z}{1 - q}$$

$$= -(\frac{zq}{w})^{1/2} \cdot \frac{z - (1 - q + z)}{z(1 - q)}$$

$$= (\frac{q}{zw})^{1/2}.$$

Hence with new variable l and m we have the desired result.   □

Lemma 4.6.5.

$$P_{L_1 \# L_2}(l,m) = P_{L_1}(l,m) \cdot P_{L_2}(l,m).$$

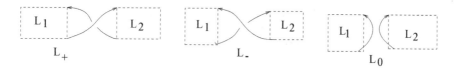

Fig. 4.4   Skein relation for the composite link

Proof: As in Figure 4.4, $L_+ = L_1 \# L_2$, $L_- = L_1 \# L_2$ and $L_0 = L_1 \coprod L_2'$.
The skein relation gives us the following identity:

$$lP_{L_1 \# L_2}(l,m) + l^{-1}P_{L_1 \# L_2}(l,m) + mP_{L_1 \coprod L_2'}(l,m) = 0.$$

Hence by Lemma 4.6.1 and Lemma 4.6.3,

$$P_{L_1 \# L_2}(l,m) = -\frac{m}{l + l^{-1}}P_{L_1 \coprod L_2'}(l,m)$$

$$= -\frac{m}{l + l^{-1}}P_{L_1 \coprod L_2}(l,m)$$

$$= -\frac{m}{l + l^{-1}}(-\frac{l + l^{-1}}{m}P_{L_1}(l,m) \cdot P_{L_2}(l,m))$$

$$= P_{L_1}(l,m) \cdot P_{L_2}(l,m).$$

$\square$

Remark 4.6.6. One can use the braid description to prove Lemma 4.6.5 as in [Jones, 1987, Example 6.7]. Note that if $L_1 = \bar{\alpha}_1, \alpha_1 \in B_n$ and $L_2 = \bar{\alpha}_2, \alpha_2 \in B_m$, then $L_1 \# L_2 = \alpha_1 \Sigma^{n-1}\alpha_2 \in B_{n+m}$, where Σ is the shift map on the inductive limit of B_n's, $\Sigma(\sigma_i) = \sigma_{i+1}$.

Exercise 4.6.7. Using the braid description, prove that $V_{L_1 \# L_2}(q,z) = V_{L_1}(q,z) \cdot V_{L_2}(q,z)$.

Lemma 4.6.8. Let $\Delta_L(t) = P_L(i, i(t^{1/2} - t^{-1/2}))$. Then $\Delta_L(t)$ satisfies the skein invariance (i) $\Delta_O(t) = 1$ and (ii)

$$\Delta_{L_+}(t) - \Delta_{L_-}(t) + (t^{1/2} - t^{-1/2})\Delta_{L_0}(t) = 0,$$

which characterizes the Alexander polynomial as normalized by Conway.

Proof: Let $l = i$ and $m = i(t^{1/2} - t^{-1/2})$. The first property $\Delta_O(t) = 1$ follows from the normalization of skein invariant. Now we have the skein relation

$$iP_{L_+}(i, i(t^{1/2} - t^{-1/2})) - iP_{L_-}(i, i(t^{1/2} - t^{-1/2}))$$

$$+i(t^{1/2} - t^{-1/2})P_{L_0}(i, i(t^{1/2} - t^{-1/2})) = 0.$$

Using the definition of $\Delta_L(t)$ and dividing i out, we have

$$\Delta_{L_+}(t) - \Delta_{L_-}(t) + (t^{1/2} - t^{-1/2})\Delta_{L_0}(t) = 0.$$

□

Lemma 4.6.9. *Let $V_L(t) = P_L(it^{-1}, -i(t^{1/2}-t^{-1/2}))$. Then $V_L(t)$ satisfies the skein invariance (i) $V_O(t) = 1$ and (ii)*

$$t^{-1}V_{L_+}(t) - tV_{L_-}(t) = (t^{1/2} - t^{-1/2})V_{L_0}(t)$$

which characterizes the one-variable Jones polynomial in [Jones, 1987, (11.1)].

Proof: Let $l = it^{-1}$ and $m = -i(t^{1/2} - t^{-1/2})$. The proof is exactly the same as in Lemma 4.6.8. □

Exercise 4.6.10. Prove Lemma 4.6.9.

In the relations (4.1) of Hecke algebra $H_{n+1}(q)$ in Definition 4.1.1, there is a $*$-algebra $A_n(q)$ with generators $1, e_1, \cdots, e_n$ and the relations:

$$\begin{cases} e_i^* = e_i, \quad e_i^2 = e_i, \\ e_i e_{i\pm1} e_i = \tau e_i, \\ e_i e_j = e_j e_i, \quad |i - j| \geq 2, \end{cases} \tag{4.11}$$

where τ is a real number related to the $\prod_1$ factors. These relations (4.11) are not for all τ, a presentation of A_n. There is a trace uniquely determined by the following Markov property:

$$tr(xe_{n+1}) = \tau tr(x), \quad x \in A_n. \tag{4.12}$$

Let t be such that $\tau^{-1} = 2 + t + t^{-1}$ and $g_i = te_i - (1 - e_i)$. Then $e_i = (1+t)^{-1}(g_i + 1)$.

$$\begin{aligned} e_i^2 &= (1+t)^{-2}(g_i + 1)^2 \\ &= (1+t)^{-2}(g_i^2 + 2g_i + 1) = (1+t)^{-2}(q + (q-1)g_i + 2g_i + 1) \\ &= (1+t)^{-2}(1+q)(g_i + 1) = (1+t)^{-1}(g_i + 1) = e_i, \end{aligned}$$

is equivalent to $(1 + q) = (1 + t)$ $(q = t)$, i.e., $g_i^2 = t + (t - 1)g_i$.

$$\begin{aligned} e_i e_{i+1} e_i &= (1+t)^{-3}(g_i + 1)(g_{i+1} + 1)(g_i + 1) \\ &= (1+t)^{-3}(g_i g_{i+1} g_i + g_{i+1} g_i + g_i g_{i+1} + g_{i+1} + (g_i^2 + 2g_i) + 1) \\ &= (1+t)^{-3}(g_i g_{i+1} g_i + g_{i+1} g_i + g_i g_{i+1} + g_{i+1} + (t + 1)(g_i + 1)). \\ \tau e_i &= (2 + t + t^{-1})(1+t)^{-1}(g_i + 1). \end{aligned}$$

Hence $e_i e_{i+1} e_i = \tau e_i$ is equivalent to

$$(1+t)^{-3}(g_i g_{i+1} g_i + g_{i+1} g_i + g_i g_{i+1} + g_{i+1} + (t+1)(g_i+1))$$

$$= (2+t+t^{-1})(1+t)^{-1}(g_i+1).$$

Combining (g_i+1) terms together, we get

$$(2+t+t^{-1})(g_i g_{i+1} g_i + g_{i+1} g_i + g_i g_{i+1} + g_{i+1}) = -(2+t+t^{-1})(g_i+1).$$

For $t \neq -1$, the relation $e_i e_{i+1} e_i = \tau e_i$ is equivalent to

$$g_i g_{i+1} g_i + g_{i+1} g_i + g_i g_{i+1} + g_{i+1} + g_i + 1 = 0. \tag{4.13}$$

Thus g_i satisfies (4.1) with $q = t$, and (4.13) does not follow from (4.1). Thus $A_n(t) = H_{n+1}(t)/(4.13)$. For example, $t = 1$, $\tau = 1/4$. Thus $A_n(1)$ is generated by $\{1, e_1, \cdots, e_n\}$ with (4.11). $A_n(1)$ is equivalent to $H_{n+1}(1)/(4.13)$ which is generated by $\{1, g_1, \cdots, g_n\}$ such that

$$g_i^2 = (t-1)g_i + t = 1, \quad g_i g_j = g_j g_i \quad (|i-j| \geq 2),$$

$$g_i g_{i+1} g_i = g_{i+1} g_i g_{i+1}, \quad (1 \leq i \leq n-1);$$

$$g_i g_{i+1} g_i + g_{i+1} g_i + g_i g_{i+1} + g_{i+1} + g_i + 1 = 0.$$

The first three relations are the symmetric group Σ_{n+1} of $n+1$ letters. The irreducible representations of a finite group are in one-to-one correspondence with conjugacy classes of the group. Conjugacy classes of Σ_{n+1} are indexed by partitions of $n+1$ corresponding to the periods of the disjoint cycles of the permutation. A partition of 7 is given by $7 = 3+2+1+1$ which corresponds to an irreducible representation. The length of the rows in the diagram is made to be non-decreasing. Such a diagram is called a Young diagram. Thus irreducible representations of Σ_{n+1} are indexed by Young diagrams. Given an irreducible representation π of Σ_{n+1} with Young diagram Y, its restriction to Σ_n is the direct sum of one copy of each representation of Σ_n obtained from Y by removing a node so as to obtain a Young diagram. For instance, the representation of Σ_8 is a partition $8 = 5+2+1$, its restriction to Σ_7 is the direct sum of representations of Σ_7 with partitions $7 = 5+2$, $7 = 5+1+1$ and $7 = 4+2+1$. It is well-known that the representations of the symmetric group Σ_{n+1} which satisfy (4.13) are those representations whose Young diagrams have at most two columns. The dimension of A_n is the Catalan number $\frac{1}{n+2}\binom{2n+2}{n+1}$.

Note that $tr(xe_{n+1}) = \tau tr(x)$ is related to the Ocneanu trace such that the trace passes to the quotient A_n. Taking the trace of (4.13), one gets,

$$Tr(g_i g_{i+1} g_i) + Tr(g_i g_{i+1}) + Tr(g_{i+1} g_i) + Tr(g_{i+1}) + Tr(g_i) + 1$$

$$= Tr(g_i^2 g_{i+1}) + 2Tr(g_i g_{i+1}) + Tr(g_{i+1}) + Tr(g_i) + 1$$
$$= zTr(g_i^2) + 2zTr(g_i) + 2z + 1$$
$$= z(t + (t-1)z) + 2z^2 + 2z + 1$$
$$= (1 + z)(zt + z + 1) = 0.$$

Thus $1 + z = 0$ or $zt + z + 1 = 0$. The value $z = -1$ is of little interest. $z = -\frac{1}{t+1}$ gives

$$Tr(e_i) = (1 + t)^{-1} Tr(g_i + 1) = \frac{t}{(1+t)^2} = \tau,$$

where $\tau^{-1} = 2 + t + t^{-1}$. Therefore $V_L(t) = X_L(q, \lambda)$ with $q = t$ and $\lambda = \frac{t^2}{q} = t$ ($V_L(t) = X_L(t,t)$) satisfies the skein relation in Lemma 4.6.9. Thus using $\rho : B_{n+1} \to A_n$ given by $\rho(\sigma_i) = g_i = te_i - (1 - e_i)$, we have, for $\alpha \in B_{n+1}$,

$$V_L(t) = \left(-\frac{1+t}{\sqrt{t}}\right)^{n-1} \sqrt{t}^{e(\alpha)} Tr(\rho(\alpha)).$$

Using (4.11) and (4.13), it suffices to calculate $Tr(\rho(\alpha))$.

Since the growth of $\dim A_n = \frac{1}{n+2} \binom{2n+2}{n+1}$ is much slower than the growth of $\dim H_n(q) = (n+1)!$, computer calculation of V_L is feasible for closed braids on more strings than for P_L. For an (n, m) torus knot K with n and m relatively prime, one has

$$V_K(t) = \frac{t^{\frac{(n-1)(m-1)}{2}}}{1 - t^2}(1 - t^{m+1} - t^{n+1} + t^{n+m}).$$

There are many other natural representations which are not necessarily irreducible, one prominent role in the algebraic theory of the Jones polynomial was first discovered by Temperley and Lieb with connection to the Potts and ice-type models of statistical mechanics. Let $\otimes_{i=1}^{\infty} M_2(\mathbb{C})$ be the infinite tensor product of 2×2 matrices with a shift endomorphism

$$\sigma(x \otimes 1 \otimes 1 \cdots) = 1 \otimes x \otimes 1 \cdots.$$

Define

$$e = \left\{ \frac{t}{1+t} e_{11} \otimes e_{22} + \frac{\sqrt{t}}{1+t}(e_{21} \otimes e_{12} + e_{12} \otimes e_{21}) + \frac{1}{1+t} e_{22} \otimes e_{11} \right\} \otimes 1 \otimes 1 \cdots,$$

where e_{ij}'s are the matrix units for $M_2(\mathbb{C})$, and $e_i = \sigma^i(e)$. One can verify that e_i's satisfies relations (4.11) for $t \in \mathbb{R}^+$ with $*$ as the usual conjugate transpose. The ensuing representation $\theta : A_n(q) \to \otimes_{i=1}^{\infty} M_2(\mathbb{C})$ is faithful for all n and generic t.

The trace tr on $A_n(q)$ is not given by the restriction of the normalized trace to $\theta(A_n(q))$. Define the power state ϕ_t on $\otimes_{i=1}^{\infty} M_2(\mathbb{C})$ by the following.

$$\phi_t(x_1 \otimes x_2 \otimes \cdots \otimes x_n \otimes 1 \otimes 1 \cdots) = TR((h \otimes h \otimes \cdots \otimes h)(x_1 \otimes x_2 \cdots \otimes x_n)),$$

where TR is the non-normalized trace on $M_{2^n}(\mathbb{C})$ and h is the 2×2 matrix $h = \begin{pmatrix} \frac{1}{1+t} & 0 \\ 0 & \frac{t}{1+t} \end{pmatrix}$ with $\text{trace}(h) = 1$. One can check the following properties (leave as exercises):

$$\phi_t(xy) = \phi_t(yx),$$
$$\phi_t(\theta(xe_{n+1})) = \frac{t}{(1+t)^2}\phi_t(\theta(x)).$$

This representation of $A_{n+1}(q)$ (or $H_n(q), B_n$) is called the PPTL representation (Pimsner, Popa, Temperley and Lieb).

For a Young diagram Y, let tr_Y be the trace on the Hecke algebra obtained by evaluating the usual trace (sum of the diagonal entries) on the image of a Hecke algebra element in the representation $\rho(\alpha)$,

$$Tr(\rho(\alpha)) = \sum_Y \frac{S_Y(q,z)}{Q_Y(q)} tr_Y(\rho(\alpha)),$$

where $S_Y(q,z)$ is the product of the terms covered by the Young diagram Y, with filling in the hook lengths as if calculating the dimension of irreducible representation of Σ_n corresponding to Y and replacing each integer m that appears by $1 - t^m$, $Q_Y(q) = Q_Y(t)$ is the product of these terms by substituting m with $1 - t^m$. Then

$$\Delta_L(t) = (-1)^{n-1}(\frac{1}{t})^{(e-n+1)/2}\frac{1-t}{1-t^n}\sum_{\beta=0}^{n-1}(-1)^\beta tr_Y(\rho(\alpha)),$$

is given in [Jones, 1987, §7]. The Alexander polynomial is given by

$$\Delta_L(t) = (-\frac{1}{\sqrt{t}})^{e(\alpha)-n(\alpha)+1}(\frac{1-t}{1-t^n}) \cdot \det(Id - m_{B_n}(\alpha)). \qquad (4.14)$$

Thus (4.14) fixes the normalizer $(-\frac{1}{\sqrt{t}})^{e(\alpha)-n(\alpha)+1}$ in Theorem 3.7.12 and verifies over a few well-known examples.

Proposition 4.6.11.

(1) $V_L(1) = (-2)^{\mu(L)-1}$.
(2) $\frac{d}{dt}V_L(1) = -3(-2)^{\mu(L)-2}$.
(3) $\frac{d^2}{dt^2}V_L(1) = -3\Delta''_L(1)$.
(4) $V_L(-1) = \Delta_L(-1)$.
(5) $V_L(e^{\pi i/2}) = \begin{cases} 0 & \text{unless } arf(L) \text{ exists} \\ (-2^{3/2})^{c-1}(-1)^{are(L)} & \text{if } arf(L) \text{ exists} \end{cases}$
(6) $V_L(e^{\pi i/3}) = \pm i^{\mu(L)-1}(i\sqrt{3})^{n(L)}$, *where* $n(L) = \dim H_1(D_L; \mathbb{Z}_3)$ *for the double branched cover of* S^3 *branched along* L.

Proof: (1) For $t = 1$, $tr(e_i) = \tau = 1/4$ and $tr(g_i) = 2tr(e_i) - 1 = -1/2$. Thus $V_L(1) = (-2)^{\mu(L)-1}$ by the above definition of $V_L(t)$. Differentiating the skein relation in Lemma 4.6.9,

$$t^{-1}V'_{L_+}(t) - t^{-2}V_{L_+}(t) - tV'_{L_-}(t) - V_{L_-}(t)$$

$$= \frac{1}{2}(t^{-1/2} + t^{-3/2})V_{L_0}(t) + (t^{1/2} - t^{-1/2})V'_{L_0}(t).$$

Hence, putting $t = 1$, we obtain

$$V'_{L_+}(1) - V'_{L_-}(1) = V_{L_+}(1) + V_{L_-}(1) + V_{L_0}(1).$$

If the crossing involves only one component of the link, then

$$V_{L_+}(1) = V_{L_-}(1) = -1/2V_{L_0}(1),$$

by [Jones, 1987, (12.1)]. Thus $V'_{L_+}(1) - V'_{L_-}(1) = 0$.
   If the crossing involves two components of the link,

$$V_{L_+}(1) = V_{L_-}(1) = -2V_{L_0}(1) = (-2)^{\mu(L)-1}.$$

So that $V'_{L_+}(1) - V'_{L_-}(1) = -3(-2)^{\mu(L)-1}$. The result (2) follows by induction on $\mu(L)$.
   (3) Changing a positive crossing of a knot to a negative one will change $\frac{d^2}{dt^2}V_L(1)$ by six times the linking number of the two-component link formed by eliminating the crossing, and hence (3) and (4) follows by skein relations with higher derivatives and comparing with the Alexander polynomial.
   (5) This has been shown by Lickorish-Millet [1986] to apply a braid projection of a knot.
   (6) Note that $V_L(t) = P_L(it^{-1}, -i(t^{1/2} - t^{-1/2}))$. So $V_L(e^{\pi i/3}) = P_L(e^{\pi i/6}, 1)$. Let $L_\pm, L_0$ be the triple of oriented links in the skein relation. Let $W_L = i^{(1-\mu(L))}V_L(e^{\pi i/3})$. Thus $V_L(e^{\pi i/3})^2$ is independent of the orientation of L by evaluations $t^{1/2} - t^{-1/2} = i$ and $t^3 = -1$ for $t = e^{\pi i/3}$.

With the sign ambiquity depending on whether or not $\mu(L_+) > \mu(L_0)$, the defining formula for V_L leads to

$$e^{-\pi i/3}W_{L_+} - e^{\pi i/3}W_{L_-} = \pm W_{L_0}.$$

Subtracting the square of this second equation $e^{-\pi i/6}W_{L_+} - e^{\pi i/6}W_{L_-} = \pm i W_{L_\infty}$ from the square of the above and using $e^{2\pi i/3} - e^{\pi i/3} = -1$, we get

$$W_{L_+}^2 + W_{L_-}^2 + W_{L_0}^2 + W_{L_\infty}^2 = 0.$$

The relation $Q_L = W_L^2$ follows from the identical defining formulae $Q_{L_+} + Q_{L_-} = x(Q_{L_0} + Q_{L_\infty})$ and $Q_U = 1$ for the unknot U. The usual induction argument on the crossing number of a link presentation shows that $W_L^2 = Q_L(-1) = (-3)^{n(L)}$. □

For $t = 1$, the PPTL representation of the braid group B_n factors through the symmetric group and is given, up to sign, by the permutations of the tensor product components of $\mathbb{C}^2 \otimes \mathbb{C}^2 \otimes \cdots$. The power state ϕ_1 is the normalized trace. Also $\frac{\sqrt{t}+1}{\sqrt{t}} = 2$ and $tr(g_i) = 2$ in this case $t = 1$. Hence $V_L(1) = (-2)^{\mu(L)-1}$. Warning on the value $t = 1$ cannot be evaluated in the two variable polynomial. $V_L(1) = \lim_{(q,\lambda) \to (1,1), q=\lambda} X_L(q, \lambda)$ the limit is only along the curve $q = \lambda$. It is a good exercise to see if the limit $\lim_{(q,\lambda) \to (1,1)} X_L(q, \lambda)$ exists.

For $t = e^{2\pi i/3}$ (as a solution of the quadratic equation $1 + t + t^2 = 0$), $\tau^{-1} = t + t^{-1} + 2 = 1$ and $Tr(e_i) = (1+t)^{-1}Tr(g_i+1) = (1+t)^{-1}(t+1) = 1$. Also $\rho(\sqrt{t}\sigma_i) = \sqrt{t}g_i$ has $tr(\rho(\sqrt{t}\sigma_i)) = \sqrt{t}tr(g_i) = t^{3/2} = 1$. Therefore

$$V_L(t) = (-\frac{1+t}{\sqrt{t}})^{n-1}(\sqrt{t})^{e(\alpha)}tr(\rho(\alpha(\sigma_1, \cdots, \sigma_{n-1})))$$

$$= (t^{3/2})^{n-1}tr(\rho\alpha(\sqrt{t}\sigma_1, \cdots, \sqrt{t}\sigma_{n-1}))$$

$$= (-1)^{\mu(L)-1}.$$

If K is a knot, then $V_K(e^{2\pi i/3}) = 1$ and $V_K(1) = 1$. Therefore $V_K(t) - 1$ can be factored by $(t - 1)$ and $t^3 - 1$. So we have

$$V_K(t) - 1 = (t - 1)(t^3 - 1)W_K(t),$$

where $W_K(t)$ is a Laurent polynomial of the knot K [Jones, 1987, Proposition 12.5]. See [Jones, 1987, §12 and §14] for more discussions on the special values of the Jones polynomial.

For $t = e^{\pm\pi i/5}$, $\rho(B_n)$ is finite when $n \leq 3$ but infinite otherwise. For example, $\rho(\sigma_1\sigma_2\sigma_3^{-1})$ has infinite order. Thus $V_L(e^{\pi i/5})$ can be used as a test for being a 3-braid (all possible values for closed 3-braids is finite).

A question asked before is to determine if every $\alpha \in B_{n+1}$ is conjugate to an braid element $\alpha_1 \sigma_n \alpha_2 \sigma_n^{-1}$ with $\alpha_1, \alpha_2 \in B_n$. If $\alpha \in B_{n+1}$ is conjugate to $\alpha_1 \sigma_n \alpha_2 \sigma_n^{-1}$ with $\alpha_1, \alpha_2 \in B_n$, then

$$\rho_0(\alpha) = \rho_0(\alpha_1)((t+1)e_n - 1)\rho_0(\alpha_2)((t^{-1}+1)e_n - 1),$$

and $tr(\rho_0(\alpha)) = 1/\tau tr(e_n \rho_0(\alpha_1)e_n \rho_0(\alpha_2))$. Let $n = 3$ and $t = e^{\pi i/5}$. The set of V_L values of elements of the above form is finite. To answer the question in the negative, it suffices to exhibit infinitely many values of $V_L(e^{\pi i/5})$ with $L = \overline{\alpha}$ for $\alpha \in B_4$. Jones showed that the Burau matrix of $\sigma_1 \sigma_2 \sigma_3^{-1}$ has an eigenvalue which is not a root of unity. In the representation on the extreme left of the A_3 level, $e_1 = e_3$ so that $\sigma_1 \sigma_2 \sigma_3^{-1}$ is the same as $\sigma_1 \sigma_2 \sigma_1^{-1}$ which has finite order. In the weighted sum version of the trace, the contributions of the first and third representations on the A_3 line are periodic whereas that of the second is aperiodic. So infinitely many powers of $\sigma_1 \sigma_2 \sigma_3^{-1}$ are not Markov equivalent to 4-braids of the special form.

For each t (not root of unity), $\{|V_L(t)| : L$ is a closed 3-braid$\}$ is dense in the interval $[0, 4\cos^2(\frac{\pi}{n})]$.

4.7 Kauffman bracket and polynomial

Definition 4.7.1. Let K be an unoriented knot or link diagram. Let $\langle K \rangle$ be the element of the ring $Z[A, B, d]$ defined by

(1) $\langle O \rangle = 1$ for the unknot O,
(2) $\langle O \cup K \rangle = d\langle K \rangle$ for nonempty K,
(3) $\langle K \rangle = A\langle K_0 \rangle + B\langle K_\infty \rangle$ as in Figure 4.5.

The bracket $\langle K \rangle$ is called the Kauffman bracket. See [Kauffman, 1987a] for more details.

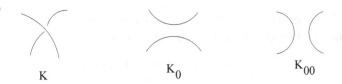

$$K \qquad\qquad K_0 \qquad\qquad K_{00}$$

Fig. 4.5   Links among Kauffman bracket relations

Remark 4.7.2. (i) The bracket $\langle \cdot \rangle$ takes values 1 on a single unknotted circle diagram; (ii) Definition 4.7.1 (2) states that $\langle K \rangle$ is multiplied by the factor d in the presence of a disjoint circular component; (iii) Definition 4.7.1 (3) entails $\langle K_- \rangle = B \langle K_0 \rangle + A \langle K_\infty \rangle$.

Proposition 4.7.3. *(1) The bracket $\langle \cdot \rangle$ under the Reidemeister move of type II satisfies*

$$\langle K_\mp \rangle = AB \langle K_\infty \rangle + (ABd + A^2 + B^2) \langle K_0 \rangle.$$

Therefore the bracket is invariant under the Reidemeister move of type II if

$$AB = 1, \qquad ABd + A^2 + B^2 = 0.$$

(2) If the bracket $\langle \cdot \rangle$ is invariant under the Reidemeister move of type II, then the bracket $\langle \cdot \rangle$ is invariant under the Reidemeister move of type III.

(3) Under the condition of (2), we have the following relations under the Reidemeister move of type I:

$$\langle K_{-0} \rangle = (-A^{-3}) \langle K_- \rangle, \quad \langle K_{+0} \rangle = (-A^3) \langle K_- \rangle.$$

See Figure 4.6 for related notations.

Proof: (1) Following Definition 4.7.1, we have

$$\langle K_\mp \rangle = AB \langle K_\infty \rangle + (ABd + A^2 + B^2) \langle K_0 \rangle.$$

See Figure 4.6 for the relations changed.

(2) is shown in Figure 4.7.

(3) follows from the calculations:

$$\begin{aligned} \langle K_{-0} \rangle &= B \langle K_- \cup O \rangle + A \langle K_- \rangle \\ &= (Bd + A) \langle K_- \rangle. \end{aligned}$$

From $d = -A^2 - A^{-2}$ and $B = A^{-1}$,

$$Bd + A = A^{-1}(-A^2 - A^{-2}) + A = (-A^{-3}).$$

Similarly $\langle K_{+0} \rangle = (Ad + B) \langle K_- \rangle$, where $Ad + B = A(-A^2 - A^{-2}) + A^{-1} = (-A^3)$. $\qquad \square$

Theorem 4.7.4. *Let $f_K(A) = (-A)^{-3w(K)} \langle K \rangle \in Z[A, A^{-1}]$ be the Kauffman polynomial, where $w(K)$ is the writhe of the diagram K. The polynomial $f_K(A)$ is an invariant of the knot K.*

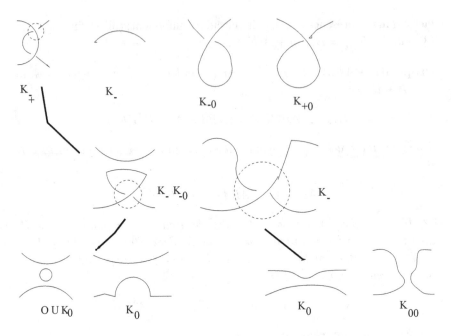

Fig. 4.6   Links in Proposition 4.7.3

Proof: The writhe of a cross is defined in Figure 3.9. The writhe is invariant under the Reidemeister moves of type II and type III by Figure 3.7 and Figure 3.8. Under the Reidemeister move of type I, $w(K) = w(K') - 1$ no matter how we orient the diagram K. Thus

$$f_K(A) = (-A)^{-3w(K)}\langle K \rangle$$
$$= (-A)^{-3(w(K')-1)}(-A^{-3})\langle K' \rangle \quad \text{by Proposition 4.7.3 (3)}$$
$$= (-A)^{-3w(K')}\langle K' \rangle$$
$$= f_{K'}(A).$$

$\square$

Note that switching all crossings results in the replacement of every appearance of A by its inverse A^{-1} in the expansion of the bracket. Let $K!$ be the mirror image of K by reversing all the crossings of K. Then $f_K(A) = f_{K!}(A^{-1})$.

Theorem 4.7.5. *The Jones polynomial and the Kauffman polynomial are related by*

$$V_K(t) = f_K(t^{-1/4}), \qquad A = t^{-1/4}.$$

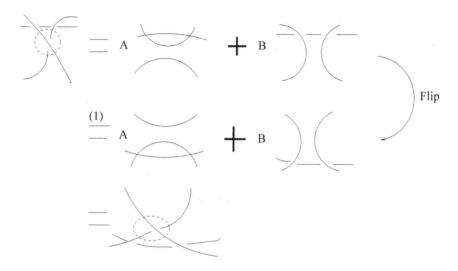

Fig. 4.7 Invariance under the Reidemeister move of type II

Proof: The identification follows from Lemma 4.6.9 and the uniqueness. Note that $\langle K_+ \rangle = A\langle K_0 \rangle + A^{-1}\langle K_\infty \rangle$ and $\langle K_- \rangle = A^{-1}\langle K_0 \rangle + A\langle K_\infty \rangle$. So

$$A\langle K_+ \rangle - A^{-1}\langle K_- \rangle = (A^2 - A^{-2})\langle K_0 \rangle.$$

Multiplying $A^3(-A)^{-3w(K_+)}$ into the above equality,

$$A^4(-A)^{-3w(K_+)}\langle K_+ \rangle - A^2(-A)^{-3w(K_+)}\langle K_- \rangle$$

$$= (A^2 - A^{-2})A^3(-A)^{-3w(K_+)}\langle K_0 \rangle.$$

Note that $w(K_+) = w(K_-) + 2$ and $w(K_+) = w(K_0) + 1$.

$$A^2(-A)^{-3w(K_+)} = A^2(-A)^{-3w(K_-)-6} = A^{-4}(-A)^{-3w(K_-)};$$

$$(A^2 - A^{-2})A^3(-A)^{-3w(K_+)} = (A^2 - A^{-2})A^3(-A)^{-3w(K_0)-3}$$

$$= -(A^2 - A^{-2})(-A)^{-3w(K_0)}.$$

Hence $A^4 = t^{-1}$ and $f_K(A) = f_K(t^{-1/4})$ satisfies Lemma 4.6.9. Then the result follows from the uniqueness of the one-variable Jones polynomial. $\square$

4.8 The Tait conjectures

We follow [Murasugi, 1987] to present the powerful technique of the Jones polynomial and Kauffman bracket which proved the Tait conjecture I and II.

Let $d_{max}V_L(t)$ and $d_{min}V_L(t)$ denote the maximal and minimal degrees of $V_L(t)$. Define

$$\text{Span}\, V_L(t) = d_{max}V_L(t) - d_{min}V_L(t). \tag{4.15}$$

Since $V_L(t) = (-t)^{3w(L)/4}\langle L \rangle|_{A=t^{-1/4}}$ by Theorem 4.7.5,

$$\begin{aligned}
\text{Span}\, V_L(t) &= \text{Span}\,\langle L\rangle|_{A=t^{-1/4}} \\
&= \text{Span}\,\langle L\rangle|_{A=t^{1/4}} \\
&= \frac{1}{4}\text{Span}\,\langle L\rangle|_{A=t} \\
&= \frac{1}{4}\text{Span}\,\langle L\rangle(A).
\end{aligned}$$

Let $c(\tilde{L})$ be the number of double points in a projection $\tilde{L}$ of the link L in S^3. For a tame link L in S^3, we take a proper connected projection $\tilde{L}$ of L in S^2. Then $\tilde{L}$ divides S^2 into finitely many domains. Let Γ be the graph of the link projection $\tilde{L}$ such that each vertex of Γ corresponds to an unshaded domain and each edge of Γ corresponds to a double point of $\tilde{L}$. The shaded domain of $\tilde{L}$ arise from the Seifert surface spanned by L. We call an edge e of Γ *positive or negative* according to the crossing as L_+ or L_- as in Figure 3.15.

Note that $\tilde{L}$ is alternating if and only if all the edges of Γ are positive (or negative). A graph Γ is called oriented if every edge is assigned with either positive or negative. Let p be the number of positive edges in Γ, and n be the number of negative edges in Γ. Thus we have

$$p + n = c(\tilde{L}).$$

Smoothing a double point on an edge e to L_0 is called *parallel*, smoothing a double point on an edge e to L_∞ is called *transverse*. See Figure 4.8.

By applying parallel or transverse smoothing on every edge in Γ, we end up with a trivial link of several components, the Kauffman bracket $\langle \tilde{L}\rangle(A)$ is the sum of these links multiplied by A^k for some integer k. Let Γ^* be the dual graph of Γ which is oriented in such a way that an edge e^* in Γ^* is positive (or negative) if and only if e^* intersects a positive (or negative) edge in Γ.

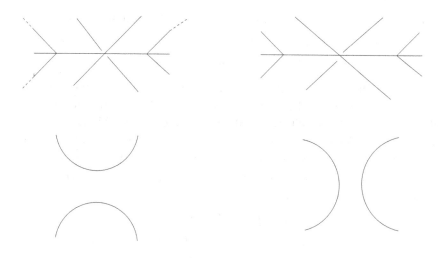

Fig. 4.8   Positive (negative) edge and parallel (transverse) smoothing

Let $\Gamma_\pm$ be a subgraph of Γ that consists of all $\pm$ edges and their vertices. Then for $0 \le a \le p, 0 \le b \le n$, $\tilde{\mathcal{L}}_\Gamma(a, b)$ is the link projection from the link projection with Γ as its oriented graph by applying parallel smoothness with exactly a many positive edges and b many negative edges, and by applying transverse smoothness on $p - a$ positive edges and $n - b$ negative edges in Γ. Note that $\tilde{\mathcal{L}}_\Gamma(a, b)$ is called *a state* in [Kauffman, 1987a]. The $\tilde{\mathcal{L}}_\Gamma(a, b)$ is a trivial link. For each pair (a, b), let $S(a, b)$ denote the collection of $\binom{p}{a}\binom{n}{b}$ many trivial links from the choices of a positive edges from p and b negative edges from n.

Define $\mu_\Gamma(a, b)$ to be the maximal number of components of links in $S(a, b)$. For $a = 0, p$ and $b = 0, n$, $S(a, b)$ contains one link only since $\binom{p}{a}\binom{n}{b} = 1$ in this case. Then $\mu_\Gamma(0, 0)$ is the number of vertices of Γ and $\mu_\Gamma(p, n)$ is the number of vertices of Γ^*.

By the classical Euler theorem $v(\Gamma) - e(\Gamma) + r(\Gamma) = 2$ in S^2 and the regions of the graph Γ $r(\Gamma) = v(\Gamma^*)$, we have

$$e(\Gamma) = v(\Gamma) + r(\Gamma) - 2 = \mu_\Gamma(0, 0) + \mu_\Gamma(p, n) - 2.$$

Lemma 4.8.1. *For any integers a and b with $0 \leq a \leq p$ and $0 \leq b \leq n$,*

(1) $\mu_\Gamma(a, b) + a + b$ is an increasing function of a and b.
(2) $\mu_\Gamma(a, b) - a - b$ is a decreasing function of a and b.

Proof: If we change a transverse smoothing to a parallel smoothing, then we apply $p - a - 1$ positive edges and $n - b$ negative edges in Γ, and the number of components of the resulting trivial link increases by one. Similarly, if one changes a parallel smoothing to a transverse smoothing, then the number of components of the resulting trivial link decreases by one. That is

$$-1 \leq \mu_\Gamma(a + 1, b) - \mu_\Gamma(a, b) \leq 1, \quad -1 \leq \mu_\Gamma(a, b + 1) - \mu_\Gamma(a, b) \leq 1.$$

Thus we obtain the following:

$$\mu_\Gamma(a+1, b) + (a+1) + b - (\mu_\Gamma(a, b) + a + b) = (\mu_\Gamma(a, b) + 1) + 1 - \mu_\Gamma(a, b) \geq 2;$$

$$\mu_\Gamma(a, b+1) + a + (b+1) - (\mu_\Gamma(a, b) + a + b) \geq (\mu_\Gamma(a, b) - 1) + 1 - \mu_\Gamma(a, b) = 0.$$

$$\mu_\Gamma(a+1, b) - (a+1) - b - (\mu_\Gamma(a, b) - a - b) \leq (\mu_\Gamma(a, b) + 1) - 1 - \mu_\Gamma(a, b) = 0;$$

$$\mu_\Gamma(a, b+1) - a - (b+1) - (\mu_\Gamma(a, b) - a - b) \leq (\mu_\Gamma(a, b) - 1) - 1 - \mu_\Gamma(a, b) = -2.$$

The result follows from an easy induction. $\quad\square$

By Lemma 4.8.1, we have, for $0 \leq a \leq p$,

$$\mu_\Gamma(0, b) + b \leq \mu_\Gamma(a, b) + a + b \leq \mu_\Gamma(p, b) + p + b.$$

Hence for all $0 \leq a \leq p, 0 \leq b \leq n$,

$$\mu_\Gamma(0, b) \leq \mu_\Gamma(a, b) + a, \quad \mu_\Gamma(a, b) \leq \mu_\Gamma(p, b) + p - a.$$

Similarly, we apply to the function $\mu_\Gamma(a, b) - a - b$:

$$\mu_\Gamma(a, 0) - a \geq \mu_\Gamma(a, b) - a - b \geq \mu_\Gamma(a, n) - a - n.$$

Therefore, one obtains for all $0 \leq a \leq p, 0 \leq b \leq n$,

$$\mu_\Gamma(a, b) \leq \mu_\Gamma(a, 0) + b, \quad \mu_\Gamma(a, n) \leq \mu_\Gamma(a, b) + n - b.$$

Note that these two functions are constant for different variables.

$$\mu_\Gamma(a, b) - a - b \leq \mu_\Gamma(0, b) - b, \quad \mu_\Gamma(0, b) + b \leq \mu_\Gamma(0, n) + n.$$

For $0 \leq a \leq p$ and $0 \leq b \leq n$,

$$\mu_\Gamma(a, b) \leq \mu_\Gamma(0, n) + a + (n - b).$$

$$\mu_\Gamma(a, b) - a + b - n \leq \mu_\Gamma(0, n). \tag{4.16}$$

From the above $\mu_\Gamma(a, b) \leq \mu_\Gamma(a, 0) + b$, and

$$\mu_\Gamma(a, 0) + a \leq \mu_\Gamma(p, 0) + p,$$

one gets the relation

$$\mu_\Gamma(a, b) + a - b - p \leq \mu_\Gamma(p, 0). \tag{4.17}$$

Lemma 4.8.2. *Let $b_i(\Gamma)$ be the i-th Betti number of a graph Γ as a 1-complex. Then we have the following:*

(1) $\mu_\Gamma(p,0) = b_1(\Gamma_+) + b_1(\Gamma_-^*) + 1$.
(2) $\mu_\Gamma(0,n) = b_1(\Gamma_-) + b_1(\Gamma_+^*) + 1$.
(3) $\mu_\Gamma(p,0) + \mu_\Gamma(0,n) \le c(\tilde{L}) + 2$.

Proof: (1) Let $\Gamma_{+,0}$ be a graph obtained by removing all negative edges from Γ (do nothing on vertices of Γ). Then $\Gamma_{+,0}$ has $\mu_\Gamma(p,0)$-many disconnected components. Let $b_i(\Gamma_{+,0}) = rankH_i(\Gamma_0)$, where Γ_0 is the resulting (possible disconnected) planar subgraph of Γ ($i = 0,1$). We have

$$\mu_\Gamma(p,0) = b_0(\Gamma_{+,0}) + b_1(\Gamma_{+,0}).$$

Note that $b_0(\Gamma_{+,0}) = b_1(S^2 - \Gamma_{+,0}) + 1 = b_1(\Gamma_-^*) + 1$ and $b_1(\Gamma_{+,0}) = b_1(\Gamma_+)$. So

$$\mu_\Gamma(p,0) = b_0(\Gamma_{+,0}) + b_1(\Gamma_{+,0})$$
$$= b_1(\Gamma_-^*) + 1 + b_1(\Gamma_+).$$

(2) Note that $\mu_\Gamma(p,0) = \mu_{\Gamma^*}(0,n)$ and $\mu_\Gamma(0,n) = \mu_{\Gamma^*}(p,0)$. Therefore one gets, by (1),

$$\mu_\Gamma(0,n) = \mu_{\Gamma^*}(p,0)$$
$$= b_1(\Gamma_+^*) + b_1(\Gamma_-^{**}) + 1$$
$$= b_1(\Gamma_+^*) + b_1(\Gamma_-) + 1.$$

(3) Note that $\Gamma = \Gamma_+ \cup \Gamma_- \subset S^2$. $\mu_\Gamma(p,n)$ is the number of boundary components of a regular neighborhood of Γ in S^2. Since $b_1(\Gamma_+ \cap \Gamma_-) = 0$ and $b_0(\Gamma) = 1$ (the graph Γ is connected), one obtains

$$b_1(\Gamma_+) + b_1(\Gamma_-) \le b_1(\Gamma).$$

We also have $b_1(\Gamma) = \mu_\Gamma(p,n) - 1$ (minus the outside component of Γ in S^2). Thus $b_0(\Gamma) + b_1(\Gamma) = \mu_\Gamma(p,n)$. Thus

$$\mu_\Gamma(p,n) \ge 1 + b_1(\Gamma_+) + b_1(\Gamma_-).$$

Similarly, we apply it to Γ^*:

$$\mu_{\Gamma^*}(p,n) \ge 1 + b_1(\Gamma_+^*) + b_1(\Gamma_-^*),$$

which is equivalent to

$$\mu_\Gamma(0,0) = \mu_{\Gamma^*}(p,n) \ge 1 + b_1(\Gamma_+^*) + b_1(\Gamma_-^*).$$

Therefore, by (1) and (2) above,

$$\mu_\Gamma(p,0) + \mu_\Gamma(0,n) = (b_1(\Gamma_+) + b_1(\Gamma_-^*) + 1) + b_1(\Gamma_+^*) + b_1(\Gamma_-) + 1$$
$$= (b_1(\Gamma_+) + b_1(\Gamma_-) + 1) + (b_1(\Gamma_+^*) + b_1(\Gamma_-^*) + 1)$$
$$\le \mu_\Gamma(p,n) + \mu_\Gamma(0,0)$$
$$= 2 + e(\Gamma) = 2 + c(\tilde{L}).$$

$\square$

Proposition 4.8.3. *For any connected projection $\tilde{L}$ of a link L,*

$$Span \langle L \rangle(A) \leq 4c(\tilde{L}).$$

Proof: For a disjoint union $\tilde{L} = \tilde{L}_1 \coprod \tilde{L}_2$, the Kauffman bracket satisfies

$$\langle \tilde{L} \rangle(A) = -(A^2 + A^{-2})\langle \tilde{L}_1 \rangle(A) \cdot \langle \tilde{L}_2 \rangle(A).$$

By definition, $\langle \tilde{L} \rangle(A) = A\langle \tilde{L}_0 \rangle(A) + A^{-1}\langle \tilde{L}_\infty \rangle(A)$ for identical projection except at a small neighborhood of one crossing in $\tilde{L}$.

By applying parallel smoothness on exactly a positive crossings and b negative crossings, each positive crossing contributes A^{-1} for a parallel smoothing. Hence for p positive crossings, A^{-a} from parallel smoothness and A^{p-a} from transverse smoothness contribute from $\tilde{\mathcal{L}}(a,b)$. For a negative crossing, A for parallel smoothing and A^{-1} for the transverse smoothing contribute into the bracket. Hence for n negative crossings, A^b from parallel smoothing and $A^{-(n-b)}$ from transverse smoothing contribute from $\tilde{\mathcal{L}}(a,b)$. The $\tilde{\mathcal{L}}(a,b)$ is trivial link with components $|\tilde{\mathcal{L}}(a,b)|$, and

$$\langle \tilde{\mathcal{L}}(a,b) \rangle(A) = \{-(A^2 + A^{-2})\}^{|\tilde{\mathcal{L}}(a,b)|-1}.$$

Thus we obtain the summation for the Kauffman bracket $\langle \tilde{L} \rangle(A) =$

$$\sum_{\tilde{\mathcal{L}}(a,b)\in S(a,b), 0\leq a\leq p, 0\leq b\leq n} A^{-a+(p-a)} A^{b-(n-b)} \{-(A^2 + A^{-2})\}^{|\tilde{\mathcal{L}}(a,b)|-1}.$$

$$(4.18)$$

By definition of $\mu_\Gamma(a,b)$, we have $|\tilde{\mathcal{L}}(a,b)| \leq \mu_\Gamma(a,b)$. In order to estimate the highest and lowest degree, we get

$$d_{max}\langle \tilde{L} \rangle(A) \leq \max_{a,b}\{p - 2a + 2b - n + 2\mu_\Gamma(a,b) - 2\},$$

$$d_{min}\langle \tilde{L} \rangle(A) \geq \min_{a,b}\{p - 2a + 2b - n - 2\mu_\Gamma(a,b) + 2\}.$$

By (4.16), $\mu_\Gamma(a,b) - a + b - n \leq \mu_\Gamma(0,n)$,

$$p - n - 2a + 2b + 2\mu_\Gamma(a,b) - 2$$
$$= p + n + 2(\mu_\Gamma(a,b) - a + b - n) - 2$$
$$\leq p + n + 2\mu_\Gamma(0,n) - 2.$$

By (4.17), $\mu_\Gamma(a,b) + a - b - p \leq \mu_\Gamma(p,0)$,

$$p - 2a + 2b - n - 2\mu_\Gamma(a,b) + 2$$
$$\geq p - n - 2(\mu_\Gamma(p,0) + p) + 2$$
$$= -2\mu_\Gamma(p,0) - p - n + 2.$$

By Lemma 4.8.2, we get

$$\text{Span}\langle \tilde{L}\rangle(A) = d_{max}\langle \tilde{L}\rangle(A) - d_{min}\langle \tilde{L}\rangle(A)$$
$$\leq p + n + 2\mu_\Gamma(0,n) - 2 - (-2\mu_\Gamma(p,0) - p - n + 2)$$
$$\leq 2(\mu_\Gamma(0,n) + \mu_\Gamma(p,0)) + 2(p+n) - 4$$
$$\leq 2(c(\tilde{L}) + 2) + 2c(\tilde{L}) - 4$$
$$= 4c(\tilde{L}).$$

$\square$

Remark 4.8.4. The proof of Proposition 4.8.3 follows from [Murasugi, 1987]. For the alternating link, Kauffman [1987a] gives a little simpler proof.

Theorem 4.8.5. *For any projection $\tilde{L}$ of a link L,*

$$Span\, V_L(t) \leq c(\tilde{L}) + \lambda - 1,$$

where λ is the number of connected components of $\tilde{L}$. If L has λ split components, then

$$Span\, V_L(t) \leq c(L)\lambda - 1,$$

where $c(L)$ is the crossing number of the link L.

Proof: By applying Proposition 4.8.3 for each connected component $\tilde{L}_i$,

$$\text{Span}\, V_L(t) = \text{Span}\,(\prod_{i=1}^{\lambda} V_{L_i}(t)) \cdot (t^{1/2} + t^{-1/2})^{\lambda-1}$$
$$= \sum_{i=1}^{\lambda} \text{Span}\, V_{L_i}(t) + (\lambda - 1)$$
$$= \sum_{i=1}^{\lambda} \frac{1}{4}\text{Span}\,\langle \tilde{L}_i\rangle(A) + (\lambda - 1)$$
$$\leq \sum_{i=1}^{\lambda} \frac{1}{4}4c(\tilde{L}_i) + (\lambda - 1)$$
$$= c(\tilde{L}) + (\lambda - 1).$$

Thus the result follows. $\square$

Theorem 4.8.6. *If $\tilde{L}$ is a connected proper alternating projection of an alternating link L, then*

$$Span\, V_L(t) = c(\tilde{L}).$$

Proof: Let L be a non-split alternating link and $\tilde{L}$ be its proper connected alternating projection. Thus all crossings of $\tilde{L}$ (same as all edges of the graph Γ of $\tilde{L}$) are positive or negative.

Without loss of generality, let $p = c(\tilde{L}), n = 0, b = 0$. Thus we have $\Gamma = \Gamma_+$ and $\Gamma_- = \emptyset$.

$$\text{Span } V_L(t) = \frac{1}{4}\text{Span } \langle \tilde{L} \rangle (A)$$

$$= \frac{1}{4}\text{Span } \{ \sum_{\tilde{\mathcal{L}}(a,0) \in S(a,0), 0 \leq a \leq p} A^{-a+(p-a)} \{-(A^2 + A^{-2})\}^{|\tilde{\mathcal{L}}(a,0)|-1} \}.$$

We obtain that $d_{max}\langle \tilde{L} \rangle (A) = \max_a \{p - 2a + 2\mu_\Gamma(a,0) - 2\}$. Note that $\mu_\Gamma(a,0) - a = \mu_\Gamma(0,0)$. Therefore

$$d_{max}\langle \tilde{L} \rangle (A) \leq p + 2\mu_\Gamma(0,0) - 2.$$

We need to show that $d_{max}\langle \tilde{L} \rangle (A) = p + 2\mu_\Gamma(0,0) - 2$ since $A^{p+2\mu_\Gamma(0,0)-2}$ appears as $\tilde{L}(0,0) \in S(0,0)$ where $S(0,0)$ has only one link $\tilde{L}(0,0)$. For any $0 < a \leq p$,

$$p - 2a + 2\mu_\Gamma(a,0) - 2 < p + 2\mu_\Gamma(0,0) - 2,$$

from the decreasing function $\mu_\Gamma(a,0) - a$ since $\mu_\Gamma(1,0) = v(\Gamma) - 1$, $\mu_\Gamma(0,0) = v(\Gamma)$ and $\mu_\Gamma(0,0) > \mu_\Gamma(1,0) - 1$ strictly decreasing. Therefore

$$d_{max}\langle \tilde{L} \rangle (A) = p + 2\mu_\Gamma(0,0) - 2.$$

Similarly, $d_{min}\langle \tilde{L} \rangle (A) = \min_a \{p - 2a - 2\mu_\Gamma(a,0) + 2\}$. Due to $\mu_\Gamma(a,0) + a \leq \mu_\Gamma(p,0) + p$,

$$d_{min}\langle \tilde{L} \rangle (A) \geq p - 2\mu_\Gamma(p,0) - 2p + 2$$
$$= -2\mu_\Gamma(p,0) - p + 2.$$

Since there is only one $\tilde{\mathcal{L}}(p,0)$ in $S(p,0)$ and $\mu_\Gamma(p,0) = \mu_\Gamma(p-1,0) + 1$, for $0 \leq a < p$,

$$\mu_\Gamma(a,0) + a < \mu_\Gamma(p,0) + p,$$

gives $p - 2a - 2\mu_\Gamma(a,0) + 2 > -p - 2\mu_\Gamma(p,0) + 2$. Thus $d_{min}\langle \tilde{L} \rangle (A) =$

$-2\mu_\Gamma(p,0) - p + 2$. Thus one gets the following identities:

$$\begin{aligned}
\text{Span}\, V_L(t) &= \frac{1}{4}(d_{max}\langle \tilde{L}\rangle(A) - d_{min}\langle \tilde{L}\rangle(A)) \\
&= \frac{1}{4}(p + 2\mu_\Gamma(0,0) - 2 - (-2\mu_\Gamma(p,0) - p + 2)) \\
&= \frac{1}{2}(\mu_\Gamma(0,0) + \mu_\Gamma(p,0) + p - 2) \\
&= \frac{1}{2}(v(\Gamma) + v(\Gamma^*) + c(\tilde{L}) - 2) \\
&= \frac{1}{2}((c(\tilde{L}) + 2) + c(\tilde{L}) - 2) \\
&= c(\tilde{L}).
\end{aligned}$$

$\square$

Remark 4.8.7. The coefficients of the maximal and minimal degree terms are ± 1.

Corollary 4.8.8. *For any alternating link L with λ split components,*
$$Span\, V_L(t) = c(L) + \lambda - 1.$$

Proof: By Theorem 4.8.5, $\text{Span}\, V_L(t) \le c(L) + \lambda - 1$. By Theorem 4.8.6, $\text{Span}\, V_L(t) = c(\tilde{L}) + \lambda - 1$, and $c(\tilde{L}) \ge c(L)$, we have $\text{Span}\, V_L(t) = c(L) + \lambda - 1$. $\square$

Theorem 4.8.9. *Let L be a prime link. Then for any non-alternating projection $\tilde{L}$ of L,*
$$Span\, V_L(t) < c(\tilde{L}).$$

Proof: Assume $\tilde{L}$ is connected and proper. Let Γ be the graph of $\tilde{L}$ with a cut vertex (deleting the cut-vertex results in two disconnected graphs). So $\Gamma = \Gamma_1 \vee \Gamma_2$ and $\tilde{L}_i$ is a link projection with graph Γ_i $(i = 1, 2)$ and $L = L_1 \# L_2$. If L is prime, then one of L_i, say L_1, is unknotted, and $V_L(t) = V_{L_2}(t)$. By Theorem 4.8.5,

$$\begin{aligned}
\text{Span}\, V_L(t) &= \text{Span}\, V_{L_2}(t) \\
&\le c(\tilde{L}_2) \\
&< c(\tilde{L}_1) + c(\tilde{L}_2) = c(\tilde{L}),
\end{aligned}$$

since $c(\tilde{L}_1) > 0$. Therefore Γ does not contain any cut-vertex. If $\text{Span}\, V_L(t) = c(\tilde{L})$, then $\text{Span}\, \langle \tilde{L}\rangle(A) = 4c(\tilde{L})$. From the proof of Theorem 4.8.5, we have

$$4c(\tilde{L}) = \text{Span}\, \langle \tilde{L}\rangle(A) \le 2c(\tilde{L}) + 2\mu_\Gamma(p,0) + 2\mu_\Gamma(0,n) - 4,$$

$$c(\tilde{L}) + 2 \leq \mu_\Gamma(p, 0) + 2\mu_\Gamma(0, n).$$

By Lemma 4.8.2 (3), $\mu_\Gamma(p, 0) + 2\mu_\Gamma(0, n) = c(\tilde{L}) + 2$. By Lemma 4.8.2 (1) and (2),

$$(b_1(\Gamma_+) + b_1(\Gamma_-) + 1) + (b_1(\Gamma_+^*) + b_1(\Gamma_-^*) + 1) = c(\tilde{L}) + 2$$

$$b_1(\Gamma_+) + b_1(\Gamma_-) + 1 = \mu_\Gamma(p, n) = b_1(\Gamma) + b_0(\Gamma)$$

$$b_1(\Gamma_+^*) + b_1(\Gamma_-^*) + 1 = \mu_{\Gamma^*}(p, n) = \mu_\Gamma(0, 0) = b_1(\Gamma^*) + b_0(\Gamma^*).$$

Hence we obtain

$$b_1(\Gamma) = b_1(\Gamma_+) + b_1(\Gamma_-), \quad b_1(\Gamma^*) = b_1(\Gamma_+^*) + b_1(\Gamma_-^*).$$

This is only possible when Γ is a positive or negative graph. Therefore $\tilde{L}$ must be an alternating projection of the link. □

Remark 4.8.10. The link $L = 4_1 \# 4_1$ has a minimal non-alternating projection, and $L = 4_1 \# 4_1$ is alternating. So Span $V_L(t) = c(\tilde{L})$. But $4_1 \# 4_1$ is not prime. The primeness of L is necessary in Theorem 4.8.9.

Theorem 4.8.11. *Let L be a non-split link. Then Span $V_L(t) = c(\tilde{L})$ if and only if L is the connected sum of alternating links.*

Proof: If a non-split link L is the connected sum of alternating links L_i, then L has a connected proper projection $\tilde{L}$ such that each L_i has a connected proper alternating projection $\tilde{L}_i$ with $c(\tilde{L}) = \sum_i c(\tilde{L}_i)$. Since $V_L(t) = \prod_i V_{L_i}(t)$ and $\langle \tilde{L} \rangle(A) = \prod_i \langle \tilde{L}_i \rangle(A)$, by Theorem 4.8.6,

$$\text{Span } V_L(t) = \sum_i \text{Span } V_{L_i}(t) = \sum_i c(\tilde{L}_i) = c(\tilde{L}).$$

If Span $V_L(t) = c(\tilde{L})$ or Span $\langle \tilde{L} \rangle(A) = 4c(\tilde{L})$ for some connected proper projection $\tilde{L}$ of a link L, from the proof of Theorem 4.8.9 we have

$$\mu_\Gamma(p, 0) + \mu_\Gamma(0, n) = c(\tilde{L}) + 2.$$

Thus Γ must be either positive or negative graph; or Γ has cut-vertices $v_1, \cdots, v_r$ such that $\Gamma = \Gamma_1 \vee_{v_1} \Gamma_2 \vee_{v_2} \cdots \vee_{v_r} \Gamma_{r+1}$ separate into positive and/or negative graphs. Thus L is the connected sum of (positive or negative) alternating links. □

Tait Conjecture I: Reduced alternating diagram has minimal crossing number. The Tait conjecture I follows from the following theorem.

Theorem 4.8.12. *The minimal projection of an alternating link is alternating.*

Proof: By Theorem 4.8.6, $\tilde{L}$ is a connecting proper alternating projection of an alternating link L, then $\operatorname{Span} V_L(t) = c(\tilde{L})$.

By Theorem 4.8.5, $c(\tilde{L}) \le c(L)$ if L has a split component. So $c(\tilde{L}) = c(L)$.

If L is a non-split alternating link, then by Theorem 4.8.11, $\operatorname{Span} V_L(t) = c(\tilde{L})$ if and only if L is a connected sum of alternating links. $\quad\square$

An alternating link always has an alternating projection with minimum number of double points among all projections. Furthermore, by Theorem 4.8.9, a non-alternating projection of a prime alternating link cannot be minimal.

Theorem 4.8.13. *(Tait Conjecture) Two alternating projections of an alternating link have the same number of double points.*

Proof: Since $\operatorname{Span} V_L(t) = c(\tilde{L})$ by Theorem 4.8.11, for another alternating projection $\tilde{L}_1$ of L, then $\operatorname{Span} V_L(t) = c(\tilde{L}_1)$. Therefore

$$c(\tilde{L}) = \operatorname{Span} V_L(t) = c(\tilde{L}_1).$$

$\quad\square$

Corollary 4.8.14. *If K is an alternating and amphicheiral knot, then any proper alternating projection $\tilde{K}$ has an even number of double points.*

Proof: Since $K = K^*$ and $V_K(t) = V_{K^*}(t^{-1}) = V_K(t^{-1})$, the Jones polynomial $V_k(t)$ is symmetric, hence $\operatorname{Span} V_K(t)$ is even. By Theorem 4.8.6,

$$c(\tilde{K}) = \operatorname{Span} V_K(t) = 2d_{max} V_K(t).$$

$\quad\square$

Corollary 4.8.15. *If L_i is alternating link, then $c(L_1 \# L_2) = c(L_1) + c(L_2)$.*

Proof: By Corollary 4.8.8,

$$\operatorname{Span} V_{L_1 \# L_2}(t) = \operatorname{Span} V_{L_1}(t) + \operatorname{Span} V_{L_2}(t) = c(\tilde{L}_1) + c(\tilde{L}_2).$$

Taking $\tilde{L}_i$ as the minimal projection, one gets

$$\begin{aligned}
c(\tilde{L}_1 \# \tilde{L}_2) &= c(L_1) + c(L_2) \\
&\ge c(L_1 \# L_2) = \operatorname{Span} V_{L_1 \# L_2}(t) \\
&= c(\tilde{L}_1) + c(\tilde{L}_2) \\
&\ge c(L_1) + c(L_2).
\end{aligned}$$

Hence $c(L_1 \# L_2) = c(L_1) + c(L_2)$. $\quad\square$

So Conjecture 3.4.5 holds for alternating links.

Define $w(\tilde{L}) = \sum_{v \in \tilde{L}} \varepsilon_v$, the writhe number of the projection $\tilde{L}$.

Theorem 4.8.16. *(Tait Conjecture II) Any two reduced alternating diagrams of a given knot have equal writhe.*

Proof: Since L is a special alternating link, any proper alternating projection of the link L must be proper alternating. Let $\tilde{L}_i$ $(i = 1, 2)$ be the proper alternating projection of the special alternating link L. Then $w(\tilde{L}_i) = \pm c(\tilde{L}_i), i = 1, 2$. Note that $w(\tilde{L}_1) = c(\tilde{L}_1) = c(L)$ or $-c(\tilde{L}_1) = -c(L)$. Similarly $w(\tilde{L}_2) = \pm c(L)$. Therefore we obtain $w(\tilde{L}_1) = \pm w(\tilde{L}_2)$. The sign of the signature of L is determined by the sign of $w(\tilde{L})$ of any special alternating projection $\tilde{L}$ of L. Therefore $w(\tilde{L}_1) = w(\tilde{L}_2)$. $\square$

Remark 4.8.17. $w(\tilde{L})$ is a knot type invariant for special alternating knots and their proper alternating projections.

4.9　The plat approach to Jones polynomials

Jones [1987] also defined his invariant through the plat constructions. Birman [1974] has proved that the analogue of Markov theorem is true for the plats.

Let $D^3 = B^3(2) \subset \mathbb{R}^3$ be a 3-ball with radius 2 centered at the origin in the standard oriented 3-space $\mathbb{R}^3$. Let $A = A_1 \coprod \cdots \coprod A_m$ be a collection of m unknotted and unlinked arcs which are embedded in D^3 with $Q_{2m} = \partial A$ a set of $2m$ points on $S^2 = \partial D^3$. Let $t : \mathbb{R}^3 \to \mathbb{R}^3$ be a translation, and $\tilde{D}_3 = t(D^3)$ and $\tilde{A} = t(A)$. Suppose that $f : (S^2, \partial A) \to (S^2, \partial A)$ is any orientation-preserving homeomorphism which keeps the set ∂A as a set. Define an identification space $(D^3, A) \cup_f (\tilde{D}_3, \tilde{A}) = (D^3, A) \cup_f (t(D^3), t(A))$ by pasting $\partial(D^3, A) = (S^2, Q_{2m})$ to $\partial t(D^3, A) = (S^2, Q_{2m})$ by fixing $tf(p) = p \in S^2$. This is a 3-sphere S^3 with a genus 0 Heegaard splitting. The subset $V = A \cup_f A$ is a collection of $\mu \leq m$ disjoint simple closed curves in $S^3 = D^3 \cup_f \tilde{D}^3$, a link L. If a link L is represented in this way, then L is displayed as a plat on $2m$ strings.

Any isotopic deformation of the surface mapping f which leaves the point set Q_{2m} fixed can be extended to an isotopy of $(D^3, A) \cup_f (\tilde{D}^3, \tilde{A})$. If f and f^* are isotopic maps of (S^2, Q_{2m}), then f and f^* are isotopic maps of $(D^3, A) \cup_f (\tilde{D}^3, \tilde{A})$ and $(D^3, A) \cup_{f^*} (t^* D^3, t^* A)$ such that they define equivalent plats $P = A \cup_f t(A)$ and $P^* = A \cup_{f^*} t^*(A)$. Hence they correspond to a link type with an element in the mapping class group

$M(0, 2m)$ of the punctured surface $S^2 \setminus Q_{2m}$.

For the plat construction, the essential feature of the link is concealed in the surface mapping f which is used to glue the boundary ∂A with $\partial \tilde{A}$, since both $A = A_1 \coprod \cdots \coprod A_m$ and $\tilde{A} = \tilde{A}_1 \coprod \cdots \coprod \tilde{A}_m$ are unknotted and unlinked. [Birman, 1974, Figure 22] illustrates a method to turn the plat construction into a braid and the link in S^3.

Theorem 4.9.1. *Every link type can be represented by a plat.*

Proof: By the Alexander theorem, every link is combinatorially equivalent to a closed braid, and every link type can be represented by a closed braid, say m strings in B_m. If a link type L is realized by a closed braid obtained from a geometric braid $\beta \in B_m$, then L is also realized by the $2m$-plat which is associated with the geometric $2m$-braid $\beta_0 \beta \beta_0^{-1}$, where β_0 is a $2m$ braid given by $\beta_0 = (\sigma_2 \sigma_3 \cdots \sigma_{m-1})(\sigma_3 \sigma_4 \cdots \sigma_{m-2}) \cdots (\sigma_m \sigma_{m+1})$. Then $\hat{\beta} = L$ is obtained by adding trivial braided strings. $\qquad \square$

Theorem 4.9.1 is Theorem 5.1 of [Birman, 1974] with its proof. Similarly, there is a similar link group presentation given by the plat representation. We only state in the following and refer readers to [Birman, 1974, Corollary 5.11].

Proposition 4.9.2. *Let $\beta \in B_{2m}$. Suppose that the braid action β on the free group F_{2m} is given by the characterization of braid representation in §3.7 with $n = 2m$. Let L be a link type which is realized by the $2m$ plat defined by the braid β. Then the link group $\pi_1(S^3 \setminus L)$ admits the presentation: for $i = 1, \cdots, m$;*

$$\text{generators: } z_1, \cdots, z_{2m},$$
$$\text{relations: } \quad z_{2i-1} = z_{2i}^{-1}$$

$$A_{2i-1}(z_1, \cdots, z_{2m}) z_{\mu_{2i-1}} A_{2i-1}^{-1}(z_1, \cdots, z_{2m})$$

$$= A_{2i}(z_1, \cdots, z_{2m}) z_{\mu_{2i}} A_{2i}^{-1}(z_1, \cdots, z_{2m}).$$

Furthermore, every link group admits a presentation of this type for some (non-unique) braid automorphism $\beta \in B_{2m}$.

Theorem 4.9.3. *[Birman, 1976] For each m, let C_m be the subgroup of B_{2m} generated by the set $X \cup Y \cup Z$, where*

$$X = \{\sigma_{2i-1} : i = 1, \cdots, m\},$$
$$Y = \{\sigma_{2i} \sigma_{2i-1} \sigma_{2i+1} \sigma_{2i} : i = 1, \cdots, m-1\},$$
$$Z = \{\sigma_{2i} \sigma_{2i-1} \sigma_{2i+1}^{-1} \sigma_{2i}^{-1} : i = 1, \cdots, m-1\}.$$

Let $S_k : B_{2k} \to B_{2k+2}$ be a map defined by $S_k(\beta) = \beta\sigma_{2k}$. Then

(1) If $\beta \in B_{2k}$, the link type $\overline{x\beta y}$ is isotopic to $\overline{\beta}$ for $x, y \in C_m$ and $\overline{S_m(\beta)}$ is isotopic to $\overline{\beta}$.

(2) The equivalent relation on $\coprod_m B_{2m}$ generated by the two moves ((i) $\beta \mapsto x\beta y$ and (ii) $\beta \mapsto S_m(\beta)$ for $x, y \in C_m$) is the same as the equivalence relation given by isotopy of the plats.

It is not obvious to tie the plat closure with the Hecke algebra. Theorem 4.9.1 shows that the link given by a plat is really closely related to the braid closure with trivial braided strings, and the plat construction is really dependent upon the mapping class group of a punctured two sphere. The plat closure of $\beta \in B_{2m}$ really only dependent on the image of β in the mapping class group of the punctured 2-sphere $S^2 \setminus D_{2m}$.

Jones [1987] showed that the idempotent $p_m = e_1 e_3 \cdots e_{2m-1}$ in A_{2m-1} defined in subsection 5.6 is minimal, i.e., $p_m A_{2m-1} p_m \subset \mathbb{C}p_m$. If ρ_Y is the irreducible representation of A_{2m-1} corresponding to the Young diagram with at most two columns Y, then $\rho_Y(p_m) \neq 0$ if and only if the Young diagram Y is rectangular with two columns.

For the 2×2 square Young diagram Y_2, $\rho_{Y_2}(e_1) = \rho_{Y_2}(e_3)$, since both $\rho_{Y_2}(e_1)$ and $\rho_{Y_2}(e_3)$ are minimal idempotent in the two dimensional representation ρ_{Y_2} and both dominate the minimal idempotent $\rho_{Y_2}(e_1 e_3) = \rho_{Y_2}(p_2)$.

In general, the number of nodes in the second (smaller) column of the Young diagram Y is the largest m for which the product of the commuting subset of size m of the e_i's is non-zero in the corresponding representation of A_n.

Definition 4.9.4. The linear functional $\phi : \cup_m A_{2m-1} \to \mathbb{C}$ is defined to be

$$\phi(x)p_m = p_m x p_m, \quad x \in A_{2m-1}.$$

Lemma 4.9.5. (1) The linear functional ϕ given in Definition 4.9.4 is well-defined.

(2) $\phi(x) = \frac{1}{\tau^m} tr(p_m x p_m)$ for $x \in A_{2m-1}$.

Proof: (1) Suppose that

$$p_{m+1} x p_{m+1} = \lambda p_{m+1}, \quad p_m x p_m = \mu p_m.$$

By definition of p_m, we have $p_{m+1} = p_m e_{2m+1}$ and $\lambda = \mu$. Hence, the linear functional ϕ is well-defined.

(2) $tr(\phi(x)p_m) = tr(p_m x p_m)$ and $tr(p_m) = \tau^m$. Hence $\tau^m tr(\phi(x)) = tr(p_m x p_m)$. $\qquad \square$

By Lemma 4.9.5 (1), the linear functional ϕ can also be defined by

$$\lim_{m \to \infty} (p_m x p_m - \phi(x)p_m) = 0.$$

Lemma 4.9.6. *Let* $g_i = te_i - (1 - e_i) \in A_n$. *Then*

(i) $p_m g_{2i-1}^{\pm 1} = g_{2i-1}^{\pm 1} p_m = t^{\pm 1} p_m$ *for* $i = 1, \cdots, m$.

(ii) $p_m(g_{2i}g_{2i-1}g_{2i+1}g_{2i}) = (g_{2i}g_{2i-1}g_{2i+1}g_{2i})p_m = t p_m$ *for* $i = 1, 2, \cdots, m - 1$.

(iii) $p_m(g_{2i}g_{2i-1}g_{2i+1}^{-1}g_{2i}^{-1}) = (g_{2i}g_{2i-1}g_{2i+1}^{-1}g_{2i}^{-1})p_m = p_m$ *for* $i = 1, 2, \cdots, m - 1$.

Proof: (i) is trivial from the definition. For (ii) and (iii), it suffices to consider the case $m = 2$ since all the e_i's except two are irrelevant to the calculation. Let f be the minimal central idempotent of A_3 corresponding to the Young diagram of Y_2 without lower right square.

(ii) $(g_2 g_1 g_3 g_2)g_1(g_2^{-1}g_1^{-1}g_3^{-1}g_2^{-1}) = g_3$ follows from the braid group relations. Thus the element $(g_2 g_1 g_3 g_2)$ commutes with $e_1 e_3 = p_3$. Hence,

$$p_3(g_2 g_1 g_3 g_2) = p_3 g_2 f g_1 g_3 g_2 = p_3 g_2 g_1^2 g_2 p_3 = \phi(g_2 g_1^2 g_2)p_3.$$

We need to calculate $\phi(g_2 g_1^2 g_2) = \frac{1}{\tau^2}tr(p_3 g_2 g_1^2 g_2) = \frac{1}{\tau}tr(e_1 g_2 g_1^2 g_2)$. Since $g_1^2 = (t-1)g_1 + t$, we obtain

$$\phi(g_2 g_1^2 g_2) = \frac{t-1}{\tau}tr(e_1 g_2 g_1 g_2)$$

$$= \frac{t}{\tau}tr(e_1 g_2^2)$$

$$= (t-1)t^2 tr(g_2) + t \cdot tr((t^2 - 1)e_2 + 1)$$

$$= \frac{-t^3 + t^2 + t^3 + t}{1 + t} = t.$$

(iii) By the definition of f, $p_3 = p_3 f$ and

$$p_3 g_2 g_1 g_3^{-1} g_2^{-1} = p_3 f g_2 g_1 g_3^{-1} g_2^{-2} = p_3,$$

by $\rho_{Y_2}(e_1) = \rho_{Y_2}(e_3)$. The result follows. $\qquad \square$

Lemma 4.9.6 builds the connection of ϕ with plats. For $\beta \in B_{2m}$ and $x, y \in C_m$,

$$\phi(x\beta y) = t^k \phi(\beta), \quad \text{for some } k \in \mathbb{Z}.$$

Proposition 4.9.7. *(1) For* $x \in A_{2m-1}$, $\phi(xg_{2m}) = -\frac{1}{1+t}\phi(x)$;

(2) If $\overline{\alpha} = \overline{\beta}$ *for* $\alpha \in B_{2m}$ *and* $\beta \in B_{2n}$, *then there exists* $k \in \mathbb{Z}$ *such that*

$$(-(t+1))^{m-1}\phi(\pi_0(\alpha)) = t^k(-(t+1))^{n-1}\phi(\pi_0(\beta)).$$

Proof: (1) By Lemma 4.9.5,

$$\phi(xg_{2m}) = \frac{1}{\tau^{m+1}} tr(xg_{2m}p_m e_{2m+1})$$

$$= \frac{1}{\tau^m} tr(g_{2m}) tr(xp_m)$$

$$= -\frac{1}{1+t}\phi(x).$$

(2) By the relation $\phi(x\beta y) = t^k\phi(\beta)$ and Theorem 4.9.3, it suffices to check that

$$(-(t+1))^{m-1}\phi(\pi_0(\alpha)) = t^k(-(t+1))^m\phi(\pi_0(S_m(\alpha))),$$

for the second move $\alpha \mapsto S_m(\alpha)$. This follows from (1) of this theorem. $\square$

Now a closed braid is a special kind of plat, and there is a potential new invariant for unoriented links defined from the construction of plats.

Theorem 4.9.8. *(1) Let $\psi : \cup_n A_n \to \mathbb{C}$ be a linear functional with $\psi(1) = 1$ and $\psi(xe_{n+1}y) = \tau\psi(xy)$ for $x, y \in A_n$. Then $\psi = tr$.*

(2) Let $L = \bar{\beta}$ be an unoriented link with $\beta \in B_{2m}$. Then there exists a $2k \in \mathbb{Z}$ such that

$$V_L(t) = t^k(-(t+1))^{m-1}\phi(\pi_0(\beta)).$$

Proof: (1) By induction assume that $\psi = tr$ on A_n. Then $\psi = tr$ on the subspace of A_{n+1} spanned by A_n and all elements of the form $xe_{n+1}y$ for $x, y \in A_n$. Then the result follows from

$$\psi(xe_{n+1}y) = \tau\psi(xy) = \tau tr(xy) = tr(xe_{n+1}y).$$

(2) Let

$$\Omega_n = (\sigma_2\sigma_3\cdots\sigma_{2n-1})(\sigma_3\sigma_4\cdots\sigma_{2n-2})\cdots(\sigma_n\sigma_{n+1}) \in B_{2n}.$$

We have $\overline{\Omega_n\beta\Omega^{-1}} = \bar{\beta} = \bar{\alpha}$ for $\alpha \in B_n$ and some integer n as isotopic unoriented links. By Proposition 4.9.7 and the definition of $V_{\bar{\alpha}}$, it suffices to prove that

$$\phi(\pi_0(\Omega_n\beta\Omega^{-1})) = tr(\pi_0(\alpha)).$$

Define $\psi : \cup_n A_n \to \mathbb{C}$ by the following, for $x \in A_{n-1}$,

$$\psi(x) = \phi(\pi_0(\Omega_n)x\pi_0(\Omega_n)^{-1}).$$

Then we have $\psi = tr$: ψ is well-defined, since for $x \in A_{n-1}$,

$$\phi(\pi_0(\Omega_{n+1})x\pi_0(\Omega_{n+1})^{-1}) = \phi(\pi_0(\Omega_n)x\pi_0(\Omega_n)^{-1}).$$

By a straightforward check in the braid group, $\Omega_p \sigma_i \Omega_p^{-1} = \sigma_{2i-1}^{-1} \sigma_{2i} \sigma_{2i-1}$ for $1 \leq i \leq p - 1$. Hence,

$$\pi_0(\Omega_p) \sigma_i \pi_0(\Omega_p^{-1}) = g_{2i-1}^{-1} g_{2i} g_{2i-1}.$$

The conjugation by the element $\pi_0(\Omega_{n+1})$ has the same effect on A_{n-1} as the conjugation by $\pi_0(\Omega_n)$. Therefore the map ψ is well-defined. We have $\psi(1) = 1$ by the definition, and all we need to show is that $\psi(x e_{n+1} y) = \tau \psi(xy)$ for $x, y \in A_{n-1}$. By the induction,

$$\begin{aligned}
&\psi(x e_{n+1} y) \\
&= \phi(\pi_0(\Omega_{n+1}) x e_n y \pi_0(\Omega_{n+1})^{-1}) \\
&= \phi(\pi_0(\Omega_n) x \pi_0(\Omega_n)^{-1}) g_{2n-1}^{-1} e_{2n} g_{2n-1} \pi_0(\Omega_n) y \pi_0(\Omega_n)^{-1}) \\
&= \tau^{-n-1} tr(p_{n+1} \pi_0(\Omega_n) x \pi_0(\Omega_n)^{-1} g_{2n-1}^{-1} e_{2n} g_{2n-1} \pi_0(\Omega_n) y \pi_0(\Omega_n)^{-1}) \\
&= \tau^{-n+1} tr(p_n \pi_0(\Omega_n) x y \pi_0(\Omega_n)^{-1}) \\
&= \tau \left(\tau^{-n} \phi(\pi_0(\Omega_n) x y \pi_0(\Omega_n)^{-1}) \right) \\
&= \tau \psi(xy).
\end{aligned}$$

$\square$

Theorem 4.9.8 gives another interpretation of the Jones polynomial from the plat construction point of view. If L_1 and L_2 are two oriented links which are isotopic as unoriented links, then there is a $k \in \mathbb{Z}$ such that

$$V_{L_1}(t) = t^k V_{L_2}(t).$$

The integer $k = k(L_1, L_2)$ can be determined from the linking numbers in the special values of the Jones polynomial.

Corollary 4.9.9. *There exists $2k_1, 2k_2 \in \mathbb{Z}$ such that*

$$V_{L_+} - t^{k_1} V_{L_-} = t^{k_2} (1 - t) V_\infty.$$

Proof: Changing a link into a plat is easy, being achieved simply by threading local maxima and minima through the link. Thus there are braids $\beta_1, \beta_2 \in B_{2m}$ such that $L_- = \beta_1 \sigma_k \beta_2$, $L_+ = \beta_1 \sigma_k^{-1} \beta_2$ and $L_0 = \beta_1 \beta_2$. By the relation g_i^2 from the Hecke algebra, we have

$$\pi_0(\beta_1 \sigma_k \beta_2) - t\pi_0(\beta_1 \sigma_k^{-1} \beta_2) = (t - 1)\pi_0(\beta_1 \beta_2).$$

Then the result follows from taking ϕ on both sides. $\square$

If L_+, L_- and L_∞ are knots, then we have

$$V'_{L_+}(1) - k_1 V_{L_-}(1) - V'_{L_-}(1) = -V_{L_\infty}(1).$$

Therefore,

$$-3(-2)^{\mu(L_+)-2} - k_1(-2)^{\mu(L_-)-1} + 3(-2)^{\mu(L_-)-2} = -(-2)^{\mu(L_\infty)-1},$$

determines $k_1 = 1$. Taking second derivatives, we have

$$V_{L_+}''(1) - V_{L_-}''(1) = -2k_2,$$

and $V_{L_+}''(1) - V_{L_-}''(1)$ is (-6) times the linking number $lk(L_0)$ of the oriented link L_0 formed by eliminating the crossing according to the orientation. Hence, $k_2 = 3lk(L_0)$. In this case with all knots,

$$V_{L_+} - tV_{L_-} = t^{3lk(L_0)}(1-t)V_\infty.$$

This can be further checked by the right-handed trefoil, where both L_- and L_∞ are unknots.

By induction, the Jones polynomial V_L is determined by certain linking numbers of two component links obtained by successively eliminating crossings of L so as to obtain knots.

There is a remarkable connection with statistical mechanics. In fact, the algebra A_n has been used by Temperley and Lieb to partially solve a statistical mechanical model known as the Potts model. The linear functional ϕ defined above is essentially the partition function in the statistic mechanics. A solution of the Potts model would be an explicit expression for $f(x,y)$, where $f(x,y) =$

$$\lim_{n \to \infty} \frac{1}{n^2} \log \left(\phi[(1_x e_1)(1 + x e_3) \cdots (1 + x e_{n-1})(y + e_2)(y + e_4) \cdots (y + e_n)]^n \right).$$

The Potts model is defined for a system of atoms arrayed on the vertices of a regular lattice in $\mathbb{R}^2$. This is how the Potts model relate to closed braids and regular knot projections. The algebra A_n are a computational device known as transfer matrices.

4.10 Homological definition of the Jones polynomial

It is natural to ask if there is a more topological definition of the Jones polynomial, rather than the algebraic properties of a braid. A partial answer was provided by Witten [1989] in his groundbreaking paper from the topological quantum field theory approach. Reshitikhin and Turaev [1991] gave a mathematically rigorous formulation of this topological quantum field theory via the quantum groups instead of the Feynman path integral. Lawrence [1990] gave a topological interpretation of the representation of B_{2n} in the construction of the Jones polynomial [Jones, 1987], and interpreted the Jones polynomial V_L as an intersection pairing between a certain element of cohomology and the image under a braid $\beta \in B_{2n}$ of a certain element of homology in [Lawrence, 1993]. We present the approach in [Bigelow, 2001b] with a more explicit description of the relevant elements of homology and cohomology classes.

This interpretation of V_L may lead to progress toward a true topological definition of the Jones polynomial. However, it has a flaw in common with many definitions since it is first defined for links with a special form, then shown to be an invariant under certain moves. The special form in the Bigelow approach is the plat closure of a braid and equivalent moves are given in [Birman, 1976] (see also Theorem 4.9.3).

Let $\overline{C}$ be the set of ordered m-tuples of distinct points in $D_{2m} = D^2 \setminus Q_{2m}$, and $C = \overline{C}/S_m$ be the quotient of $\overline{C}$ by the symmetric group S_m, where C is the set of unordered m-tuples of distinct points in D_{2m}. We would like to define a homomorphism $\Phi : \pi_1(C) \to \langle q \rangle \oplus \langle t \rangle$ from $\pi_1(C)$ to an abelian group $\langle q \rangle \oplus \langle t \rangle$ with a universal property that any map from $\pi_1(C)$ to an abelian group which is invariant under the action of B_{2m} must factor uniquely through the homomorphism Φ.

Let $\alpha : I \to C$ be a loop in C, where the loop α of unordered n-tuples in the disk is considered to be a braid in B_m. Let b be the image of this braid under the usual abelianization map from B_m to $\mathbb{Z}$ by taking each generator σ_i to 1. The map $s \mapsto \{p_1, \cdots, p_{2m}\} \cup \alpha(s)$ determines a braid in B_{3m}. Let b' be the image of this braid under the usual abelianzation map from B_{3m} to $\mathbb{Z}$. Both b and b' have the same parity, equal to the parity of the image of the braid α in the symmetric group S_m. Define $a = \frac{b'-b}{2}$.

Define $\Phi(\alpha) = q^{a(\alpha)} t^{b(\alpha)}$. For each loop in C represented by $\alpha : I \to C$, the $a(\alpha) = \frac{b'(\alpha)-b(\alpha)}{2}$ is the total winding number of these arcs $\alpha(s) = (\alpha_1(s), \cdots, \alpha_m(s))\}$ in D_{2m} around the puncture points Q_{2m}, and $b(\alpha)$ is twice the winding number of these arcs around each other. Let $\tilde{C}$ be the covering space of C corresponding to Φ. The group of covering transformation of $\tilde{C}$ is $\langle q \rangle \oplus \langle t \rangle$.

Definition 4.10.1. Suppose A and B are immersed m-manifolds in C such that at least one of them is closed, and the other intersects every compact subset of C in a compact set. Let $\tilde{A}$ and $\tilde{B}$ be the lifts of A and B respectively in $\tilde{C}$. Let $q^a t^b \tilde{A}$ be the image of $\tilde{A}$ under the covering transformation $q^a t^b$. There is a well-defined algebraic intersection number $(q^a t^b \tilde{A}, \tilde{B}) \in \mathbb{Z}$. Define the pairing,

$$\langle \tilde{A}, \tilde{B} \rangle = \sum_{a,b \in \mathbb{Z}} (q^a t^b \tilde{A}, \tilde{B}) q^a t^b \in \mathbb{Z}[q^{\pm 1}, t^{\pm 1}].$$

The algebraic intersection number is non-zero for finitely many terms of $(a, b) \in \mathbb{Z} \times \mathbb{Z}$ since the number of non-zero terms is at most the geometric intersection number of A and B in C.

Let $d_1, \cdots, d_m$ be distinct points on ∂D which lies in the lower half plane, ordered from left to right. Let $c = \{d_1, \cdots, d_m\} \in C$ and fix a choice $\tilde{c} \in \tilde{C}$ in the covering space.

A fork diagram in D_{2m} consists of maps

$$E_1, \cdots, E_m : I \to D,$$

as tine edges, and maps

$$E_1', \cdots, E_m' : I \to D,$$

as handles, subject to the following conditions.

- The tine edges are disjoint embeddings of the interior of I into D_{2m}, and map the end points of I to the puncture points in D (not necessarily injectively here);
- The handles are disjoint embeddings of I into D_{2m};
- E_i' is the path from d_i to a point in the interior of E_i.

Such a fork diagram determines an immersed open m-ball $\tilde{B}$ in $\tilde{C}$, since $E_1 \times \cdots \times E_m$ maps the interior of $I \times \cdots \times I$ into $\tilde{C}$ with its projection B into C. Let γ be the path in C given by

$$\gamma(s) = \{E_1'(s), \cdots, E_m'(s)\}.$$

Lift the path γ to $\tilde{\gamma}$ with starting $\tilde{c}$ in $\tilde{C}$. The $\tilde{B}$ as the lift of B containing $\tilde{\gamma}(1)$ is an oriented open m-ball in $\tilde{C}$.

Let $\tilde{S}$ be the open m-ball in $\tilde{C}$ corresponding to the standard fork diagram shown [Bigelow, 2001b, Figure 1], where all tine edges are oriented from left to right. For each tine edge E_i in the standard fork diagram, let F_i be the map from S^1 to the figure-eight as shown in [Bigelow, 2001b, Figure 2]. Then $F_1 \times \cdots \times F_m$ is an immersion from the m-torus into $\tilde{C}$. Let T be the projection of this map into C, and specify a lift $\tilde{T}$ of T to $\tilde{C}$.

If $[\beta] \in B_{2m} = M(0, m)$ in the mapping class group of punctured 2-sphere, then β is a homemorphism from D_{2m} to itself. Let β' be the induced map from C to itself. This can be lifted from $\tilde{C}$ to itself. Let $\tilde{\beta}'$ be the lift of β' which starts at $\tilde{c}$.

An oriented braid is a braid $\beta \in B_{2m}$ together with a choice of orientation for each of the $2m$ strands such that the orientations match up correctly when we take the plat closure of β.

Definition 4.10.2. Suppose $\beta \in B_{2m}$ is an oriented braid. Let w be the writhe of β (the number of right handed crossings minus the number of left handed crossings). Let $e(\beta)$ be the sum of the exponents of the generators in any word representing β. We define

$$V'_\beta = \lambda \langle \tilde{S}, \beta \tilde{T} \rangle|_{t=-q^{-1}},$$

where $\lambda = (-q^{1/2} - q^{-1/2})^{-1}(-q^{3/4})^w(q^{-1/4})^{e(\beta)+2m}$.

Define K_{2m} be a subgroup of B_{2m} generated by

$$\sigma_1, \quad \sigma_2\sigma_1^2\sigma_2, \quad Y = \{\sigma_{2i}\sigma_{2i-1}\sigma_{2i+1}\sigma_{2i} : i = 1, 2, \cdots, m-1\}.$$

There is a modified result of Birman for oriented braids in the following.

Proposition 4.10.3. *The equivalent relation on $\coprod_m B_{2m}$ generated by the two moves (1) $\beta \mapsto x\beta y$ for $x, y \in K_{2m}$ and (2) $\beta \mapsto S_m(\beta)$, is the same equivalence relation for oriented braids with same isotopic plat closure.*

Birman [1976] showed the result for same plat closure only for knots by using the subgroup C_{2m}. The proof is required to have the commutativity between two moves. We do not need to commute two moves, the result in Proposition 4.10.3 [Bigelow, 2001b, Lemma 5.2] works for links. There is a small technical issue on the orientation. Proposition 4.10.3 really specify the effect on the oriented braids with same plat closure, where this orientation is not considered in [Birman, 1976].

Proposition 4.10.4. V'_β *defined in Definition 4.10.2 is invariant for oriented braids.*

Proof: By Proposition 4.10.3, this amounts to checking that V'_β is invariant under the two moves given in Proposition 4.10.3.

(1) For an oriented braid $\beta \in B_{2m}$, we show that $V'_{x\beta} = V'_\beta$ and $V'_{\beta y} = V'_\beta$ for $x, y \in K_{2m}$.

Note that σ_1 contributes a left-handed crossing to $\sigma_1\beta$, so the writhe $w(\sigma_1\beta) = w(\beta) - 1$ For other generators in K_{2m}, we have $w(\times\beta) = w(\beta)$. For $x = \sigma_1$, this is to show that

$$(-q)^{-1}\langle \tilde{S}, (\sigma_1\beta)\tilde{T} \rangle = \langle \tilde{S}, \beta\tilde{T} \rangle,$$

and for $x = \sigma_2\sigma_1^2\sigma_2$ or $x = \sigma_{2i}\sigma_{2i-1}\sigma_{2i+1}\sigma_{2i}$,

$$q^{-1}\langle \tilde{S}, (x\beta)\tilde{T} \rangle = \langle \tilde{S}, \beta\tilde{T} \rangle.$$

This is equivalent to $\langle \delta^{-1}\tilde{S}, x\beta\tilde{T} \rangle = \langle x\tilde{S}, x\beta\tilde{T} \rangle$ for $\delta = -q^{-1}$ if $x = \sigma_1$ and $\delta = q^{-1}$ otherwise. Thus, it reduces to verifying that $x\tilde{S} = \delta^{-1}\tilde{S}$ which is done in [Bigelow, 2001b, Lemma 4.1]. The result $V'_{x\beta} = V'_\beta$ follows.

Note that $\tilde{T}$ can be isotopes equal to $(1-q)^n \tilde{S}$ except in an arbitrarily small neighborhood of the puncture points. Thus

$$
\begin{aligned}
\langle \tilde{S}, \beta\tilde{T} \rangle &= (1-q)^{-n}\langle \tilde{T}, \beta\tilde{T} \rangle \\
&= \langle (1-q)^{-n}\beta^{-1}\tilde{T}, \tilde{T} \rangle \\
&= \langle \beta^{-1}\tilde{T}, (1-q)^{-n}\tilde{T} \rangle \\
&= \langle \beta^{-1}\tilde{T}, \tilde{S} \rangle.
\end{aligned}
$$

Similarly, we have

$$
\langle \tilde{S}, \beta y\tilde{T} \rangle = \langle \beta^{-1}\tilde{S}, \tilde{T} \rangle.
$$

Therefore $\langle \beta^{-1}\tilde{S}, \tilde{T} \rangle = \delta^{-1}\langle \beta^{-1}\tilde{T}, \tilde{S} \rangle$ follows from previous discussion, and both sides gives the identifications of $V'_{\beta y} = V'_{\beta}$.

(2) For an oriented braid $\beta \in B_{2m}$, we show that $V'_{\beta} = V'_{S_m(\beta)}$ for the second move of the oriented braid for the same plat closure.

Denote β' as the natural image of $\beta \in B_{2m}$ in $B_{2(m+1)}$ and $S_m(\beta) = S_m(\beta')$. Define $S_m^{-1}(\beta') = \sigma_{2m}^{-1}\beta'$. Then we get the following relation.

$$
S_m(\beta')\sigma_{2m+1}^2 = \sigma_{2m}\sigma_{2m+1}^2\sigma_{2m}S_m^{-1}(\beta').
$$

We have that both σ_{2m+1} and $\sigma_{2m}\sigma_{2m+1}^2\sigma_{2m}$ lie in $K_{2(m+1)}$. By the first part of the proposition (1),

$$
V'_{S_m(\beta')} = V'_{S_m^{-1}(\beta')}.
$$

By a special skein relation of Lemma 5.5 of [Bigelow, 2001b], we have

$$
q^{-1}V'_{S_m(\beta')} - qV'_{S_m^{-1}(\beta')} = (q^{1/2} - q^{-1/2})V'_{\beta'}.
$$

Thus we have

$$
(-q^{1/2} - q^{-1/2})V'_{S_m(\beta')} = V'_{\beta'}.
$$

In order to show that $V'_{S_m(\beta')} = V'_{\beta}$, it suffices to prove that

$$
V'_{\beta'} = (-q^{1/2} - q^{-1/2})V'_{\beta}.
$$

Note that $w(\beta') = w(\beta)$ and $e(\beta') = e(\beta)$. The oriented braids β and β' have $2m$ and $2(m+1)$ strands respectively. The coefficient λ for β' is $q^{-1/2}$ times the coefficient λ for β. Hence, it amounts to verifying

$$
\langle \tilde{S}', \beta\tilde{T}' \rangle = q^{-1/2}(-q^{1/2} - q^{-1/2})\langle \tilde{S}, \beta\tilde{T} \rangle,
$$

where $\tilde{S}'$ is the open $(m+1)$-ball corresponding to the standard fork diagram in $D_{2(m+1)}$, and $\tilde{T}'$ is the corresponding $(m+1)$-torus.

The open $(m + 1)$-ball $\tilde{S}'$ is the product of the open m-ball $\tilde{S}$ with an edge, and the $(m+1)$-torus $\tilde{T}'$ is the product of the m-torus $\tilde{T}$ with a circle. This edge meets this circle at two points in the disk. These two points contribute -1 and $-q$ times $\langle \tilde{S}, \beta \tilde{T} \rangle$. Therefore $V'_\beta = V'_{S_m(\beta)}$ follows. ☐

Proposition 4.10.4 shows that V'_β for the oriented braid β is invariant under two moves and gives an isotopy invariant of the plat closure link L of β. Now we are going to verify that this invariant is indeed the Jones polynomial invariant constructed in [Jones, 1987].

Theorem 4.10.5. *The invariant V'_β of an oriented braid $\beta \in B_{2m}$ is the Jones polynomial of the plat closure link L, i.e., $V'_\beta = V_L(t)$.*

Moreover, the simpler formula $q^{-(e(\beta)+2m)/4} \langle \tilde{S}, \beta \tilde{T} \rangle |_{t=-q^{-1}}$ is the Kauffman bracket of the plat closure link L of $\beta \in B_{2m}$, normalized to equal one for the empty diagram.

Proof: This is done by showing the skein relation for V'_β. Since V'_β is an isotopy invariant, there are good representations we can choose for those $L_\pm$, L_0 in the skein relation as plat closures (see some choices in previous section). We can move the ball on which the three links differ to the top right of the diagram. Then we can isotope the rest of the diagram to be a plat closure. It reduces to the situation proved in [Bigelow, 2001b, Lemma 5.5].

First, a straightforward computation gives the one for the unknot. Then the skein relation completes the proof that $V'_\beta = V_L(t)$. ☐

There is an algebraic way to compute the Jones polynomial V_L for the plat closure of a braid $\beta \in B_{2m}$, and a geometric way to evaluate the invariant $V'_\beta = V_L$ by Theorem 4.10.5 for an oriented braid $\beta \in B_{2m}$. Bigelow [2001b] further described the method to compute the invariant V'_β. We follow the method given in [Bigelow, 2001b] to illustrate the computational method.

Let $E_1, \cdots, E_m$ and $E'_1, \cdots, E'_m$ be the tine edges and handles of the standard fork diagram. Let $F_1, \cdots, F_m$ be the corresponding figure-eights to all E_i's. We can assume that βF_i and E_j intersect transversely without triple points by applying isotopies.

The intersection $S \cap \beta T$ consists of points $\mathbf{e} = \{e_1, \cdots, e_m\} \in C$ such that $e_i \in E_i \cap F_{\pi(i)}$, where π is a permutation of $\{1, \cdots, m\}$. Each intersection point e_i contributes a monomial $\pm q^a t^b$ to $\langle \tilde{S}, \beta \tilde{T} \rangle$.

The sign $\pm$ in front of the monomial $q^a t^b$ is the sign of the intersection of S with βT at $\mathbf{e}$. Let $\tilde{\mathbf{e}}$ be the lift of $\mathbf{e}$ which lies in $\tilde{S}$ with fixed starting points in **c**. The integer pair $(a, b) \in \mathbb{Z} \times \mathbb{Z}$ is determined to have $q^a t^b \tilde{\mathbf{e}}$ lie in $\beta \tilde{T}$. Let m be the number of points e_i such that the sign of the intersection of E_i with $F_{\pi(i)}$ at e_i is negative. The sign for our desired monomial is then $(-1)^m$ times the parity of the permutation π.

Let $\gamma(s) = \{E_1'(s), \cdots, E_m'(s)\}$ be the path in C, and $\varepsilon(s) = \{\varepsilon_1(s), \cdots, \varepsilon_m(s)\}$ be the path such that $\varepsilon_i(s)$ is a segment of E_i starting from $E_i'(1)$ and ending at e_i. Then the composition $\gamma \circ \varepsilon$ lifts to a path from $\tilde{\mathbf{c}}$ to $\tilde{\mathbf{e}}$.

Let $\phi(s) = \{\phi_1(s), \cdots, \phi_m(s)\}$ be the path such that $\phi_i(s)$ is a segment of $\beta F_{\pi(i)}$ starting from $\beta E_{\pi(i)}'(1)$ and ending at e_i. Then the composition $\beta(\gamma) \circ \phi$ lifts to a path from $\tilde{\mathbf{c}}$ to $q^a t^b \tilde{\mathbf{e}} \in \beta \tilde{T}$. Now the path $\alpha = \beta(\gamma) \circ \phi \circ (\gamma \circ \varepsilon)^{-1}$ lifts to a path from $\tilde{\mathbf{c}}$ to $q^a t^b \tilde{\mathbf{c}}$. Thus, we have

$$\Phi(\alpha) = q^{a(\alpha)} t^{b(\alpha)}.$$

The sum of these monomial terms $\pm q^a t^b$ over all **e** gives the required intersection pairing $\langle \tilde{S}, \beta \tilde{T} \rangle$. Evaluating $t = -q^{-1}$ into the intersection pairing $\langle \tilde{S}, \beta \tilde{T} \rangle$, we have another method to calculate the Jones polynomial of a knot or a link, other than the method given in the previous section.

Given a knot diagram, change every sequence of consecutive undercrossings with the figure-eight F_i, and attach handles E_i''s in any convenient way to create a fork diagram. Then we can proceed with the above method to compute the correct factor λ or simply to determine the value of the Jones polynomial up to sign and multiplication by a power of $q^{1/2}$.

Chapter 5

Casson type invariants

5.1 The $SU(2)$ group

The group $SU(2)$ is defined by

$$SU(2) = \{A \in M_{2\times 2}(\mathbb{C}) : A\overline{A}^T = Id, \quad \det A = 1\}.$$

Let $A = \begin{pmatrix} a & b \\ c & d \end{pmatrix}$ be a 2×2 complex matrix. Then we have

$$A^{-1} = \frac{AdA}{\det A} = \begin{pmatrix} d & -b \\ -c & a \end{pmatrix}$$

$$\overline{A}^T = \begin{pmatrix} \overline{a} & \overline{c} \\ \overline{b} & \overline{d} \end{pmatrix}.$$

Hence $a = \overline{d}$ and $b = -\overline{c}$ follows from $A^{-1} = \overline{A}^T$. We can identify the group $SU(2)$ to the matrix group under the multiplication:

$$SU(2) = \{\begin{pmatrix} a & b \\ -\overline{b} & \overline{a} \end{pmatrix} : a\overline{a} + b\overline{b} = 1\} \cong S^3.$$

As real variables, one can see that

$$SU(2) = \{\begin{pmatrix} x_1 + ix_2 & x_3 + ix_4 \\ -x_3 + ix_4 & x_1 - ix_2 \end{pmatrix} : x_1^2 + x_2^2 + x_3^2 + x_4^2 = 1\},$$

where $a = x_1 + ix_2$ and $b = x_3 + ix_4$. One can use the trace function $Tr : SU(2) \to \mathbb{R}, A \mapsto Tr(A) = 2Re(a) = 2x_1$ as a Morse function [Akbulut and McCarthy, 1990]. Hence x_1 as height has only two critical points (one maximal and one minimal). Therefore $H^*(SU(2), \mathbb{Z}) = H^*(S^3, \mathbb{Z})$.

As quaternion variables, $SU(2) \cong Sp(1)$ via $\mathbb{C}^2 \stackrel{\cong}{\to} \mathbb{H}$ (identified by $(a, b) \mapsto a + bj$). Let $q = x + yi + zj + wk \in \mathbb{H}$ and $\overline{q} = x - yi - zj - wk$ with $|q|^2 = q\overline{q}$.

$$Sp(1) = \{q \in \mathbb{H} : |q|^2 = 1\},$$

167

again is a 3-sphere. The identification between $SU(2)$ and $Sp(1)$ is

$$A_q = \begin{pmatrix} a & b \\ -\bar{b} & \bar{a} \end{pmatrix} \in SU(2) \leftrightarrows q = a + bj \in \mathbb{H}.$$

The identification preserves the multiplication rule as well, and it provides a group isomorphism. Typically, we have

$$1 \in \mathbb{H} \leftrightarrows \begin{pmatrix} 1 & 0 \\ 0 & 1 \end{pmatrix} \in SU(2), \quad i \in \mathbb{H} \leftrightarrows \begin{pmatrix} i & 0 \\ 0 & -i \end{pmatrix} \in SU(2),$$

$$j \in \mathbb{H} \leftrightarrows \begin{pmatrix} 0 & 1 \\ -1 & 0 \end{pmatrix} \in SU(2), \quad k \in \mathbb{H} \leftrightarrows \begin{pmatrix} 0 & i \\ i & 0 \end{pmatrix} \in SU(2).$$

Note that the maximal commutative subgroup of $SU(2)$ is $U(1) = \{\begin{pmatrix} e^{i\phi} & 0 \\ 0 & e^{-i\phi} \end{pmatrix} : \phi \in [0, 2\pi]\}$. From the quaternion point of view, $q = a + bj$ is not commutative unless $b = 0$. Thus $a \in \mathbb{C}$ with $a\bar{a} = 1$. All other maximal commutative subgroups in $SU(2)$ are $C^{-1} \cdot U(1) \cdot C$ for $C \in SU(2)$.

Lemma 5.1.1. *Two elements A and A' in $SU(2)$ are conjugate if and only if $Tr(A) = Tr(A')$.*

Proof: If A and A' in $SU(2)$ are conjugate, then there exists an element $C \in SU(2)$ such that $A = C^{-1}A'C$. Hence $Tr(A) = Tr(A')$ follows from the invariance under the conjugacy.

Let $A : \mathbb{C}^2 \to \mathbb{C}^2$ be a linear transformation with eigenvalue $\lambda \in \mathbb{C}$ and eigenvector $\psi = (x, y)$. So $A\psi = \lambda\psi$. Rescaling to have a unit eigenvector ψ, $C = \begin{pmatrix} x & -\bar{y} \\ y & \bar{x} \end{pmatrix}$ satisfies $C\begin{pmatrix} 1 \\ 0 \end{pmatrix} = \begin{pmatrix} x \\ y \end{pmatrix}$, and $C^{-1} = \begin{pmatrix} \bar{x} & \bar{y} \\ -y & x \end{pmatrix}$. Therefore

$$C^{-1}AC\begin{pmatrix} 1 \\ 0 \end{pmatrix} = C^{-1}A\begin{pmatrix} x \\ y \end{pmatrix} = \lambda C^{-1}\begin{pmatrix} x \\ y \end{pmatrix} = \lambda\begin{pmatrix} 1 \\ 0 \end{pmatrix}.$$

So $C^{-1}AC = \begin{pmatrix} \lambda & \alpha \\ 0 & \beta \end{pmatrix} \in SU(2)$ which implies $\alpha = 0$ and $\beta = \bar{\lambda}$ and $\lambda = e^{i\phi}$. So any element A is conjugate in $SU(2)$ to $\begin{pmatrix} e^{i\phi} & 0 \\ 0 & e^{-i\phi} \end{pmatrix}$ with $Tr(A) = 2\cos\phi$. The trace is uniquely defined by the angle ϕ and vice versa, up to ± 1.

$$\begin{pmatrix} e^{i\phi} & 0 \\ 0 & e^{-i\phi} \end{pmatrix} = \begin{pmatrix} 0 & -1 \\ 1 & 0 \end{pmatrix} \begin{pmatrix} e^{i\phi} & 0 \\ 0 & e^{-i\phi} \end{pmatrix} \begin{pmatrix} 0 & 1 \\ -1 & 0 \end{pmatrix}.$$

$\square$

Conjugacy classes in $SU(2)$ are in one-to-one correspondence with $Tr^{-1}(c)$ for $-2 \leq c \leq 2$.

$$Tr \begin{pmatrix} a & b \\ -\bar{b} & \bar{a} \end{pmatrix} = a + \bar{a} = 2Re(a) = c.$$

(1) The maximal value $c = 2$ ($a = 1$), we have $Tr^{-1}(2) = \{Id\}$.
(2) The minimal value $c = -2$ ($a = -1$), we have $Tr^{-1}(-2) = \{-Id\}$.
(3) $-2 < c < 2$, we have $Tr^{-1}(c) = \{A \in SU(2) : Re(a) = c/2\} = S^2$
   which corresponds to $\frac{c^2}{4} + x_2^2 + x_3^2 + x_4^2 = 1$ the 2-sphere with radius $\sqrt{4 - c^2}/2$.

From the identification of conjugacy classes, $c = Tr(A) = 2\cos\phi$ and the argument $\phi = \cos^{-1}(Tr(A)/2)$ defines a surjective map $arg(A) = \phi$ from $SU(2)$ to $[0, \pi]$.

$$arg(A) = 0 \leftrightarrows A = Id; \quad arg(A) = \pi \leftrightarrows A = -Id.$$

Define a natural splitting (Heegaard decomposition of the standard 3-sphere):

$$D_+^3 = \{A \in SU(2) : 0 \leq arg(A) \leq \pi/2\} = \{A \in SU(2) : 0 \leq Tr(A) \leq 2\}$$

$$D_-^3 = \{A \in SU(2) : \pi/2 \leq arg(A) \leq \pi\} = \{A \in SU(2) : -2 \leq Tr(A) \leq 0\}$$

$$S^2 = D_+^3 \cap D_-^3 = \{A \in SU(2) : Tr(A) = 0\}.$$

The 2-sphere from the matrix representative can be characterized as $Re(a) = 0$, $a = it$ and $a + bj = it + bj$ a pure quaternion number with norm one.

There is a natural involution of the decomposition

$$i : (S^3, D_+^3, D_-^3, S^2) \to (S^3, D_+^3, D_-^3, S^2), \quad i(A) = -A^{-1}.$$

It is easy to see $Tr(A) = Tr(A^{-1})$ and $Tr(-A^{-1}) = -Tr(A)$. So

$$arg(-A^{-1}) = \cos^{-1}(Tr(-A^{-1})/2) = \pi - \cos^{-1}(Tr(A)/2) = \pi - arg(A).$$

Note that $i(A) = A$ if and only if $Tr(A) = 0$ or $arg(A) = \pi/2$.

The tangent space $T_{Id}SU(2) \cong \mathfrak{su}(2)$ is the Lie algebra which can be identified as the following.

$$T_A SU(2) = \{(A, U) : q_A \cdot q_U = 0\}(\text{as tangent line to a quaternion circle})$$
$$= \{(A, U) : a\bar{u} + ovau + b\bar{v} + \bar{b}v = 0\},$$

for $U = \begin{pmatrix} u & v \\ -\bar{v} & \bar{u} \end{pmatrix}$. For $A = Id$ ($a = 1, b = 0$), the Lie algebra $\mathfrak{su}(2) = \{(Id, U) : \bar{u} + u = 0\}$. Let $u = is$ and $v = t_1 + t_2 i \in \mathbb{C}$.

$$\mathfrak{su}(2) = \left\{ \begin{pmatrix} is & t_1 + t_2 i \\ -t_1 + t_2 i & -is \end{pmatrix} : (s, t_1, t_2) \in \mathbb{R}^3 \right\}$$

$$= \left\{ s \begin{pmatrix} i & 0 \\ 0 & -i \end{pmatrix} + t_1 \begin{pmatrix} 0 & 1 \\ -1 & 0 \end{pmatrix} + t_2 \begin{pmatrix} 0 & i \\ i & 0 \end{pmatrix} : (s, t_1, t_2) \in \mathbb{R}^3 \right\}$$

$$= \{ si + t_1 j + t_2 k = si + vj \in \mathbb{H} \}.$$

The Lie algebra $\mathfrak{su}(2)$ can be identified with the pure quaternion numbers.

Let $B_\pi(0)$ be an open ball of radius π centered at the origin in $\mathfrak{su}(2)$. Note that the standard Riemannian metric on S^3 has injective radius π. Define $\exp : (-i\pi, i\pi) \to (S^1 \setminus \{-1\})$ to be $\exp(i\phi) = e^{i\phi}$. Let $Exp(g\alpha g^{-1}) = g\exp(\alpha)g^{-1}$ be the defined map: $B_\pi(0) \subset \mathfrak{su}(2) \to SU(2) \setminus \{-Id\}$. Since $\exp$ is diffeomorphism, the map Exp is diffeomorphism as well.

Define an adjoint action on $SU(2)$:

$$Ad_A : SU(2) \to SU(2); \qquad B \mapsto ABA^{-1},$$

and the isotropy group

$$\Gamma_A = \{ B \in SU(2) : Ad_A(B) = ABA^{-1} = B \}.$$

For $U(1) \subset SU(2)$,

$$Ad_C(U(1)) = \left\{ C \begin{pmatrix} e^{i\phi} & 0 \\ 0 & e^{-i\phi} \end{pmatrix} C^{-1} : C \in SU(2) \right\}.$$

For $\{\pm Id\} = Z(SU(2))$ the center of $SU(2)$, $\Gamma_A = SU(2)$.

Lemma 5.1.2. *For $A \in Ad_C(U(1) \setminus Z(SU(2)))$, $\Gamma_A = Ad_C(U(1))$.*

Proof: For $A \in Ad_C(U(1) \setminus Z(SU(2)))$, we have $A = C \begin{pmatrix} e^{i\phi} & 0 \\ 0 & e^{-i\phi} \end{pmatrix} C^{-1}$ for some $C \in SU(2)$. Thus any element in $Ad_C(U(1))$ commutes with A since

$$AB = C \begin{pmatrix} e^{i\phi} & 0 \\ 0 & e^{-i\phi} \end{pmatrix} C^{-1} C \begin{pmatrix} e^{i\phi_1} & 0 \\ 0 & e^{-i\phi_1} \end{pmatrix} C^{-1}$$

$$= C \begin{pmatrix} e^{i\phi_1} & 0 \\ 0 & e^{-i\phi_1} \end{pmatrix} C^{-1} C \begin{pmatrix} e^{i\phi} & 0 \\ 0 & e^{-i\phi} \end{pmatrix} C^{-1} = BA.$$

Therefore $Ad_C(U(1)) \subset \Gamma_A$.

If $B \in \Gamma_A$ ($BAB^{-1} = A$), then $C^{-1}AC = \begin{pmatrix} e^{i\phi} & 0 \\ 0 & e^{-i\phi} \end{pmatrix}$ and $C^{-1}BAB^{-1}C = C^{-1}AC$ is equivalent to

$$(C^{-1}BC)(C^{-1}AC)(C^{-1}BC)^{-1} = C^{-1}AC$$

$$(C^{-1}BC) \begin{pmatrix} e^{i\phi} & 0 \\ 0 & e^{-i\phi} \end{pmatrix} (C^{-1}BC)^{-1} = \begin{pmatrix} e^{i\phi} & 0 \\ 0 & e^{-i\phi} \end{pmatrix}.$$

So $B \in Ad_C(U(1))$. Hence we obtain $\Gamma_A \subset Ad_C(U(1))$. $\square$

Note that this gives a standard stratification of S^3 (see [Akbulut and McCarthy, 1990] for more details).

Computing $DAd_A : T_{Id}SU(2) \to T_{Id}SU(2)$, we have the linear term of ε from the expansion:

$$Ad_A(Id + \varepsilon\alpha) = A(Id + \varepsilon\alpha)A^{-1}) = Id + \varepsilon A\alpha A^{-1}.$$

Thus $DAd_A : \mathfrak{su}(2) = T_{Id}SU(2) \to \mathfrak{su}(2)$ is given by $DAd_A(\alpha) = A\alpha A^{-1}$. Then one obtains a map $DAd : SU(2) \to Aut(\mathfrak{su}(2))$ given by $A \mapsto DAd_A$. Note that $DAd_{Id} = Id$. Thus we have

$$D(DAd) = ad : T_{Id}SU(2) \to T_{Id}Aut(\mathfrak{su}(2)).$$

The map ad is the linear term of DAD for $A = Id + \varepsilon\alpha$:

$$\begin{aligned} DAD_{(Id+\varepsilon\alpha)}(\beta) &= (Id + \varepsilon\alpha)\beta(Id + \varepsilon\alpha)^{-1} \\ &= (\beta + \varepsilon\alpha\beta)(Id - \varepsilon\alpha + o(\varepsilon^2)) \\ &= \beta + \varepsilon(\alpha\beta - \beta\alpha) + o(\varepsilon^2). \end{aligned}$$

Thus $ad_\alpha = [\alpha, \cdot] : \mathfrak{su}(2) \to \mathfrak{su}(2)$ and $ad_\alpha(\beta) = [\alpha, \beta]$.

Lemma 5.1.3. *The map* $DAd : SU(2) \to Aut(\mathfrak{su}(2))$ *via* $DAd(A) = DAd_A : \mathfrak{su}(2) \to \mathfrak{su}(2)$ *is the well-known double cover map* $SU(2) \to SO(3)$.

Proof: $DAd(AB) = DAd_{AB}$ and

$$\begin{aligned} DAd_{AB}(\alpha) &= AB\alpha(AB)^{-1} \\ &= A(B\alpha B^{-1})A^{-1} \\ &= A(DAd_B(\alpha))A^{-1} \\ &= DAd_A(DAd_B(\alpha)) \\ &= DAd_A \circ DAd_B(\alpha). \end{aligned}$$

Thus $DAd : SU(2) \to Aut(\mathfrak{su}(2))$ is a group homomorphism. From the above identification, we have $\mathfrak{su}(2) \cong \mathbb{R}^3$. For $U, V \in \mathfrak{su}(2)$, $-\frac{1}{2}Tr(UV) = q_U \cdot q_V$. In particular,

$$(si + vj) \cdot (-si - \bar{v}j) + (-si - vj) \cdot (si + vj) = -2(s^2 + v\bar{v}).$$

Thus the Euclidean inner product of pure quaternion numbers is

$$\langle q_1, q_2 \rangle_E = -\frac{1}{2} \langle q_1, q_2 \rangle_{\mathbb{H}}.$$

Now we check that

$$\langle DAd_A\alpha, DAd_A\beta \rangle_E = -\frac{1}{2} \langle DAd_A\alpha, DAd_A\beta \rangle_{\mathbb{H}}$$

$$= -\frac{1}{2} \langle A\alpha A^{-1}, A\beta A^{-1} \rangle_{\mathbb{H}}$$

$$= -\frac{1}{2} \langle \alpha, \beta \rangle_{\mathbb{H}}$$

$$= \langle \alpha, \beta \rangle_E.$$

Thus DAd_A preserves the Euclidean inner product and $DAd_A \in O(3)$. Since $SU(2)$ is connected, $DAd_A \in SO(3)$.

Now we show that the map DAd is a surjective map onto $SO(3)$. Note that $SO(3)$ is the product of rotations about the coordinate axes. For the rotation about the x-axis, the rotation matrix is given by

$$R_x = \begin{pmatrix} 1 & 0 & 0 \\ 0 & \cos\psi & -\sin\psi \\ 0 & \sin\psi & \cos\psi \end{pmatrix}, \quad \phi = \psi/2, \quad A = \begin{pmatrix} e^{i\phi} & 0 \\ 0 & e^{-i\phi} \end{pmatrix}.$$

We have $DAd_A(i) = e^{i\phi} i e^{-i\phi} = i$ and

$$DAd_A(j) = e^{i\phi} j e^{-i\phi} = e^{2i\phi} j = \cos\psi j + \sin\psi k;$$

$$DAd_A(k) = e^{i\phi} k e^{-i\phi} = e^{2i\phi} k = \cos\psi k - \sin\psi j.$$

Thus $DAd_A = R_x$. Similarly one can get the images of R_y and R_z. Note that $Ad_A = Ad_b$ if and only if $BA^{-1} \in \{\pm Id\}$. Hence our result follows. $\square$

5.2 The $SU(2)$ representations

Let G be a group and $R(G) = Hom(G, SU(2))$ be the $SU(2)$-representations of the group G.

$$R(G) = \{\phi : G \to SU(2) : \phi(e_G) = Id_{2\times 2}, \ \phi(g_1 g_2) = \phi(g_1)\phi(g_2)\},$$

where e_G is the identity of the group G. We use the compact open topology on $R(G)$ with discrete topology on G. For a compact subset $K \subset G$ (must be finite from the discrete topology), define $U_{K,V} = \{f : K \to SU(2) : f(K) \subset V\}$ the basis for $R(G)$ for an open subset V in $SU(2)$. Thus $U_{g,V} = \{f : G \to SU(2) : f(g) \subset V\}$. The representation $\phi_n \in R(G)$

converge to $\phi \in R(G)$ if and only if $\phi_n(g)$ converges to $\phi(g)$ for each $g \in G$. In particular, this convergence only needs to be checked for those generators in G.

If $h : G_1 \rightarrow G_2$, then there is an associated map $R(h) : G(G_2) \rightarrow R(G_1)$ given by $\phi \mapsto \phi \circ h$. Note that $R(h_1 \circ h_2) = R(h_2) \circ R(h_1)$ and $R(Id) = Id$. Thus R is a contravariant functor from the category of groups to the category of topological $SU(2)$-representation spaces.

The adjoint map $Ad : SO(3) \times SU(2) \rightarrow SU(2)$ induces an action on the space $R(G)$: $SO(3) \times R(G) \rightarrow R(G)$ given by $(A, \phi) \mapsto (Ad_A(\phi))$, where $Ad_A(\phi) : G \rightarrow SU(2)$ is defined by $g \mapsto A\phi(g)A^{-1}$.

Exercise 5.2.1. Check that the map $Ad \circ (Id \times R(h)) : SO(3) \times R(G_2) \rightarrow R(G_1)$ agrees with $R(h) \circ Ad : SO(3) \times R(G_2) \rightarrow R(G_1)$.

Definition 5.2.2. Let $S_0(G) = \{\phi \in R(G) : \phi : G \rightarrow \{\pm Id\}\}$, $S_1(G) = \{\phi \in R(G) : \phi : G \rightarrow U(1)\}$ and $S(G) = \{\phi \in R(G) : \phi : G \rightarrow SU(2), \text{reducible}\}$. A representation $\phi \in R(G)$ is reducible if there is a proper non-zero $\mathbb{C}$-linear subspace $U \subset \mathbb{C}^2$ such that $\phi(g)U \subset U$ for all $g \in G$. Let $R^*(G) = R(G) \setminus S(G)$ be the set of irreducible $SU(2)$-representations.

Note that $SO(3)$ action on $R^*(G)$ is free.

Exercise 5.2.3. Prove that $S_0(G) \subset S_1(G) \subset S(G) \subset R(G)$.

Lemma 5.2.4. *We have* $S(G) = \cup\{Ad_C S_1(G) : C \in SU(2)\}$.

Proof: Since the conjugate relation of a reducible representation ϕ has the following property:
$$\phi(g)U \subset U \Rightarrow Ad_C(\phi)U = C\phi(g)C^{-1}U \subset U,$$
the representation $Ad_C(\phi)$ is again reducible. Thus $\{Ad_C S_1(G) : C \in SU(2)\} \subset S(G)$.

Every reducible representation is conjugate to a diagonal representation and $\phi : G \rightarrow SU(2)$ has an invariant subspace $U \subset \mathbb{C}^2$. Pick a unit vector $(x, y) \in U$ and $C = \begin{pmatrix} x & -\overline{y} \\ y & \overline{x} \end{pmatrix}$. Therefore $C^{-1}\phi(g)C \in S_1(G)$ for all $g \in G$. That is $Ad_{C^{-1}}(\phi) \in S_1(G)$ and $\phi \in Ad_C S_1(G)$. $\square$

For a group homomorphism $h : G_1 \rightarrow G_2$, there is a commutative diagram

$$\begin{array}{ccccccc} S_0(G_2) & \subset & S_1(G_2) & \subset & S(G_2) & \subset & R(G_2) \\ \downarrow R(h) & & \downarrow R(h) & & \downarrow R(h) & & \downarrow R(h) \\ S_0(G_1) & \subset & S_1(G_1) & \subset & S(G_1) & \subset & R(G_1) \end{array}$$

Define $\Gamma_\phi = \{A \in SU(2) : A\phi(g)A^{-1} = \phi(g), g \in G\}$ to be the isotropy group of ϕ in $SU(2)$ for all $g \in G$.

Exercise 5.2.5. (i) If $\phi \in S_0(G)$, then $\Gamma_\phi = SU(2)$; (ii) If $\phi \in S_1(G)$, then $\Gamma_\phi = CU(1)C^{-1}$; (iii) If $\phi \in R^*(G)$, then $\Gamma_\phi = \{\pm Id\}$.

Example 5.2.6. If $G = \pi_1(M^3) = \{g_1, \cdots, g_n | r_1, r_2, \cdots, r_m\}$ is the fundamental group of a closed 3-manifold M^3 with generators $g_1, \cdots, g_n$ and relations $r_1, \cdots, r_m$, then the $SU(2)$-representation space $R(G)$ can be described as

$$R(G) = \{(\phi(g_1), \cdots, \phi(g_n)) \in SU(2)^n | r_1(\phi(g_i)) = Id, \cdots, r_m(\phi(g_i)) = Id\}.$$

Each generator $g_i \in G$ gives an element of $SU(2)$ through the representation ϕ, and the n-tuple $SU(2)$ matrices $(\phi(g_i))_{1 \le i \le n}$ satisfies the m relations. The relations give a system of polynomials of n-tuples matrices. Hence $R(G)$ is a real algebraic variety defined by real polynomials from the real algebraic group $SU(2)$. Since $SU(2)^n$ is compact and $R(G)$ is closed in $SU(2)^n$, the $SU(2)$ representation space $R(G)$ is always compact of dimension $3(n - m)$. Note that if M^3 is an integral homology 3-sphere (i.e., $H_*(M^3, \mathbb{Z}) = H_*(S^3, \mathbb{Z})$, then $S(\pi(M^3)) = \{\pm Id\}$. Every $U(1)$-reducible representation $\phi : \pi_1(M^3) \to U(1)$ factors through the abelianization of $\pi_1(M^3)$ so that

$$\phi : \pi_1(M^3) \to \pi_1(M^3)/[\pi_1(M^3), \pi_1(M^3)] = H_1(M^3, \mathbb{Z}) \to U(1) \subset SU(2).$$

With $H_1(M^3, \mathbb{Z}) = H_1(S^3, \mathbb{Z}) = 0$, the representation $\phi : \pi_1(M^3) \to \{\pm Id\} \subset SU(2)$ maps into the center of the group $SU(2)$. Therefore $S(\pi(M^3)) = S_0(\pi_1(M^3))$ for integral homology 3-spheres M^3.

There is a canonical action of $Aut(G) \times SU(2)$ on $R(G)$:

$$Aut(G) \times SU(2) \times R(G) \to R(G); \quad (\sigma, A, \phi) \mapsto \phi_{(\sigma, A)},$$

where $\phi_{(\sigma, A)} : G \to SU(2)$ the $SU(2)$-representation is given by $\phi_{(\sigma, A)}(g) = A\phi(\sigma(g))A^{-1}$.

If $\sigma_y \in Inn(G)$ is the inner automorphism of G induced by $y \in G$, then $\sigma_y(g) = y^{-1}gy$ and $\phi(\sigma_y(g)) = \phi^{-1}(y)\phi(g)\phi(y)$,

$$\phi_{(\sigma, A)} : G \to SU(2), \quad g \mapsto \phi(y)\phi(\sigma_y(g))\phi^{-1}(y) = \phi(g).$$

Thus $Inn(G)$ acts on $R(G)$ trivially. Thus we pass to the group $Out(G) = Aut(G)/Inn(G)$ acting on $R(G)$.

Let $\mathcal{R}(G) = Hom(G, SU(2))/SU(2)$ be the equivalent classes of $SU(2)$ representations under the conjugacy relation. So we have the action

$Out(G) \times \mathcal{R}(G) \to \mathcal{R}(G)$ induced from the action $Out(G) \times SU(2) \times R(G) \to R(G)$. The outmorphisms act on $\mathcal{R}(G)$ effectively. If $G = \pi_1(\Sigma_g)$ is the fundamental group of a closed oriented surface with genus g, then $Out(G)$ is the mapping class group of the surface Σ_g.

Lemma 5.2.7. *The Zariski tangent space $T_\phi Hom(G, SU(2))$ is the space $Z^1(G, ad\phi)$ of 1-cocycles with values in $ad\phi$, where $ad\phi$ is the G-module obtained by the composition $G \xrightarrow{\phi} SU(2) \xrightarrow{DAd} Aut(\mathfrak{su}(2))$.*

Proof: Let $\phi_t \in Hom(G, SU(2))$ be a path dependent differentiably on the real variable t and $\phi_t(g) = e^{tu(g)+o(t^2)} \cdot \phi(g)$ for all $g \in G$. Since ϕ_t is a representation $\phi_t(g_1 \cdot g_2) = \phi_t(g_1) \cdot \phi_t(g_2)$, we have the following:

$$e^{tu(g_1 g_2)+o(t^2)}\phi(g_1 g_2) = e^{tu(g_1)+o(t^2)}\phi(g_1)e^{tu(g_2)+o(t^2)}\phi(g_2).$$

Thus, by expanding $e^x = 1 + x + o(x^2)$,

$$\phi(g_1 g_2) + tu(g_1 g_2)\phi(g_1 g_2) + o(t^2) =$$

$$(\phi(g_1) + tu(g_1)\phi(g_1) + o(t^2))(\phi(g_2) + tu(g_2)\phi(g_2) + o(t^2)).$$

By using $\phi(g_1 g_2) = \phi(g_1)\phi(g_2)$ and dividing t,

$$u(g_1 g_2)\phi(g_1)\phi(g_2) + o(t) = u(g_1)\phi(g_1)\phi(g_2) + \phi(g_1)u(g_2)\phi(g_2) + o(t).$$

Let $t \to 0$. We get

$$u(g_1 g_2) = u(g_1) + \phi(g_1)u(g_2)\phi(g_1)^{-1} = u(g_1) + Ad_{\phi(g_1)}(u(g_2)), \quad (5.1)$$

which is equivalent to the 1-cocycle condition for $u \in Z^1(G, ad\phi)$.

Let u be a 1-cocycle in $Z^1(G, ad\phi)$. Thus u satisfies (5.1). So any $\phi_t(g) = e^{tu(g)+o(t^2)}\phi(g)$ satisfies the homomorphism condition up to the first order

$$\lim_{t \to 0} \frac{\phi_t(gh) - \phi_t(g)\phi_t(h)}{t} = 0.$$

$\square$

Let $Z(\phi)$ be the centralizer of $\phi(G)$ in $SU(2)$. The group $Z(SU(2)) = \{\pm Id\} = \mathbb{Z}_2$ is the center of $SU(2)$. Note that $\pi_1(\Sigma_g) = \{a_1, b_1, \cdots, a_g, b_g | \prod_{i=1}^g [a_i, b_i] = 1\}$. If $G = \pi_1(\Sigma_g)$, then one can compute

$$\dim Z^1(G, ad\phi) = \dim T_\phi Hom(G, SU(2))$$
$$= (2g-1) \cdot \dim SU(2) + \dim Z(\phi)$$
$$= (6g-3) + \dim Z(\phi).$$

Definition 5.2.8. A representation ϕ is a simple point of $Hom(G, SU(2))$ if and only if $\dim Z(\phi)/Z(SU(2)) = 0$, if and only if $\phi \in R^*(G) = R(G) \setminus S(G)$, and if and only if ϕ is irreducible.

A representation $\phi \in Hom(G, SU(2))$ is a singular point of $Hom(G, SU(2))$ if and only if $\dim Z(\phi)/Z(SU(2)) = 1$, and if and only if ϕ is reducible.

For a simple point $\phi \in R(G)$ (or irreducible representation), $\dim Z^1(G, ad\phi) = 6g - 3$. For a singular point $\phi \in S(G) \setminus S_0(G)$, there is a natural submanifold of $Hom(G, SU(2))$ containing ϕ such that ϕ is a simple point of $Hom(G, U(1))$, where $U(1) = Z_{SU(2)}(Z(\phi))$ is the centralizer of the centralizer of $\phi(G)$. Thus $\dim Z(\phi)/Z_{SU(2)}(Z(\phi)) = 0$.

$$\dim Z^1(G, ad\phi) = \dim T_\phi Hom(G, U(1))$$
$$= (2g - 1)\dim U(1) + \dim Z(\phi)$$
$$= 2g.$$

Lemma 5.2.9. *The tangent space of the orbit $\phi \in Hom(G, SU(2))$ under the conjugate action is the space $B^1(G, ad\phi)$ of 1-coboundaries.*

Proof: Let $\phi_t(g) = y_t^{-1}\phi(g)y_t$ be a path in $Hom(G, SU(2))$ under a conjugate action from a path $y_t \in SU(2)$. If $y_t = e^{tu_0 + o(t^2)}$, then $\phi_t(g) = e^{tu(g) + o(t^2)}\phi(g)$ corresponding to the cocycle $u : G \to ad\phi$ as the expansion:

$$(1 + tu(g) + o(t^2))\phi(g) = (1 - tu_0 + o(t^2))\phi(g)(1 + tu_0 + o(t^2)).$$

By comparing both sides, one has

$$u(g)\phi(g) = -u_0\phi(g) + \phi(g)u_0,$$

from the first order in t. Thus

$$u(g) = \phi(g)u_0\phi(g)^{-1} - u_0 = Ad_{\phi(g)}u_0 - u_0 = \delta u_0, \qquad (5.2)$$

is the coboundary. Similarly, the 1-coboundary $u : G \to ad\phi$ gives the path $\phi_t(g) = e^{tu(g) + o(t^2)}\phi(g)$ satisfies $\phi_t(g) = y_t^{-1}\phi(g)y_t$. $\square$

Let $\mathfrak{z}(\phi)$ be the Lie algebra of $Z(\phi)$. Then $B^1(G, ad\phi)$ is isomorphic as a vector space to the quotient $\mathfrak{su}(2)/\mathfrak{z}(\phi)$. Thus

$$\dim G \cdot \phi = \dim B^1(G, ad\phi)$$

$$= \dim \mathfrak{su}(2) - \dim \mathfrak{z}(\phi) = \dim SU(2) - \dim Z(\phi).$$

For a simple point $\phi \in R^*(G)$, $\dim B^1(G, ad\phi) = \dim SU(2) - \dim Z(\phi) = 3$; for a singular point $\phi \in S(G) \setminus S_0(G)$, $\dim B^1(G, ad\phi) = \dim U(1) - \dim Z(\phi) = 0$.

Remark 5.2.10. (1) The $SU(2)$-action on $R(G)$ is locally free precisely on $R^*(G) = R(G) \setminus S(G)$ (simple points).

(2) The $SU(2)$-action on the representation space $Hom(G, SU(2))$ induces a quotient space $Hom(G, SU(2))/SU(2)$ which is Hausdorff. For the irreducible representations,

$$\dim R^*(G) = \dim T_\phi Hom(G, SU(2)) - \dim G \cdot \phi = (6g - 3) - 3 = 6g - 6.$$

(3)

$$T_\phi \mathcal{R}^*(G) = T_\phi Hom(G, SU(2))/T_\phi SU(2)$$

$$= Z^1(G, ad\phi)/B^1(G, ad\phi) = H^1(G, ad\phi),$$

where $\mathcal{R}^*(G) = R^*(G)/SU(2)$ is the space of equivalent classes of irreducible $SU(2)$ representations under the conjugate relation.

Proposition 5.2.11. *A cohomology class $\xi \in H^1(G, ad\phi)$ is tangent to a path $\{\phi_t\} \in \mathcal{R}(G)$ if and only if $\xi_2 = [\xi, \xi] = 0$ and $\xi_k = 0$ $(k \geq 2)$, where $\xi_k, k \geq 2$ are obstruction classes aroused from solutions u_k in*

$$\phi_t(g) = e^{tu(g) + \sum_{k=2}^{\infty} u_k(g)t^k} \phi(g), \quad \phi_t(gh) = \phi_t(g)\phi_t(h).$$

Proof: Let $u \in Z^1(G, ad\phi)$ be a representative of the cohomology class $\xi \in H^1(G, ad\phi)$. If u is tangent to a differentiable path $\phi_t \in R(G)$, then $\phi_t(gh) = \phi_t(g)\phi_t(h)$. We are going to show the first obstruction class $\xi_2 = [\xi, \xi]$. Using $e^x = 1 + x + x^2/2 + o(x^3)$, we have

$$\phi_t(gh) = (1 + tu(gh) + t^2 u_2(gh) + \frac{1}{2}u(gh)u(gh) + o(t^3))\phi(g)\phi(h);$$

$$\phi_t(g) = (1 + tu(g) + t^2 u_2(g) + \frac{1}{2}u(g)u(g) + o(t^3))\phi(g).$$

Similarly for $\phi_t(h)$. The constant term from $\phi_t(gh) = \phi_t(g)\phi_t(h)$ gives $\phi(gh) = \phi(g)\phi(h)$, i.e., $\phi \in Hom(G, SU(2))$, and the t-linear term from $\phi_t(gh) = \phi_t(g)\phi_t(h)$ gives $u(gh) = u(g) + Ad_{\phi(g)}u(h)$ (the 1-cocycle condition). Comparing t^2-term, we obtain the term

$$L = (u_2(gh) + \frac{1}{2}u(gh)u(gh))\phi(g)\phi(h),$$

from $\phi_t(gh)$, and

$$R = (u_2(g) + \frac{1}{2}u(g)u(g))\phi(g)\phi(h)$$

$$+u(g)\phi(g)u(h)\phi(h) + \phi(g)((u_2(h) + \frac{1}{2}u(h)u(h))\phi(h),$$

from $\phi_t(g)\phi_t(h)$. By the 1-cocycle condition on u,

$$L(\phi(g)\phi(h))^{-1} = u_2(gh) + \frac{1}{2}(u(g) + Ad_{\phi(g)}u(h))^2$$

$$= u_2(gh) + \frac{1}{2}(u(g)^2 + Ad_{\phi(g)}u^2(h))$$

$$+ \frac{1}{2}u(g)Ad_{\phi(g)}u(h) + \frac{1}{2}Ad_{\phi(g)}u(h)u(g),$$

where $Ad_{\phi(g)}u(h) \cdot Ad_{\phi(g)}u(h) = Ad_{\phi(g)}u^2(h)$ follows from its definition.

$$R(\phi(g)\phi(h))^{-1}$$

$$= u_2(g) + \frac{1}{2}u(g)^2 + u(g)Ad_{\phi(g)}u(h) + Ad_{\phi(g)}(u_2(h) + \frac{1}{2}u(h)^2)$$

$$= u_2(g) + \frac{1}{2}u(g)^2 + u(g)Ad_{\phi(g)}u(h) + Ad_{\phi(g)}u_2(h) + \frac{1}{2}Ad_{\phi(g)}u(h)^2.$$

Thus the identity reduces into the following by canceling $\frac{1}{2}u(g)^2 + \frac{1}{2}Ad_{\phi(g)}u(h)^2$ term:

$$u_2(g) + Ad_{\phi(g)}u_2(h) - u_2(gh) = \frac{1}{2}Ad_{\phi(g)}u(h) \cdot u(g) - \frac{1}{2}u(g) \cdot Ad_{\phi(g)}u(h)$$

$$= \frac{1}{2}[Ad_{\phi(g)}u(h), u(g)].$$

For the 1-cocycle $u \in Z^1(G, ad\phi)$, the map $(g, h) \mapsto \frac{1}{2}[Ad_{\phi(g)}u(h), u(g)]$ defines a 2-cocycle in $Z^2(G, ad\phi)$, and it is just the product $[\xi, \xi]$: $H^1(G, ad\phi) \times H^1(G, ad\phi) \to H^2(G, ad\phi)$ which is the cup-product in G using the Lie product in $ad\phi$ as coefficients [Lang, 1996, Chapter IV]. Thus the tangent condition is equivalent to solving

$$u_2(g) + Ad_{\phi(g)}u_2(h) - u_2(gh) = \frac{1}{2}[Ad_{\phi(g)}u(h), u(g)]$$

for u_2. But $u_2(g) + Ad_{\phi(g)}u_2(h) - u_2(gh)$ represents a coboundary of 1-cocycle ξ_1, therefore $[\xi, \xi] = \delta\xi_1 = 0$ as a cohomology class.

Note that from the expansion of $e^{tu(g) + \sum_{k=2}^{\infty} u_k(g)t^k}\phi(g)$ there are infinite many obstructions classes ξ_k arising from solving equations for u_k as $u \in Z^1(G, ad\phi)$ is tangent to a differentiable path in $R(G)$. Each u_k defines in terms of the preceding solutions $u_{k-1}, u_{k-2}, \cdots, u_2, u$ and takes value in

$H^2(G, ad\phi)$. From the discussion above, each u_k gives the obstruction class ξ_k related to the cohomology class ξ, and the obstructions for each $k \geq 2$ vanish if and only if u can be tangent to a path in $Hom(G, SU(2))$. $\qquad\square$

Remark 5.2.12. If $H^2(G, ad\phi) = 0$, then every $\xi \in H^1(G, ad\phi)$ is tangent to a smooth path in $\mathcal{R}(G)$ since all obstruction classes $\xi_k = 0$ $(k \geq 2)$ in $H^2(G, ad\phi) = 0$.

There is a nondegenerate symmetric bilinear form $B : \sim\cong(2) \times \sim\cong(2) \to \mathbb{R}$ invariant under the adjoint representation. The bilinear form B defines an isomorphism $ad\phi \cong (ad\phi)^*$ for any $\phi \in R(G)$. Then by Poincaré duality from the cup-product,

$$H^i(G, ad\phi) \times H^{n-i}(G, ad\phi^*) \to H^n(G, ad\phi \otimes ad\phi^*),$$

with coefficients $ad\phi \otimes ad\phi^* \cong \mathbb{R}$. Therefore, for $G = \pi_1(\Sigma_g)$ and $n = 2$, we have

$$H^i(\pi_1(\Sigma_g), ad\phi) \cong H^{2-i}(\pi_1(\Sigma_g), ad\phi^*)^*, \quad i = 0, 1.$$

$$H^2(\pi_1(\Sigma_g), ad\phi) \cong H^0(\pi_1(\Sigma_g), ad\phi^*)^* \cong H^0(\pi_1(\Sigma_g), ad\phi)^*.$$

The cohomology group $H^0(\pi_1(\Sigma_g), ad\phi)$ is the space of invariants in $ad\phi$, is the infinitesimal centralizer $\mathfrak{z}(\phi)$ of $\phi(\pi_1(\Sigma_g))$ in $\mathfrak{su}(2)$. Hence,

$$\dim H^2(\pi_1(\Sigma_g), ad\phi) = \dim H^0(\pi_1(\Sigma_g), ad\phi)^*$$
$$= \dim H^0(\pi_1(\Sigma_g), ad\phi)$$
$$= \dim \mathfrak{z}(\phi).$$

Note that $H^i(\pi_1(\Sigma_g), ad\phi)$ is the twisted cohomology group which cochains are given by the ordinary cochains of Σ_g tensoring with coefficients module $ad\phi$, and the structure of the coefficients module takes place in the differential of the twisted complex. Therefore the Euler characteristic $\sum_{i=0}^2 (-1)^i \dim H^i(\pi_1(\Sigma_g), ad\phi)$ is independent of ϕ, and equal to

$$\sum_{i=0}^2 (-1)^i \dim H^i(\pi_1(\Sigma_g), \mathfrak{su}(2)) = \chi(\Sigma_g) \cdot \dim \mathfrak{su}(2),$$

for the trivial representation $\phi = Id$.

$$\dim H^0(\pi_1(\Sigma_g), ad\phi) - \dim H^1(\pi_1(\Sigma_g), ad\phi) + \dim H^2(\pi_1(\Sigma_g), ad\phi)$$

$$= (2 - 2g) \cdot \dim SU(2).$$

Hence we get

$$\dim H^1(\pi_1(\Sigma_g), ad\phi) = (2g - 2)\dim SU(2) + 2\dim 3(\phi) \equiv 0 \pmod{2}.$$

Theorem 5.2.13. *Let* $\omega_B : H^1(\pi_1(\Sigma_g), ad\phi) \times H^1(\pi_1(\Sigma_g), ad\phi) \to H^2(\pi_1(\Sigma_g), \mathbb{R}) \cong \mathbb{R}$ *is the composition of the cup-product with the bilinear form* $B : ad\phi \times ad\phi \overset{\cong}{\to} \mathbb{R}$. *Then the 2-form* ω_B *is a symplectic form on the space* $\mathcal{R}(\pi_1(\Sigma_g)) = Hom(\pi_1(\Sigma_g), SU(2))/SU(2)$.

Proof: A symplectic form is a nondegenerate skew closed 2-form. The nondegeneracy of ω_B follows from the Poincaré duality and the nondegeneracy of the bilinear form B. The skew property follows from $\alpha \cup \beta = (-1)^{\deg \alpha \deg \beta} \beta \cup \alpha$ for $\alpha, \beta \in H^1(\pi_1(\Sigma_g), ad\phi)$. The closeness of ω_B is a much deeper result. We refer the interested readers to [Goldman, 1984, §1.8 and §3.10] for a complete proof on this closeness.   □

5.3   The $SU(2)$ Casson invariant of integral homology 3-spheres

We recall briefly the definition of the $SU(2)$ Casson invariant for integral homology 3-spheres. There are good references on this subject [Akbulut and McCarthy, 1990; Walker, 1992].

Let $\mathcal{R}^*(X)$ be the irreducible $SU(2)$-representations of $\pi_1(X)$. Thus the Heegaard decomposition of $Y = H_1 \cup_\Sigma H_2$, and $\Sigma_0 = \Sigma \setminus D^2$. We have

$$\begin{array}{ccc} \pi_1(\Sigma_0) \to \pi_1(\Sigma) & \to & \pi_1(H_1) \\ \downarrow & & \downarrow \\ \pi_1(H_2) & \to & \pi_1(Y), \end{array}$$

and the corresponding representation varieties

$$\begin{array}{ccc} \mathcal{R}^*(\Sigma_0) \longleftarrow \mathcal{R}^*(\Sigma) & \longleftarrow & \mathcal{R}^*(H_1) \\ \uparrow & & \uparrow \\ \mathcal{R}^*(H_2) & \longleftarrow & \mathcal{R}^*(Y). \end{array}$$

In [Akbulut and McCarthy, 1990], the intersection $\mathcal{R}(H_1) \cap \mathcal{R}(H_2)$ in $\mathcal{R}(\Sigma)$ is transversal at the single isolated reducible representation θ_Y. The intersection of the smooth open manifolds $\mathcal{R}^*(H_1)$ and $\mathcal{R}^*(H_2)$ in $\mathcal{R}^*(\Sigma)$ is compact due to the compact group $SU(2)$. Choose any isotopy of $\mathcal{R}(\Sigma)$ with compact support around θ_Y such that $\mathcal{R}^*(H_2)$ is perturbed into $\tilde{\mathcal{R}}^*(H_2)$ with transversal intersection points $\mathcal{R}^*(H_1) \cap \tilde{\mathcal{R}}^*(H_2)$. There are orientations on spaces $\mathcal{R}(\Sigma), \mathcal{R}(H_i)$ $(i = 1, 2)$. Then the intersection point $p \in \mathcal{R}^*(H_1) \cap \tilde{\mathcal{R}}^*(H_2)$ is assigned a number $\varepsilon_p(= \pm 1)$ depending upon whether the orientation of $T_p\mathcal{R}^*(H_1) \oplus T_p\tilde{\mathcal{R}}^*(H_2)$ matches with the one of $T_p\mathcal{R}(\Sigma)$.

Definition 5.3.1. For an integral homology 3-sphere Y $(H_*(Y;\mathbb{Z}) = H_*(S^3;\mathbb{Z}))$, the Casson invariant of Y is defined to be

$$\lambda_C(Y) = \frac{(-1)^{g(\Sigma)}}{2} \sum_{p \in \mathcal{R}^*(H_1) \cap \tilde{\mathcal{R}}^*(H_2)} \varepsilon_p,$$

where $Y = H_1 \cup_\Sigma H_2$ is an Heegaard decomposition of Y.

The number $\lambda_C(Y)$ is independent of the choice of perturbations and independent of the Heegaard decomposition [Akbulut and McCarthy, 1990]. The Casson invariant gives a powerful tool to solve some topological problems. (This is beyond what we try to do in this book.) This has been extended to rational homology 3-spheres by Walker [1992]. The Casson-Walker invariant is an invariant of Q-homology 3-spheres Y via the counting of irreducible $SU(2)$-representations of $\pi_1(Y)$ and the correction term from the $U(1)$-representations of $\pi_1(Y)$. The proper counting on a stratified space is a challenge problem in general. The transversal, compactness and orientation are three key ingredients involved in the counting, and perturbations play an essential role to make it work. Due to the stratified structure , counting on the top-stratum (open manifold) depends upon the perturbation along the next stratum, mainly effected by the normal direction information.

(1) Casson looked at the Z-homology 3-spheres case with only one isolated trivial representation θ in the next stratum, then he used the compact support perturbation fixed around θ to make the counting on the top-stratum to be an invariant;

(2) Walker looked at the Q-homology 3-spheres case with the isolated $U(1)$ representations in the next stratum, then he used the compact support perturbations fixed around the isolated $U(1)$-stratum to understand the normal direction. Tangent direction is automatically transverse with topological invariant on the tangent Morse data. The Walker correction term is a kind of relative normal Morse data on the $U(1)$-stratum (Maslov index or spectral flow). The relative tangent Morse data on the $U(1)$-stratum is independent of the perturbation. Thus adding correction term from the normal data on the counting irreducible representations on the top-stratum gives an invariant.

5.4   The $SU(2)$ Casson invariant of knots

For a knot K in an integral homology 3-sphere Y, define $K_n = Y(K, \frac{1}{n})$ to be the $1/n$-Dehn surgery on K in Y. Note that K_n is again an integral homology sphere and $K_0 = Y$. Define

$$\lambda_C(K, n) = \lambda_C(K_{n+1}) - \lambda_C(K_n). \tag{5.3}$$

One of the key ingredients in Casson's invariant [Akbulut and McCarthy, 1990] is to show that $\lambda_C(K, n)$ is independent of n. By choosing preferred Heegaard decompositions for K_{n+1} and K_n and introducing a canonical isotopy, one can construct a difference cycle which is independent of the surgery coefficient n. The number $\lambda_C(K, n)$ defined in (5.3) is essentially counting the intersection between the difference cycle and the unperturbed representation space of one of the handlebodies (see [Akbulut and McCarthy, 1990] for more details). The number $\lambda_C(K) = \lambda_C(K, n)$ is called the Casson invariant of knots K.

Let F be a Seifert surface in Y for K. Let V be the linking matrix $V = (v_{ij})$, where $v_{ij} = link(e_i, e_j^+)$ for a basis $e_i \in H_1(F, \mathbb{Z})$ and e_j^+ is the pushoff of e_j for $1 \le i, j \le rank H_1(F, \mathbb{Z})$. Thus the Alexander polynomial of K is given by $\Delta_F(t) = \det(V - tV^T)$. Note that $\Delta_{F'}(t) = \pm t^l \Delta_F(t)$ for a different Seifert surface F'. Hence we define a symmetric Alexander polynomial $\Delta_K(t)$ with $\Delta_K(t) = \Delta_K(t^{-1})$ and $\Delta_K(t) = a_0 + \sum_{i>0} a_i(t^i + t^{-i})$.

Proposition 5.4.1. *Let K be a knot in an integral homology 3-sphere Y. Then there exists a knot L in S^3 such that*

(i) $\lambda_C(K) = \lambda_C(L)$ and $\Delta_K(t) = \Delta_L(t)$;

(ii) $\lambda_C(K) = \frac{1}{2}\Delta_K''(1) = \sum_{i>0} i^2 a_i$, up to the normalization $\lambda(3_1) = +1$.

The key relation is $\lambda_C(K_+) - \lambda_C(K_-) = lk(K_0)$. See [Akbulut and McCarthy, 1990, pp. 142–149] for more details.

We are going to give a combinatoric interpretation of the Casson invariant of knots through the Gauss diagram and Vassiliev invariants of links [Polyak and Viro, 1994].

Recall that a Gauss diagram is an immersing circle with the preimage of each double point connected with a chord. By orienting each chord from the upper branch to the lower branch, we recover the information of overpasses and underpasses. Each chord c is assigned the sign $\varepsilon(c)$ of the local writhe number. A based Gauss diagram is a Gauss diagram with a marked point on the circle, distinct from the end points of the chords. We show a combinatorial calculation of the Casson invariant of a knot by the work of Polyak and Viro [1994].

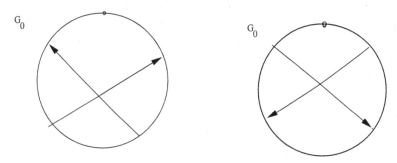

Fig. 5.1   Gauss subdiagrams G_0 and $\overline{G}_0$

Theorem 5.4.2. *If G is any Gauss diagram of a knot K, then*

$$\lambda_C(K) = \sum_{G' \subset G} \varepsilon(c_1) \cdot \varepsilon(c_2),$$

where the summation is taken from all subdiagrams of G isomorphic to G_0 (see Figure 5.1), and c_1 and c_2 are the chords of subdiagrams.

Proof: By the skein relation of the Casson invariant of knots

$$\lambda_C(K_+) - \lambda_C(K_-) = lk(K_0),$$

we present all knots by their fragments. Note that the unknot O has zero Casson invariant $\lambda_C(O) = 0$.

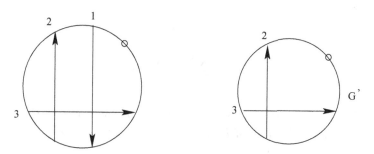

Fig. 5.2   Normalizing the value on trefoil knot

The right hand evaluated on the trefoil knot 4_1 equals to one: $\sum_{G' \subset G} \varepsilon(c_2) \cdot \varepsilon(c_3) = (-1) \cdot (-1) = 1$ in Figure 5.2. One needs to change K to the unknot by walking around the base point along the orientation and replacing an undercrossing by an overcrossing, if at the first passage through the point we walk along the undercrossing. When we pass over the whole diagram with repeating the changes, the knot represents the unknot. Each time we change a crossing at s: if $K = K_-$ changing from undercrossing to overcrossing, then

$$\lambda_C(K_+) - \lambda_C(K) = lk(K_0^s) = (-\varepsilon(s)) \cdot lk(K_0^s);$$

if $K = K_+$ changing from overcrossing to undercrossing, then

$$\lambda_C(K) - \lambda_C(K_-) = lk(K_0^s) = \varepsilon(s) \cdot lk(K_0^s).$$

Since $\lambda_C(O) = 0$, then we obtain

$$\lambda_C(K) = \sum_s \varepsilon(s) \cdot lk(K_0^s),$$

where K_0^s runs over links which smooth at double points where the cross-changing is performed.

To calculate $lk(K_0^s)$, we first collect all components containing the base point which go below the other components and sum up the signs over those crossing points. Note that these points correspond to the chords of G intersecting the chord of $c(s)$ corresponding to s and directed to the side of $c(s)$ containing the base point. Thus all arrows in the Gauss diagram with heads between the base point and the head of $c(s)$ have to be inverted. The linking number $lk(K_0^s)$ is the sum of signs of arrows crossing $c(s)$ and having heads between the tail of $c(s)$ and the base point. Therefore $lk(K_0^s) = \sum_{c_2} \varepsilon(c_2)$ with the summation over all subdiagrams of the type G_0. Hence

$$\lambda_C(K) = \sum_s \varepsilon(s) \cdot lk(K_0^s)$$

$$= \sum_{c_2, s} \varepsilon(s) \cdot \varepsilon(c_2)$$

$$= \sum_{G' \subset G} \varepsilon(c_1) \cdot \varepsilon(c_2).$$

□

Remarks: (1) Note that a rotation of a knot K by π around the x-axis results in a Gauss diagram of K with all arrows of G reversed, while their signs are preserved. Hence

$$\lambda_C(K) = \sum_{\overline{G}_0 \subset G} \varepsilon(c_1) \cdot \varepsilon(c_2),$$

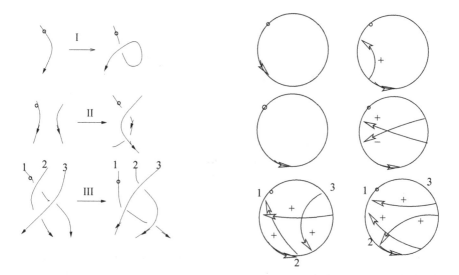

Fig. 5.3 Invariance under Reidemeister moves

with $\overline{G}_0$ is isomorphic to G_0 with arrows reversed (see Figure 5.1).

(2) The expression $\sum_{G' \subset G} \varepsilon(c_1) \cdot \varepsilon(c_2)$ gives a combinatorial method to define the Casson invariant of knots . The invariance under Reidemeister moves follows from the study of the corresponding changings of a Gauss diagram (see Figure 5.3). The first and third Reidemeister moves do not change any subdiagrams isomorphic to G_0, the second Reidemeister move creates a pair of new chords with $\pm$ sign, so any new subdiagrams cancel out in pairs.

(3) There are many methods to calculate the Casson invariant of knots. Theorem 5.4.2 essentially follows from Kauffman's [1987b] algorithm for calculation of $\Delta_K''(1)$.

Example: We calculate the Casson invariant of the figure eight knot from Theorem 5.4.2. By picking up a base point in the figure, and changing the knot into a Gauss diagram, we have only one subdiagram in the Gauss diagram which is isomorphic to G_0. Hence the result follows from the calculation of local writhe numbers of the chords.

Theorem 5.4.3. *For any knot K which admits a diagram with n crossing points, then*

$$|\lambda_C(K)| \leq [\frac{n^2}{8}],$$

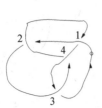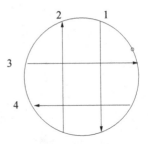

Fig. 5.4　Calculation of the figure eight knot

where $[\cdot]$ is the largest integer function.

Proof: By representing a base Gauss diagram G of K with n crossing points, we divide the set of chords into $C^+ = \{c \in G : (t(c), b, h(c)) = o(S^1)\}$ and $C^- = \{c_2 \in G : (t(c), b, h(c)) = -o(S^1)\}$ (the tail of c, the base point and the head of c gives a counterclockwise (+-orientation) or clockwise (−-orientation) on the plane). Thus $|C^+| = k$ is the cardinality of C^+ and $|C^-| = n-k$ ($0 \le k \le n$). Any subdiagram $G' \subset G$ isomorphic to G_0 (or G_0 with arrows reversed, denoted by $\overline{G}_0$) consists of a pair of $S^+ = \{(c_1, c_2) \cong G_0 \subset G\}$ and $S^- = \{(c_1, c_2) \cong \overline{G}_0 \subset G\}$. Thus $|S^+| + |S^-| \le k(n - k)$.

$$|\lambda_C(K)| = |\sum_{G' \subset G} \varepsilon(c_1) \cdot \varepsilon(c_2)|$$
$$\le |S^+|$$
$$|\lambda_C(K)| = |\sum_{\overline{G}' \subset G} \varepsilon(c_1) \cdot \varepsilon(c_2)|$$
$$\le |S^-|$$
$$\le \min\{|S^+|, |S^-|\}$$
$$\le \frac{k(n - k)}{2} \le [\frac{n^2}{8}].$$

$\square$

Remark: For $n = 4$ in the example of the figure eight knot, $|\lambda_C(4_1)| = 1$ and $[\frac{4^2}{8}] = 2$. Hence the inequality is not sharp. For torus knot $(n, 2)$ with $n \ge 3$ odd (n is also the crossing number of the knot), the Casson invariant is $\frac{n^2-1}{8}$ which equals to $[\frac{n^2}{8}]$.

5.5 Representation space of knot complements

Let (S^3, D^3_+, D^3_-, S^2) be a Heegaard decomposition of S^3 with genus 0, where

$$S^3 = D^3_+ \cup_{S^2} D^3_-, \quad \partial D^3_+ = \partial D^3_- = D^3_+ \cap D^3_- = S^2.$$

Suppose that a knot $K \subset S^3$ is in a general position with respect to this Heegaard decomposition. So $K \cap S^2 = \{x_1, \cdots, x_n, y_1, \cdots, y_n\}$, $K \cap D^3_\pm$ is a collection of unknotted, unlinked arcs $\{\gamma_1^\pm, \cdots, \gamma_n^\pm\} \subset D^3_\pm$, where $\partial \gamma_i^- = \{x_i, y_i\}$ and $\{\gamma_1^+, \cdots, \gamma_n^+\} = K \cap D^3_+$ is a braid of n strands inside D^3_+. Let β be a word in the braid group B_n. For the top end point x_i of γ_i^+, the bottom end points of $\{\gamma_1^+, \cdots, \gamma_n^+\}$ give a permutation of $\{y_1, \cdots, y_n\}$ which generates a map

$$\pi : B_n \to S_n,$$

where $\pi(\beta)$ is the permutation of $\{y_1, \cdots, y_n\}$ in the symmetric group of n letters. Let $K = \bar{\beta}$ be the closure of β. It is well-known that there is a correspondence between a knot and a braid β with $\pi(\beta)$ being a complete cycle of the n letters [Birman, 1974].

There is a corresponding Heegaard decomposition for the complement of a knot K,

$$S^3 \setminus K = (D^3_+ \setminus K) \cup_{(S^2 \setminus K)} (D^3_- \setminus K),$$

$$D^3_\pm \setminus K = D^3_\pm \setminus (D^3_\pm \cap K), \quad S^2 \setminus K = S^2 \setminus (S^2 \cap K).$$

Thus from the Seifert-Van Kampen theorem:

$$
\begin{array}{ccc}
\pi_1(S^2 \setminus K) & \to & \pi_1(D^3_+ \setminus K) \\
\downarrow & & \downarrow \\
\pi_1(D^3_- \setminus K) & \to & \pi_1(S^3 \setminus K),
\end{array}
$$

we obtain a pull-back of representation spaces

$$
\begin{array}{ccc}
\mathcal{R}(S^2 \setminus K) & \leftarrow & \mathcal{R}(D^3_+ \setminus K) \\
\uparrow & & \uparrow \\
\mathcal{R}(D^3_- \setminus K) & \leftarrow & \mathcal{R}(S^3 \setminus K),
\end{array}
\qquad (5.4)
$$

where $\mathcal{R}(X) = Hom(\pi_1(X), SU(2))/SU(2)$ for $X = S^2 \setminus K, D^3_\pm \setminus K, S^3 \setminus K$. The group $SU(2)$ admits a natural Riemannian metric which arises by translation from the inner product on the Lie algebra of S^3. The natural identification of S^3 with $SU(2)$ is an isometry of the standard metric on S^3 and the above Riemannian metric on $SU(2)$. Also note that $S^3 \setminus K$ is the Eilenberg-MacLane space $K(\pi_1(S^3 \setminus K), 1)$ for a classical knot K due to the asphericity theorem of Papakyriakopoulos.

Let $\mathcal{R}(S^3 \setminus K)^{[i]}$ be the space of $SU(2)$ representations $\rho : \pi_1(S^3 \setminus K) \to SU(2)$ such that

$$\rho([m_{x_i}]) \sim \begin{pmatrix} i & 0 \\ 0 & -i \end{pmatrix}, \quad \rho([m_{y_i}]) \sim \begin{pmatrix} i & 0 \\ 0 & -i \end{pmatrix}, \tag{5.5}$$

where $m_{x_i}, m_{y_i} (i = 1, 2, \cdots, n)$ are the meridian circles around x_i, y_i respectively, and $\sim$ is the conjugate relation. Note that $\pi_1(S^2 \setminus K)$ is generated by $m_{x_i}, m_{y_i} (i = 1, 2, \cdots, n)$ with one relation $\prod_{i=1}^n m_{x_i} = \prod_{i=1}^n m_{y_i}$. Corresponding to (5.4), we have

$$
\begin{array}{ccc}
\mathcal{R}(S^2 \setminus K)^{[i]} & \leftarrow & \mathcal{R}(D_+^3 \setminus K)^{[i]} \\
\uparrow & & \uparrow \\
\mathcal{R}(D_-^3 \setminus K)^{[i]} & \leftarrow & \mathcal{R}(S^3 \setminus K)^{[i]}.
\end{array}
\tag{5.6}
$$

The conjugacy class in $SU(2)$ is completely determined by its trace. So the condition (5.5) can be reformulated for $\rho \in \mathcal{R}(X)^{[i]}$ as

$$\mathrm{tr}\rho([m_{x_i}]) = \mathrm{tr}\rho([m_{y_i}]) = 0. \tag{5.7}$$

The space $\mathcal{R}(S^2 \setminus K)^{[i]}$ can be identified with the space of $2n$ matrices $X_1 \cdots, X_n, Y_1, \cdots, Y_n$ in $SU(2)$ satisfying

$$\mathrm{tr}(X_i) = \mathrm{tr}(Y_i) = 0, \quad \text{for } i = 1, \cdots, n, \tag{5.8}$$

$$X_1 \cdot X_2 \cdots X_n = Y_1 \cdot Y_2 \cdots Y_n. \tag{5.9}$$

Remark: Let $\tilde{S}^2$ be the branched double covering of S^2 along $K \cap S^2$. Let $\mathcal{R}^*(\tilde{S}^2)$ be the subset of $\mathcal{R}(\tilde{S}^2)$ consisting of irreducible representations. Then $\mathbb{Z}_2$ operates on $\mathcal{R}^*(\tilde{S}^2)$. The component of the fixed point set $\mathcal{R}^*(\tilde{S}^2)$ with trace zero condition coincides with $\mathcal{R}^*(S^2 \setminus K)^{[i]}$.

Lemma 5.5.1. *The space $\mathcal{R}^*(S^2 \setminus K)^{[i]}$ is a manifold of dimension $4n - 6$.*

Proof: Note that $\mathcal{R}^*(S^2 \setminus K)^{[i]} = (H_n \setminus S_n)/SU(2)$ in Lin's [1992] notation, where

$$H_n = \{(X_1, \cdots, X_n, Y_1, \cdots, Y_n) \in Q_n \times Q_n | \ X_1 \cdots X_n = Y_1 \cdots Y_n\},$$

S_n is the subspace of H_n consisting of all the reducible points which can be diagonalized by $A \in SU(2)$ simultaneously, and $H_n \setminus S_n$ is the total space of a $SU(2)$-fiber bundle over $\hat{H}_n = \mathcal{R}^*(S^2 \setminus K)^{[i]}$. It is remarkable that the points in H_n where the $SU(2)$-action is not locally free are precisely those points in S_n.

Define $f : Q_n \times Q_n \to SU(2)$ by $f(X_1, \cdots, X_n, Y_1, \cdots, Y_n) = X_1 \cdots X_n Y_n^{-1} \cdots Y_1^{-1}$. Hence $f^{-1}(Id_{2\times 2}) = H_n$. It suffices to show that $Tf : T_\rho H_n \to T_{Id} SU(2)$ is onto for $\rho \in H_n \setminus S_n$. Without loss of generality, we can arrange that $X_1 = \begin{pmatrix} i & 0 \\ 0 & -i \end{pmatrix}$ and $Y_1^{-1} = \begin{pmatrix} is & v \\ -\bar{v} & -is \end{pmatrix}$, where $s^2 + v\bar{v} = 1$, $v \neq 0$ and $Y_1 = \begin{pmatrix} -is & -v \\ \bar{v} & is \end{pmatrix}$. Thus the representation ρ corresponding with $(X_1, \cdots, Y_n)$ with those X_1 and Y_1 cannot be diagonalized simultaneously, therefore it stands for an element in $H_n \setminus S_n$.

Since $(X_1, \cdots, X_n, Y_1, \cdots, Y_n) \in f^{-1}(Id)$ which is equivalent to the identity

$$X_1 \cdots X_n Y_n^{-1} \cdots Y_1^{-1} = Id,$$

we have

$$X_2 \cdots X_n Y_n^{-1} \cdots Y_2^{-1} = X_1^{-1} Y_1 = \begin{pmatrix} -i & 0 \\ 0 & i \end{pmatrix} \cdot \begin{pmatrix} -is & -v \\ \bar{v} & is \end{pmatrix} = \begin{pmatrix} -s & iv \\ i\bar{v} & -s \end{pmatrix}.$$

Let $X_1^t = \begin{pmatrix} i\sqrt{1-t^2} & t \\ -t & -i\sqrt{1-t^2} \end{pmatrix}$ be a 1-parameter family with $\frac{d}{dt} X_1^t|_{t=0} = \begin{pmatrix} 0 & 1 \\ -1 & 0 \end{pmatrix}$.

$$\frac{d}{dt} f(X_1^t, X_2, \cdots, Y_1)|_{t=0} = \frac{d}{dt}(X_1^t X_2 \cdots Y_1^{-1})|_{t=0}$$

$$= \frac{d}{dt} X_1^t|_{t=0} X_2 \cdots X_n Y_n^{-1} \cdots Y_1^{-1}$$

$$= \frac{d}{dt} X_1^t|_{t=0} X_1^{-1}$$

$$= \begin{pmatrix} 0 & 1 \\ -1 & 0 \end{pmatrix} \cdot \begin{pmatrix} i & 0 \\ 0 & -i \end{pmatrix}$$

$$= \begin{pmatrix} 0 & i \\ i & 0 \end{pmatrix} \in \mathbf{su}(2).$$

Note that $\begin{pmatrix} 0 & i \\ -i & 0 \end{pmatrix}$ does not lie in the Lie algebra $\mathbf{su}(2)$ of $SU(2)$ (see §6.1). There are several typos in [Lin, 1992].

Let $Y_1^{-1}(t) = \begin{pmatrix} is & ve^{it} \\ -\bar{v}e^{-it} & -is \end{pmatrix}$ with $\frac{d}{dt} Y_1^{-1}(t)|_{t=0} = \begin{pmatrix} 0 & iv \\ i\bar{v} & 0 \end{pmatrix}$ and

$Y_1(0) = Y_1.$

$$\frac{d}{dt}f(X_1, X_2, \cdots, Y_1(t))|_{t=0} = \frac{d}{dt}(X_1 \cdots X_n Y_n^{-1} \cdots Y_1^{-1}(t))|_{t=0}$$

$$= X_1 \cdots X_n Y_n^{-1} \cdots Y_2^{-1} \frac{d}{dt}Y_1^{-1}(t)|_{t=0}$$

$$= Y_1(0)\frac{d}{dt}Y_1^{-1}(t)|_{t=0}$$

$$= \begin{pmatrix} -is & -v \\ \bar{v} & is \end{pmatrix} \cdot \begin{pmatrix} 0 & iv \\ i\bar{v} & 0 \end{pmatrix}$$

$$= \begin{pmatrix} -iv\bar{v} & sv \\ -s\bar{v} & iv\bar{v} \end{pmatrix}$$

$$= (-v\bar{v})\begin{pmatrix} i & 0 \\ 0 & -i \end{pmatrix} + (sv)\begin{pmatrix} 0 & 1 \\ -1 & 0 \end{pmatrix}.$$

Therefore Tf is onto from §6.1, and $H_n \setminus S_n$ is a manifold of $4n - 3$ dimension. From the previous remark on the $SU(2)$-action which is locally free on $H_n \setminus S_n$, so $\mathcal{R}^*(S^2 \setminus K)^{[i]} = (H_n \setminus S_n)/SU(2)$ is a manifold of dimension $4n - 6$. □

Given $\beta \in B_n$, we denote the graph of β in $Q_n \times Q_n$ by Γ_β:

$$\Gamma_\beta = \{(X_1, \cdots, X_n, \beta(X_1), \cdots, \beta(X_n)) \in Q_n \times Q_n\}.$$

As an automorphism of the free group $Z[m_1] * Z[m_2] * \cdots * Z[m_n]$, the element $\beta \in B_n$ preserves the word $[m_1] \cdots [m_n]$. Hence we obtain

$$X_1 \cdots X_n = \beta(X_1) \cdots \beta(X_n).$$

In other words, Γ_β is a subspace of H_n. In fact, for $\bar{\beta} = K$, the subspace Γ_β coincides with the subspace of representations $\rho : \pi_1(S^2 \setminus K) \to SU(2)$ in H_n which can be extended to $\pi_1(D_+^3 \setminus K)$. So we have $\Gamma_\beta = Hom(\pi_1(D_+^3 \setminus K), SU(2))^{[i]}$. The space $\mathcal{R}^*(D_+^3 \setminus K)^{[i]} = \Gamma_{\beta,irre}/SU(2) = \overline{\Gamma}_\beta$ is the irreducible $SU(2)$ representations with trace zero condition over $D_+^3 \setminus K$.

If $\beta = id$, then Γ_{id} represents the diagonal in $Q_n \times Q_n$:

$$\Gamma_{id} = \{(X_1, \cdots, X_n, X_1, \cdots, X_n) \in Q_n \times Q_n\}.$$

Since $K \cap D_-^3$ represents the trivial braid, this subspace $\Gamma_{id} \subset H_n$ can be identified with the subspace of representations in $Hom(\pi_1(S^2 \setminus K), SU(2))^{[i]}$ which can be extended to $\pi_1(D_-^3 \setminus K)$, i.e.

$$\Gamma_{id} = Hom(\pi_1(D_-^3 \setminus K), SU(2))^{[i]}.$$

By the Seifert-Van Kampen Theorem, the intersection $\Gamma_\beta \cap \Gamma_{id}$ is the same as the space of representations of $\pi_1(S^3 \setminus K)$ satisfying the monodromy condition $[i]$ (see (5.4)),

$$\Gamma_\beta \cap \Gamma_{id} = Hom(\pi_1(S^3 \setminus K), SU(2))^{[i]}.$$

For $\beta \in B_n$ with $\overline{\beta} = K$, there is an induced diffeomorphism (still denoted by β) from Q_n to itself. Such a diffeomorphism also induces a diffeomorphism $\phi_\beta : \mathcal{R}^*(S^2 \setminus K)^{[i]} \to \mathcal{R}^*(S^2 \setminus K)^{[i]}$.

Lemma 5.5.2. *For $\beta \in B_n$ with $\overline{\beta} = K$, the induced diffeomorphism ϕ_β has the fixed point set which can be identified with $\mathcal{R}^*(S^3 \setminus K)^{[i]}$.*

Proof: Using the above identification, we have

$$\mathcal{R}(S^3 \setminus K)^{[i]} = \{X_1, \cdots, X_n : \quad X_i = \beta(X_i), 1 \le i \le n\}.$$

So the fixed point set of the induced diffeomorphism on $\mathcal{R}^*(S^2 \setminus K)^{[i]}$ is the space of irreducible representations with the property $X_i = \beta(X_i)$ for all i. Hence the result follows. $\square$

Note that the reducible representations of $\pi_1(S^3 \setminus K)$ are conjugate to the $U(1)$ representations for which each generator X_j of $\pi_1(S^3 \setminus K)$ is represented by $\begin{pmatrix} a & 0 \\ 0 & \overline{a} \end{pmatrix}$ for $a \in \mathbb{C}$ with $|a| = 1$. Let $\mathcal{S}^*(X)^{[i]} = Hom(\pi_1(X), U(1))^{[i]}/\mathbb{Z}_2$ be the strata of $\mathcal{R}(X)$ corresponding to the representations with $U(1)$ stabilizer. Similar to (5.6), we have the following diagram:

$$\begin{array}{ccc} \mathcal{S}^*(S^3 \setminus K)^{[i]} & \to & \mathcal{S}^*(D^3_+ \setminus K)^{[i]} \\ \downarrow & & \downarrow \\ \mathcal{S}^*(D^3_- \setminus K)^{[i]} & \to & \mathcal{S}^*(S^2 \setminus K)^{[i]}. \end{array} \qquad (5.10)$$

Let $S_n = \mathcal{S}^*(S^2 \setminus K)^{[i]}$ be the subset of H_n consisting of $(X_1, \cdots, X_n, Y_1, \cdots, Y_n)$ such that there is a matrix $A \in SU(2)$ with the property

$$A^{-1} X_i A, \quad A^{-1} Y_i A \in U(1), \quad \text{for all } i = 1, 2, \cdots, n.$$

Note that $Hom(\pi_1(S^3 \setminus K), U(1)) \cong U(1)$, $\mathcal{S}^*(S^3 \setminus K) = U(1)/\mathbb{Z}_2$ the upper half unit circle in the complex plane. In general, we would like to understand the effect of the reducible representations $\mathcal{S}^*(S^2 \setminus K)^{[i]}$ to get more information about knots. It is a challenging problem to understand the Walker correction terms (see [Walker, 1992] for more details) for the family of $U(1)$ reducible representations $\mathcal{S}^*(S^3 \setminus K)$, rather than for an isolated $U(1)$ reducible representation.

There is only one reducible conjugacy class of representations $\rho : \pi_1(S^3 \setminus K) \to U(1) \hookrightarrow SU(2)$ which satisfy the trace condition $\mathrm{tr}(\rho([m_{x_i}])) = \mathrm{tr}(\rho([m_{y_i}])) = 0$. Without loss of generality, we assume that $s : \pi_1(S^3 \setminus K) \to U(1)$ is the diagonal matrix

$$s([m_{x_i}]) = s([m_{y_i}]) = \begin{bmatrix} i & 0 \\ 0 & -i \end{bmatrix}.$$

Write $s = \sigma \oplus \sigma^{-1}$, where $\sigma : \pi_1(S^3 \setminus K) \to U(1)$ is given by $\sigma([m_{x_i}]) = \sigma([m_{y_i}]) = i$. Thus

$$ad(s) = End^0(\sigma \oplus \sigma^{-1}) = \mathbb{R} \oplus \sigma^{\otimes 2}.$$

By deformation theory, any deformation in $H^1(\cdot, \mathbb{R})$-direction changes the trace condition. So only $H^1(\cdot, \sigma^{\otimes 2})$ fits into the Mayer-Vietoris sequence:

$$H^1(S^3 \setminus K, \sigma^{\otimes 2}) \to H^1(D_-^3 \setminus K, \sigma^{\otimes 2}) \oplus H^1(D_+^3 \setminus K, \sigma^{\otimes 2})$$

$$\to H^1(S^2 \setminus K, \sigma^{\otimes 2}) \to \cdots.$$

The intersections between $\mathcal{S}^*(D_+^3 \setminus K)^{[i]} = S_n \cap \Gamma_\beta$ and $\mathcal{S}^*(D_-^3 \setminus K)^{[i]} = S_n \cap \Gamma_{id}$ are transversal if and only if $H^1(S^3 \setminus K, \sigma^{\otimes 2}) = 0$. According to Milnor [1968], $H^1(S^3 \setminus K, \sigma^{\otimes 2})$ can be computed by the cohomology of the infinite cyclic covering space $\widehat{S^3 \setminus K}$. The cohomology $H^1(\widehat{S^3 \setminus K})$ is a module over the Laurent polynomial ring $\mathbb{Z}[t, t^{-1}]$:

$$H^1(\widehat{S^3 \setminus K}) \otimes_{\mathbb{Z}[t,t^{-1}]} \mathbb{C} = H^1(S^3 \setminus K, \sigma^{\otimes 2}).$$

Therefore $0 = H^1(\widehat{S^3 \setminus K}) = \mathbb{Z}[t, t^{-1}]/\Delta_K(t)$ if and only if $\Delta_K(-1) \neq 0$, which is true for any knot K. See §4.7 also.

Lemma 5.5.3. *For any* $K = \bar{\beta}$, $\mathcal{S}^*(D_+^3 \setminus K)^{[i]}$ *and* $\mathcal{S}^*(D_-^3 \setminus K)^{[i]}$ *intersect transversally in* $T_s \mathcal{S}^*(S^2 \setminus K)^{[i]}$. *Also* $\mathcal{S}^*(S^3 \setminus K)^{[i]} = \{s\}$ *is an isolated point of the intersection between* $\mathcal{S}^*(D_+^3 \setminus K)^{[i]}$ *and* $\mathcal{S}^*(D_-^3 \setminus K)^{[i]}$ *in* $\mathcal{S}^*(S^2 \setminus K)^{[i]}$.

Proof: In general, one can take $a = e^{i\theta}$ with the normalized Alexander polynomial $\Delta_K(a^2) \neq 0$ and $0 \leq \theta \leq \pi$. For our case, $\theta = \pi/2$. Note that $\Delta_K(a^2) \neq 0$ implies $\det(Id - m_{B_n}(\beta)) \neq 0$, where $m_{B_n}(\beta)$ is the reduced Burau matrix of β with parameter $t = a^2$ ($= -1$ for our case $a = i$) obtained from Burau representation in §4.6 and (4.14)

$$\Delta_K(t) = (-\frac{1}{\sqrt{t}})^{e-n+1}(\frac{1-t}{1-t^n}) \cdot \det(Id - m_{B_n}(\beta)),$$

for the exponent sum e of β. For knots, $e - n + 1$ is always an even integer. Thus the intersection of $T_{a \times a}\Gamma_{id}$ and $T_{a \times a}\Gamma_\beta$ in $T_{a \times a}\mathcal{R}(S^2 \setminus K)^{[i]}$ is of the minimal dimension 3, where $a = (X_1, \cdots, X_n)$ and $X_i = \begin{pmatrix} e^{i\theta} & 0 \\ 0 & e^{-i\theta} \end{pmatrix}$ for $1 \leq i \leq n$. In a neighborhood of $a \times a$ in $\mathcal{R}(S^2 \setminus K)^{[i]}$, $\Gamma_{id} \cap \Gamma_\beta$ is a 3-dimensional manifold. On the other hand reducible representations of $\pi_1(S^3 \setminus K)$ can be identified with the diagonal of $Hom(F_n, SU(2))$ which is diffeomorphic to $SU(2)$, where F_n is a free group of n generators. Therefore we obtain a neighborhood of a in $Hom(\pi_1(S^3 \setminus K), SU(2))$ consisting of only reducible representations. $\qquad\square$

Lemma 5.5.4. *The intersection* $\mathcal{R}^*(D_+^3 \setminus K)^{[i]} \cap \mathcal{R}^*(D_-^3 \setminus K)^{[i]}$ *is a compact subset in* $\mathcal{R}^*(S^2 \setminus K)^{[i]}$.

Proof: The result follows from Lemma 5.5.2 and Lemma 5.5.3 since there is a neighborhood U_s of the reducible representation s such that $U_s \cap \text{Fix}\,(\phi_\beta)$ consists of only s. $\qquad\square$

5.6 The $SU(2)$ Casson-Lin invariant of knots

Let $\tilde{D}^3_\pm$ be the double branched covering of $D^3_\pm$ along $D^3_\pm \cap K$. There is an induced $\mathbb{Z}_2$-action on the representation space $\mathcal{R}(\tilde{D}^3_\pm)$. The submanifolds $\overline{\Gamma}_\beta = \mathcal{R}^*(D_+^3 \setminus K)^{[i]}$ and $\overline{\Gamma}_{id} = \mathcal{R}^*(D_-^3 \setminus K)^{[i]}$ can be identified as components of the fixed point set $\mathcal{R}(\tilde{D}^3_\pm)^{\mathbb{Z}_2}$ under the $\mathbb{Z}_2$-action. It follows that $\mathcal{R}^*(D_+^3 \setminus K)^{[i]}$ and $\mathcal{R}^*(D_-^3 \setminus K)^{[i]}$ have orientations inherited from the orientation on $\mathcal{R}(D_\pm^3)^{\mathbb{Z}_2}$. These oriented submanifolds $\mathcal{R}^*(D_+^3 \setminus K)^{[i]}, \mathcal{R}^*(D_-^3 \setminus K)^{[i]}$ intersect each other in a compact subspace of $\mathcal{R}^*(S^2 \setminus K)^{[i]}$ by Lemma 5.5.4. Hence we can perturb ϕ_β (via a Hamiltonian vector field if necessary). In other words, we can perturb $\mathcal{R}^*(D_+^3 \setminus K)^{[i]} = \overline{\Gamma}_\beta$ to $\hat{\mathcal{R}}^*(D_+^3 \setminus K)^{[i]} = \hat{\Gamma}_\beta$ by a compactly support isotopy so that $\hat{\mathcal{R}}^*(D_+^3 \setminus K)^{[i]}$ intersects $\mathcal{R}^*(D_-^3 \setminus K)^{[i]}$ transversally at a finite number of points. Denote the perturbed (symplectic) diffeomorphism by $\tilde{\phi}_\beta$. So the fixed points of $\tilde{\phi}_\beta$ are all nondegenerate.

Definition 5.6.1. *The Casson-Lin invariant of a knot* $K = \overline{\beta}$ *is the algebraic intersection number of* $\hat{\mathcal{R}}^*(D_+^3 \setminus K)^{[i]}$ *and* $\mathcal{R}^*(D_-^3 \setminus K)^{[i]}$, *or the algebraic number of* $\text{Fix}(\tilde{\phi}_\beta)$:

$$\begin{aligned}
\lambda_{CL}(K) = \lambda_{CL}(\beta) &= \#\text{Fix}(\tilde{\phi}_\beta) \\
&= \#(\hat{\mathcal{R}}^*(D_+^3 \setminus K)^{[i]} \cap \mathcal{R}^*(D_-^3 \setminus K)^{[i]}) \\
&= \#(\tilde{\Gamma}_\beta \cap \overline{\Gamma}_{id}).
\end{aligned}$$

Lin [1992, Theorem 1.8] showed that $\lambda_{CL}(K) = \lambda_{CL}(\beta)$ is independent of braid representatives of the knot K, i.e. $\lambda_{CL}(\beta)$ is invariant under the Markov moves of type I and type II on β. We give the proof here from arguments in [Li, 1997; Lin, 1992].

Theorem 5.6.2. *For $\beta_1 \in B_n$ and $\beta_2 \in B_m$ with $K = \overline{\beta}_1 = \overline{\beta}_2$ as knots, then*

$$\lambda_{CL}(\beta_1) = \lambda_{CL}(\beta_2).$$

Proof: It suffices to show that $\lambda_{CL}(\beta)$ is invariant under the Markov moves of type I and type II by Theorem 2.6.4.

Suppose that we have a Markov move of type I: change β to $\xi^{-1}\beta\xi$ for some $\xi \in B_n$ and $\beta \in B_n$. Recall that B_n is generated by $\sigma_1, \cdots, \sigma_{n-1}$. For any $\sigma_i^{\pm 1}$, the induced diffeomorphism

$$\sigma_i^{\pm 1} \times \sigma_i^{\pm 1} : Q_n \times Q_n \to Q_n \times Q_n$$

is an orientation preserving diffeomorphism. After compositions of $\sigma_i^{\pm 1} \times \sigma_i^{\pm 1}$, we obtain that $\xi : Q_n \to Q_n$ is also an orientation preserving diffeomorphism. So the diffeomorphism $\xi \times \xi : Q_n \times Q_n \to Q_n \times Q_n$ commutes with the $SU(2)$-action and satisfies:

$\xi \times \xi(\mathcal{R}^*(S^2 \setminus K)^{[i]}) = \mathcal{R}^*(S^2 \setminus K)^{[i]}$ (changing variables by $\xi \times \xi$),

$\xi \times \xi(\mathcal{R}^*(D_-^3 \setminus K)^{[i]}) = \mathcal{R}^*(D_-^3 \setminus K)^{[i]}$ (in new coordinates $\xi(X_1), \cdots, \xi(X_n)$),

$\xi \times \xi(\mathcal{R}^*(D_+^3 \setminus K)^{[i]}) = \mathcal{R}^*(D_+^3 \setminus K)^{[i]}$ (in new coordinates $\xi(X_1), \cdots, \xi(X_n)$),

as oriented manifolds. Let $f_\xi : \mathcal{R}^*(S^2 \setminus K)^{[i]} \to \mathcal{R}^*(S^2 \setminus K)^{[i]}$ be the (symplectic) diffeomorphism induced from $\xi \times \xi$. We have

$$\phi_\beta = f_\xi \circ \phi_{\xi^{-1}\beta\xi} \circ f_\xi^{-1}$$

by changing variables via f_ξ. Note that $\mathrm{Fix}(\phi_{\xi^{-1}\beta\xi})$ is identified with $\mathrm{Fix}(\phi_\beta)$ under f_ξ. Since the induced map f_ξ is a diffeomorphism, there exists a perturbation with a compact support such that $\hat{\mathcal{R}}(D_+^3 \setminus K)^{[i]}$ intersects with $\hat{\mathcal{R}}(D_-^3 \setminus K)^{[i]}$ transversally at a finite number of points and $f_\xi(\hat{\mathcal{R}}(D_+^3 \setminus K)^{[i]})$ intersects with $f_\xi(\hat{\mathcal{R}}(D_-^3 \setminus K)^{[i]})$ transversally at a finite number of points simultaneously. With the orientation preserving property of f_ξ, we have

$$\lambda_{CL}(\xi^{-1}\beta\xi) = \#\mathrm{Fix}\,(\tilde{\phi}_{\xi^{-1}\beta\xi}) = \#\mathrm{Fix}\,(\tilde{\phi}_\beta) = \lambda_{CL}(\beta).$$

It is clear that the argument goes through for the inverse operation of a Markov move of type I.

Suppose that we have a Markov move of type II: change β to $\sigma_n\beta \in B_{n+1}$. Recall that $\sigma_n(x_i) = x_i, 1 \le i \le n-1, \sigma_n(x_n) = x_n x_{n+1} x_n^{-1}$ and $\sigma_n(x_{n+1}) = x_n$. We must identify the intersection from the construction in $\hat{H}_n = \mathcal{R}^*(S^2 \setminus \overline{\beta})^{[i]}$ with the one from the construction in $\hat{H}_{n+1} = \mathcal{R}^*(S^2 \setminus \overline{\sigma_n\beta})^{[i]}$. There is an embedding $g : Q_n \times Q_n \to Q_{n+1} \times Q_{n+1}$ given by

$$g(X_1, \cdots, X_n, Y_1, \cdots, Y_n) = (X_1, \cdots, X_n, Y_n, Y_1, \cdots, Y_n, Y_n).$$

Such an embedding commutes with the $SU(2)$-action and $g(H_n) \subset H_{n+1}$, and induces an embedding

$$\hat{g} : \hat{H}_n(= \mathcal{R}^*(S^2 \setminus \overline{\beta})^{[i]}) \to \hat{H}_{n+1}(= \mathcal{R}^*(S^2 \setminus \overline{\sigma_n\beta})^{[i]}).$$

Note that $\hat{g}$ is an embedding such that the space $\hat{H}_n$ is naturally the sub-manifold $\hat{g}(\hat{H}_n)$ of $\hat{H}_{n+1}$. With this embedding, $\hat{g}(\phi_\beta) : \hat{H}_{n+1} \to \hat{H}_{n+1}$ is given by

$$(X_1, \cdots, X_n, X_1, \cdots, X_n) \mapsto (X_1, \cdots, X_n, \beta(X_n), \beta(X_1), \cdots, \beta(X_n), \beta(X_n)).$$
$$(5.11)$$

The image of $\hat{g}(\phi_\beta)$ is equivariant under the operation of σ_n:

$$\hat{g}(\phi_\beta) \circ \sigma_n(X_1, \cdots, X_n, X_1, \cdots, X_n)$$
$$= \hat{g}(\phi_\beta)(X_1, \cdots, X_{n-1}, X_n X_{n+1} X_n^{-1}, X_1, \cdots, X_{n-1}, X_n X_{n+1} X_n^{-1})$$
$$= \hat{g}(X_1, \cdots, X_{n-1}, X_n X_{n+1} X_n^{-1}, \beta(X_1), \cdots, \beta(X_{n-1}), \beta(X_n)\beta(X_{n+1})\beta(X_n)^{-1})$$
$$= (X_1, \cdots, X_{n-1}, X_n X_{n+1} X_n^{-1}, \beta(X_n)\beta(X_{n+1})\beta(X_n)^{-1},$$
$$\beta(X_1), \cdots, \beta(X_{n-1}), \beta(X_n)\beta(X_{n+1})\beta(X_n)^{-1}, \beta(X_n)\beta(X_{n+1})\beta(X_n)^{-1});$$

$$\sigma_n \circ \hat{g}(\phi_\beta)(X_1, \cdots, X_n, X_1, \cdots, X_n)$$
$$= \sigma_n(X_1, \cdots, X_n, \beta(X_n), \beta(X_1), \cdots, \beta(X_n), \beta(X_n))$$
$$= \sigma_n(X_1, \cdots, X_n, Y_n, Y_1, \cdots, Y_n, Y_n)$$
$$= (X_1, \cdots, X_{n-1}, X_n X_{n+1} X_n^{-1}, Y_n Y_{n+1} Y_n^{-1},$$
$$Y_1, \cdots, Y_{n-1}, Y_n Y_{n+1} Y_n^{-1}, Y_n Y_{n+1} Y_n^{-1}),$$

where $Y_i = \beta(X_i)$, and $\hat{g}(\phi_\beta) \circ \sigma_n = \sigma_n \circ \hat{g}(\phi_\beta) : H_n \to H_{n+1}$. The corresponding diffeomorphism $\phi_{\sigma_n\beta}$ of the braid $\sigma_n\beta$ is given by

$$\phi_{\sigma_n\beta}(X_1, \cdots, X_n, X_{n+1}, X_1, \cdots, X_n, X_{n+1})$$

$$= (X_1, \cdots, X_{n+1}, \beta(X_1), \cdots, \beta(X_{n-1}), \beta(X_n)X_{n+1}\beta(X_n)^{-1}, \beta(X_n)).$$
$$(5.12)$$

Thus we obtain

$$\hat{g}(\mathcal{R}^*(D_-^3 \setminus \overline{\beta})^{[i]}) \subset \mathcal{R}^*(D_-^3 \setminus \overline{\sigma_n\beta})^{[i]}, \quad \hat{g}(\mathcal{R}^*(D_+^3 \setminus \overline{\beta})^{[i]}) \subset \mathcal{R}^*(D_+^3 \setminus \overline{\sigma_n\beta})^{[i]}.$$

The fixed points of $\phi_{\sigma_n \beta}$ are the elements

$$\beta(X_i) = X_i, 1 \le i \le n - 1; \quad \beta(X_n)X_{n+1}\beta(X_n)^{-1} = X_n, \quad \beta(X_n) = X_{n+1}.$$
$$(5.13)$$

So, from the last equation of (5.13),

$$X_n = \beta(X_n)X_{n+1}\beta(X_n)^{-1}$$
$$= X_{n+1}X_{n+1}X_{n+1}^{-1}$$
$$= X_{n+1}$$

and $\beta(X_n) = X_n$ follows from the second last equation of (5.13). Therefore (5.13) is equivalent to $\beta(X_i) = X_i, 1 \le i \le n$. So

$$\text{Fix}(\phi_{\sigma_n \beta}) = \text{Fix}(\hat{g}(\phi_\beta)) = \text{Fix}(\phi_\beta).$$

In order to identify the intersection numbers or fixed points numbers, we have to consider the orientations of various manifolds involved, i.e., we need to identify the orientation of the normal bundle of the embedded submanifold.

Let $X = (X_1, \cdots, X_n) \in Q_n$ be a fixed point of $\phi_\beta|_{Q_n}$.

$$T_{g(X,X)}Hom(\pi_1(D_-^3 \setminus \overline{\sigma_n \beta}), SU(2))^{[i]}$$

$$= dg(T_{(X,X)}Hom(\pi_1(D_-^3 \setminus \overline{\beta}), SU(2))^{[i]}) \oplus U,$$

$$T_{g(X,X)}Hom(\pi_1(D_+^3 \setminus \overline{\sigma_n \beta}), SU(2))^{[i]}$$

$$= dg(T_{(X,X)}Hom(\pi_1(D_+^3 \setminus \overline{\beta}), SU(2))^{[i]}) \oplus V,$$

as oriented vector spaces, where U and V are oriented vector spaces of the following forms:

$$U \cong \{(u, u, 0) : u \in T_{X_n}S^2\}, \quad V \cong \{(v, 0, v) : v \in T_{X_n}S^2\}.$$

We also have

$$T_{g(X,X)}(Q_{n+1} \times Q_{n+1}) = dg(T_{(X,X)}(Q_n \times Q_n)) \oplus W,$$

as oriented vector spaces and

$$W \cong \{(u + v, u, v) : u, v \in T_{X_n}S^2\} \cong U \oplus V,$$

as oriented vector spaces. The decompositions pass down to the tangent spaces of $\hat{g}(\hat{H}_n) \subset \hat{H}_{n+1}$, $\hat{g}(\mathcal{R}^*(D_-^3 \setminus \overline{\beta})^{[i]}) \subset \mathcal{R}^*(D_-^3 \setminus \overline{\sigma_n \beta})^{[i]}$, and $\hat{g}(\mathcal{R}^*(D_+^3 \setminus \overline{\beta})^{[i]}) \subset \mathcal{R}^*(D_+^3 \setminus \overline{\sigma_n \beta})^{[i]}$ at the corresponding points. Thus, with the transversal intersection point of $\mathcal{R}^*(D_-^3 \setminus \overline{\sigma_n\beta})^{[i]}$ and $\mathcal{R}^*(D_+^3 \setminus \overline{\sigma_n\beta})^{[i]}$ at $\hat{g}(X, X)$,

$$
\begin{aligned}
& T_{\hat{g}(X,X)} \mathcal{R}^*(S^2 \setminus \overline{\sigma_n\beta})^{[i]} \\
&= T_{\hat{g}(X,X)} \mathcal{R}^*(D_-^3 \setminus \overline{\sigma_n\beta})^{[i]} \oplus T_{\hat{g}(X,X)} \mathcal{R}^*(D_+^3 \setminus \overline{\sigma_n\beta})^{[i]} \\
&= d\hat{g}(T_{(X,X)} \mathcal{R}^*(D_-^3 \setminus \overline{\beta})^{[i]}) \oplus U \oplus d\hat{g}(T_{(X,X)} \mathcal{R}^*(D_+^3 \setminus \overline{\beta})^{[i]}) \oplus V \\
&= d\hat{g}(T_{(X,X)} \mathcal{R}^*(D_-^3 \setminus \overline{\beta})^{[i]} \oplus T_{(X,X)} \mathcal{R}^*(D_+^3 \setminus \overline{\beta})^{[i]}) \oplus (U \oplus V) \\
&= d\hat{g} T_{(X,X)} \mathcal{R}^*(S^2 \setminus \overline{\sigma_n\beta})^{[i]} \oplus W,
\end{aligned}
$$

as oriented vector spaces. This implies that the intersection number at (X, X) has the same sign in both $\mathcal{R}^*(S^2 \setminus \overline{\sigma_n\beta})^{[i]}$ and $\mathcal{R}^*(S^2 \setminus \overline{\beta})^{[i]}$ through the embedding $\hat{g}$. The above identifications of oriented vector spaces, especially the normal bundle $W \cong U \oplus V$, are independent of the intersection points , as well as independent of the deformations $\tilde{\mathcal{R}}^*(D_+^3 \setminus \overline{\beta})^{[i]}$. Thus

$$
\lambda_{CL}(\sigma_n\beta) = \#\hat{\mathcal{R}}^*(D_+^3 \setminus \overline{\beta})^{[i]} \cap \mathcal{R}^*(D_-^3 \setminus \overline{\beta})^{[i]}
$$

$$
= \#\hat{\mathcal{R}}^*(D_+^3 \setminus \overline{\beta})^{[i]} \cap \mathcal{R}^*(D_-^3 \setminus \overline{\beta})^{[i]} = \lambda_{CL}(\beta).
$$

It is clear that the argument goes through for the inverse operation of a Markov move of type II. Hence we have shown that $\lambda_{CL}(\beta)$ is invariant under Markov moves and defines an invariant of the knot $K = \overline{\beta}$. $\qquad\square$

5.7 Surgery relation for the $SU(2)$ Casson-Lin invariants

If K_+ is the closure of $\beta_+ = \beta_1 \sigma_k \beta_2$ and K_- is the closure of $\beta_- = \beta_1 \sigma_k^{-1} \beta_2$, then K_0 is the closure of $\beta_0 = \beta_1 \beta_2$.

$$
\begin{aligned}
\lambda_{CL}(K_+) - \lambda_{CL}(K_-) &= \lambda_{CL}(\beta_1 \sigma_k \beta_2) - \lambda_{CL}(\beta_1 \sigma_k^{-1} \beta_2) \\
&= \#\overline{\Gamma}_{id} \cap \overline{\Gamma}_{\beta_1 \sigma_k \beta_2} - \#\overline{\Gamma}_{id} \cap \overline{\Gamma}_{\beta_1 \sigma_k^{-1} \beta_2} \\
&= \#\overline{\Gamma}_{\beta_1^{-1}} \cap \overline{\Gamma}_{\sigma_k \beta_2} - \#\overline{\Gamma}_{\beta_1^{-1}} \cap \overline{\Gamma}_{\sigma_k^{-1} \beta_2} \\
&= \#\overline{\Gamma}_{\sigma_k^{-1} \beta_1^{-1}} \cap \overline{\Gamma}_{\beta_2} - \#\overline{\Gamma}_{\sigma_k \beta_1^{-1}} \cap \overline{\Gamma}_{\beta_2} \\
&= \#(\overline{\Gamma}_{\sigma_k^{-1} \beta_1^{-1}} - \overline{\Gamma}_{\sigma_k \beta_1^{-1}}) \cap \overline{\Gamma}_{\beta_2} \\
&= \#(\overline{\Gamma}_{\sigma_k^{-2} \beta_1^{-1}} - \overline{\Gamma}_{\beta_1^{-1}}) \cap \overline{\Gamma}_{\beta_2}.
\end{aligned}
$$

Without loss of generality, let $\beta_+ = \sigma_1^2 \beta$, $\beta_- = \beta$. It suffices to understand the difference cycle $\overline{\Gamma}_{\sigma_1^{-2}} - \overline{\Gamma}_{id}$. We are going to follow the discussion in [Lin, 1992] to obtain the surgery relation.

Note that the difference cycle $\overline{\Gamma}_{\sigma_1^{-2}} - \overline{\Gamma}_{id}$ is supported in the space $\overline{V}_n$, where $\overline{V}_n = (V_n \setminus V_n \cap S_n)/SU(2)$ is the space of equivalence classes of irreducible representations in

$$V_n = \{(X_1, \cdots, X_n, Y_1, \cdots, Y_n) \in H_n | X_i = Y_i, 3 \leq i \leq n\}.$$

Let $\overline{H}_2$ be the space of equivalence classes of irreducible representations in

$$H_2 = \{(X_1, X_2, Y_1, Y_2) \in Q_2 \times Q_2 | X_1 X_2 = Y_1 Y_2\}.$$

The space $\overline{V}_n$ is a submanifold of $\overline{H}_n = \mathcal{R}^*(S^2 \setminus K)^{[i]}$ whose dimension is $2(n+2) - 6 = 2n - 2$. The submanifold $\overline{V}_n$ in $\overline{H}_n$ has codimension $4n - 6 - (2n - 2) = 2n - 4$. With perturbed submanifold $\overline{\Gamma}_\beta$ of dimension $2n - 3$ through a compact support perturbation so that $\overline{V}_n$ and $\hat{\Gamma}_\beta = \hat{\mathcal{R}}^*(D_+^3 \setminus K)^{[i]}$ intersect transversally, and the intersection $\overline{V}_n \cap \hat{\Gamma}_\beta$ is a one-dimensional submanifold in $\mathcal{R}^*(S^2 \setminus K)^{[i]} = \overline{H}_n$. One can orient the one-dimensional manifold $\overline{V}_n \cap \hat{\Gamma}_\beta$ in the following way. Let $(X, Y) \in \overline{V}_n \cap \hat{\Gamma}_\beta$. Then there is an oriented subspace $P \subset T_{(X,Y)}\hat{\Gamma}_\beta$ of codimension one (i.e., $\dim P = \dim \overline{\Gamma}_\beta - 1 = 2n - 4$) such that

$$T_{(X,Y)}\overline{H}_n = T_{(X,Y)}\overline{V}_n \oplus P$$

as oriented vector spaces. Then

$$T_{(X,Y)}\hat{\Gamma}_\beta = T_{(X,Y)}(\overline{V}_n \cap \hat{\Gamma}_\beta) \oplus P$$

gives an orientation of the submanifold $\overline{V}_n \cap \hat{\Gamma}_\beta$.

Lemma 5.7.1. *Let* $\langle \overline{\Gamma}_{\sigma_1^{-2}} - \overline{\Gamma}_{id}, \hat{\Gamma}_\beta \rangle = \#(\overline{\Gamma}_{\sigma_1^{-2}} - \overline{\Gamma}_{id}) \cap \hat{\Gamma}_\beta$ *be the intersection number in* $\mathcal{R}^*(S^2 \setminus K)^{[i]}$. *The following identity holds*

$$\langle \overline{\Gamma}_{\sigma_1^{-2}} - \overline{\Gamma}_{id}, \hat{\Gamma}_\beta \rangle = \langle \overline{\Gamma}_{\sigma_1^{-2}} - \overline{\Gamma}_{id}, \overline{V}_n \cap \hat{\Gamma}_\beta \rangle_{\overline{V}_n},$$

where $\langle \cdot \rangle_{\overline{V}_n}$ *is the intersection number in* $\overline{V}_n$.

Proof: Assume that $(X, X) \in \overline{\Gamma}_{id} \cap \hat{\Gamma}_\beta$ is the positive 1 intersection point in $\overline{H}_n = \mathcal{R}^*(S^2 \setminus K)^{[i]}$. Thus we have

$$
\begin{aligned}
T_{(X,X)}\overline{H}_n &= T_{(X,X)}\overline{\Gamma}_{id} \oplus T_{(X,X)}\hat{\Gamma}_\beta \\
&= T_{(X,X)}\overline{\Gamma}_{id} \oplus T_{(X,X)}(\overline{V}_n \cap \hat{\Gamma}_\beta) \oplus P \\
&= T_{(X,X)}(\overline{\Gamma}_{id} \cap \overline{V}_n \cap \hat{\Gamma}_\beta) \oplus P,
\end{aligned}
$$

since $\dim P = 2n - 4$ is even. Therefore

$$T_{(X,X)}\overline{\Gamma}_{id} \oplus T_{(X,X)}(\overline{V}_n \cap \hat{\Gamma}_\beta) = T_{(X,X)}\overline{V}_n,$$

the intersection number of $\overline{\Gamma}_{id} \cap \overline{V}_n \cap \hat{\Gamma}_\beta$ in $\overline{V}_n$ has the sign which is determined uniformly by the normal vector subspace P. Thus $\langle \overline{\Gamma}_{id}, \overline{V}_n \cap \hat{\Gamma}_\beta \rangle_{\overline{V}_n} = \langle \overline{\Gamma}_{id}, \hat{\Gamma}_\beta \rangle$ and $\langle \overline{\Gamma}_{\sigma_1^{-2}}, \overline{V}_n \cap \hat{\Gamma}_\beta \rangle_{\overline{V}_n} = \langle \overline{\Gamma}_{\sigma_1^{-2}}, \hat{\Gamma}_\beta \rangle$. The result follows.
$\square$

Lemma 5.7.1 reduces the skein relation into the counting intersection numbers in a smaller space $\overline{V}_n$, where $\dim \overline{V}_n = \frac{1}{2} \dim \overline{H}_n - 1$. The space $\overline{V}_n$ has a natural projection to $\overline{H}_2$ by $p(X_1, \cdots, X_n, Y_1, \cdots, Y_n) = (X_1, X_2, Y_1, Y_2)$. Now the intersection number in $\overline{V}_n$ can be reduced further into a space $\overline{H}_2$ which can be computed explicitly. Note that $\dim \overline{H}_2 + \dim \overline{V}_n = 2n$ and $\dim \overline{H}_2 = 2$, $\overline{H}_2$ is a much smaller space than $\overline{V}_n$.

Lemma 5.7.2.

$$\langle \overline{\Gamma}_{\sigma_1^{-2}} - \overline{\Gamma}_{id}, \overline{V}_n \cap \hat{\Gamma}_\beta \rangle_{\overline{V}_n} = \langle \overline{\Gamma}_{\sigma_1^{-2}} - \overline{\Gamma}_2, p(\overline{V}_n \cap \hat{\Gamma}_\beta) \rangle_{\overline{H}_2},$$

where $\overline{\Gamma}_2$ *is the space* $\overline{\Gamma}_{id}$ *with* $n = 2$ *case.*

Proof: Again this is basically matching the orientations through the identification. Assume that $(X, X) \in \overline{\Gamma}_{id} \cap (\overline{V}_n \cap \hat{\Gamma}_\beta)$ has a positive one intersection number in $\overline{V}_n$. As oriented vector spaces,

$$T_{(X,X)}\overline{V}_n = p^*(T_{p(X,X)}\overline{H}_2) \oplus Q, \quad T_{(X,X)}\overline{\Gamma}_{id} = p^*(T_{p(X,X)}\overline{\Gamma}_2) \oplus Q,$$

where Q has dimension $2n - 4$. Now we have

$$\begin{aligned}
T_{(X,X)}\overline{V}_n &= T_{(X,X)}\overline{\Gamma}_{id} \oplus T_{(X,X)}(\overline{V}_n \cap \hat{\Gamma}_\beta) \\
&= \{p^*(T_{p(X,X)}\overline{\Gamma}_2) \oplus Q\} \oplus p^*(T_{p(X,X)}p(\overline{V}_n \cap \hat{\Gamma}_\beta)) \\
&= p^*(T_{p(X,X)}\overline{\Gamma}_2 \oplus T_{p(X,X)}p(\overline{V}_n \cap \hat{\Gamma}_\beta)) \oplus Q \\
&= p^*(T_{p(X,X)}\overline{H}_2) \oplus Q,
\end{aligned}$$

where the first equality follows from the assumption of the transversal intersection, the second from the above identification as oriented vector space, the third from the even dimensions of the vector spaces involved changing order, and the last equality implies that

$$T_{p(X,X)}\overline{H}_2 = T_{p(X,X)}\overline{\Gamma}_2 \oplus T_{p(X,X)}p(\overline{V}_n \cap \hat{\Gamma}_\beta). \tag{5.14}$$

Hence the intersection number of $\overline{\Gamma}_2$ with $p(\overline{V}_n \cap \hat{\Gamma}_\beta)$ at $p(X, X)$ in $\overline{H}_2$ is also a positive one. Similarly, one can show that

$$T_{p(X,X)}\overline{H}_2 = T_{p(X,X)}\overline{\Gamma}_{\sigma_1^{-2}} \oplus T_{p(X,X)}p(\overline{V}_n \cap \hat{\Gamma}_\beta). \tag{5.15}$$

The normal bundle of the intersection point used in this identification of orientations is independent of the intersection points. Hence the result follows from (5.14) and (5.15).
$\square$

Let us analyze the space H_2 in more detail. By definition, $H_2 = \{(X_1, X_2, Y_1, Y_2) \in Q_2 \times Q_2 : X_1 X_2 = Y_1 Y_2\}$. From linear algebra, we can find a conjugate matrix A such that $(AX_1 A^{-1}, AX_2 A^{-1}, AY_1 A^{-1}, AY_2 A^{-1})$ the equivalent element has the following:

$$AX_1 A^{-1} = x_1 = \begin{pmatrix} i \cos \theta_1 & \sin \theta_1 \\ -\sin \theta_1 & -i \cos \theta_1 \end{pmatrix}, \quad AX_2 A^{-1} = x_2 = \begin{pmatrix} i & 0 \\ 0 & -i \end{pmatrix},$$

for $0 \le \theta_1 \le \pi$. Let $y_1 = AY_1 A^{-1}$ and $y_2 = AY_2 A^{-1}$. Thus $x_1 x_2 = y_1 y_2$ gives $y_2 = y_1^{-1} x_1 x_2$. Hence y_2 is uniquely determined by x_1, x_2 and y_1. Since $y_1 = \begin{pmatrix} ir & b \\ -\bar{b} & -ir \end{pmatrix}$ and y_2 are of trace zero, one has

$$\begin{aligned}
y_2 &= y_1^{-1} x_1 x_2 \\
&= \begin{pmatrix} -ir & -b \\ \bar{b} & ir \end{pmatrix} \cdot \begin{pmatrix} i \cos \theta_1 & \sin \theta_1 \\ -\sin \theta_1 & -i \cos \theta_1 \end{pmatrix} \cdot \begin{pmatrix} i & 0 \\ 0 & -i \end{pmatrix} \\
&= \begin{pmatrix} ir \cos \theta_1 + b \sin \theta_1 i & -r \sin \theta_1 + b \cos \theta_1 \\ -\bar{b} \cos \theta_1 + r \sin \theta_1 & -i\bar{b} \sin \theta_1 - ir \cos \theta_1 \end{pmatrix}.
\end{aligned}$$

Hence we have $tr(y_2) = (b - \bar{b}) \sin \theta_1 i = 0$, and b is a real number with $r^2 + b^2 = 1$. So

$$y_1 = \begin{pmatrix} i \cos \theta_2 & \sin \theta_2 \\ -\sin \theta_2 & -i \cos \theta_2 \end{pmatrix}, \quad -\pi \le \theta_2 \le \pi.$$

From the parametrization, we have H_2 characterized by $(\theta_1, \theta_2) \in [0, \pi] \times [-\pi, \pi]$. The reducible representations in H_2 correspond to the diagonalized matrices for all x_1, x_2, y_1, y_2. Hence $(\theta_1, \theta_2) = (0, 0), (0, \pm\pi), (\pi, \pm\pi)$ and $(\pi, 0)$ are four different reducible representations. Note that $(0, \pi)$ and $(0, -\pi)$ correspond to two reducible representations which are equivalent by the element $A = \begin{pmatrix} 0 & 1 \\ -1 & 0 \end{pmatrix}$. See Figure 5.5 for the space $\overline{H}_2$.

Identifying the corresponding sides AC, BD and CD, we fold the rectangle into a pillow case. Note that $\overline{H}_2$ is homeomorphic to $S^2 \{A, B, C, D\}$ the 2-sphere with four punctured points.

Exercise 5.7.3. Identify the subspaces $\overline{\Gamma}_2$ with $\{(\theta_1, \theta_2) \in H_2 : \theta_1 = \theta_2, 0 \le \theta_1 \le \pi\}$ and $\overline{\Gamma}_{\sigma_1^{-2}}$ with $\{(\theta_1, \theta_2) \in H_2 : \theta_1 = -\theta_2, 0 \le \theta_1 \le \pi\}$.

Now the difference cycle $\overline{\Gamma}_{\sigma_1^{-2}} - \overline{\Gamma}_2$ can be viewed in Figure 5.6.

The next task is to determine the intersection number of $p(\overline{V}_n \cap \hat{\Gamma}_\beta)$ with the difference cycle $\overline{\Gamma}_{\sigma_1^{-2}} - \overline{\Gamma}_2$. This amounts to seeing how a smooth curve

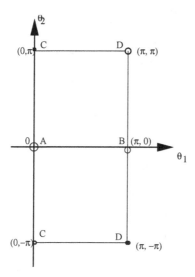

Fig. 5.5   Parametrization of the space $\overline{H}_2$

in $\overline{H}_2$ intersects with $\overline{\Gamma}_{\sigma_1^{-2}} - \overline{\Gamma}_2$. Let $c(t)$ be a smooth curve in $\overline{H}_2$ with parametrization $(\theta_1(t), \theta_2(t))$ for $-\infty < t < \infty$. Here $A = \lim_{t \to -\infty} c(t)$ and $c'(-\infty) = \lim_{t \to -\infty} c'(t) = \lim_{t \to -\infty}(\theta_1'(t), \theta_2'(t))$. Assume that $c'(-\infty) = (\theta_1^0, \theta_2^0)$ has a well-defined derivative. Hence the slope of the curve $c(t)$ at A can be defined by

$$ s = \frac{\theta_2^0}{\theta_1^0}, $$

and $s = \infty$ is understood if $\theta_1^0 = 0$.

Let $U_n = \{(X_1, X_2, X_3, \cdots, X_n, Y_1, Y_2, X_3, \cdots, X_n) \in Q_n \times Q_n\}$ be a submanifold of $Q_n \times Q_n$ of dimension $2n + 4$. Recall $\Gamma_\beta = \{(X_1, \cdots, X_n, \beta(X_1), \cdots, \beta(X_n) \in Q_n \times Q_n\}$ is a submanifold of $Q_n \times Q_n$ of dimension $2n$. The minimal dimension of $T_s U_n \cap T_s \Gamma_\beta$ is 4, where s stands for the unique reducible representation of the knot group by Lemma 5.5.3, i.e., $X_1 = \cdots = X_n = Y_1 = \cdots = Y_n = \begin{pmatrix} i & 0 \\ 0 & -i \end{pmatrix}$. Its image under the projection p gives $A = p(s)$.

Proposition 5.7.4. *Let n be an odd number ≥ 3. The curve $p(\overline{V}_n \cap \hat{\Gamma}_\beta)$ approaches to A in a neighborhood of A.*

Proof: We are going to show that the intersection $U_n \cap \Gamma_\beta$ is a submanifold of dimension 4 in a neighborhood of s. Then under the conjugacy equivalent

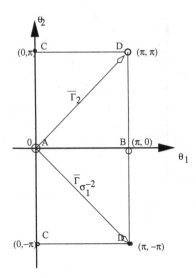

Fig. 5.6   The difference cycle in the space $\overline{H}_2$

relation the space $p(\overline{V}_n \cap \hat{\Gamma}_\beta)$ represents a 1-dimensional manifold in a neighborhood of A.

Note that the tangent map $T\beta|_{Q_n}$ under the standard basis is given by the Burau matrix of β with parameter equal to -1. Denote it by the matrix $\Psi_\beta = \begin{pmatrix} a & b \\ c & d \end{pmatrix}$, where a is a 2×2 matrix and d is a $(n-2) \times (n-2)$ matrix, the sizes of matrices b and c are clear from the ones of a and d. It suffices to show that the real solution space of $\Psi v^T = w^T$ has dimension 2 for $v = (v_1, v_2, \cdots, v_n)$ and $w = (w_1, w_2, v_3, \cdots, v_n)$. Then $\dim T_s U_n \cap T_s \Gamma_\beta = 2 + 2 = 4$ as desired.

(i) For any knot K, we have $\Delta_K(-1) \neq 0$ and $\beta|_{Q_n}$ leaves the whole diagonal of Q_n fixed. The fixed vector of Ψ_β is a scalar multiplication of the vector $(1, 1, \cdots, 1)^T$;

(ii) let $\{v\} = \sum_{j=1}^n (-1)^j v_j$. Then $\{\Psi_\beta \cdot v^T\} = \{v\}$ for any vector v since β preserves the product $X_1 \cdots X_n$ for any $(X_1, \cdots, X_n) \in Q_n$.

The n is assumed to be odd ≥ 3 so that $\Delta_K(-1) \neq 0$ for $K = \overline{\beta}$ implies that $\det(Id_{n \times n} - m_{B_n}(\beta)) \neq 0$. We split this into cases of $\det(Id_{(n-2) \times (n-2)} - d) \neq 0, = 0$.

(1) If $\det(Id_{(n-2) \times (n-2)} - d) \neq 0$, then the solutions of $\Psi_\beta v^T = w^T$ are

given by

$$\begin{bmatrix} v_3 \\ \vdots \\ v_n \end{bmatrix} = (Id_{(n-2)\times(n-2)} - d)^{-1}c \begin{bmatrix} v_1 \\ v_2 \end{bmatrix}.$$

Taking $\begin{bmatrix} v_1 \\ v_2 \end{bmatrix} = \begin{bmatrix} 1 \\ 0 \end{bmatrix}$, we obtain

$$\Psi_\beta \begin{bmatrix} 1 \\ 0 \\ v_3 \\ \vdots \\ v_n \end{bmatrix} = \begin{bmatrix} s \\ s-1 \\ v_3 \\ \vdots \\ v_n \end{bmatrix}, \tag{5.16}$$

where $s \neq 0$ is the slope of $p(\overline{V}_n \cap \hat{\Gamma}_\beta)$ at A, and the solution space of $\Psi_\beta v^T = w^T$ has dimension 2.

(2) If $\det(Id_{(n-2)\times(n-2)} - d) = 0$, then there is a non-zero vector $v = (v_3, \cdots, v_n)$ such that

$$\Psi_\beta \begin{bmatrix} 0 \\ 0 \\ v_3 \\ \vdots \\ v_n \end{bmatrix} = \begin{bmatrix} k \\ k \\ v_3 \\ \vdots \\ v_n \end{bmatrix},$$

for some $k \neq 0$. The existence of the solution v is, in fact, unique. If there is another non-zero vector v' with k' which satisfies the equation, then we obtain

$$\Psi_\beta \begin{bmatrix} 0 \\ 0 \\ v^T - \frac{k}{k'}(v')^T \end{bmatrix} = \begin{bmatrix} k \\ k \\ v_3 \\ \vdots \\ v_n \end{bmatrix} - \frac{k}{k'} \begin{bmatrix} k' \\ k' \\ v_3 \\ \vdots \\ v_n \end{bmatrix} = \begin{bmatrix} 0 \\ 0 \\ v^T - \frac{k}{k'}(v')^T \end{bmatrix}.$$

Hence $(0, 0, v - \frac{k}{k'}(v'))^T$ is also fixed by Ψ_β which is not a scalar of $(1, 1, \cdots, 1)$. This leads to a contradiction. Suppose there is a vector $w = (w_3, \cdots, w_n)$ such that

$$\Psi_\beta \begin{bmatrix} 1 \\ 0 \\ w^T \end{bmatrix} = \begin{bmatrix} s \\ s-1 \\ w^T \end{bmatrix}.$$

Since the vector $(s, s-1)$ and (k, k) are linearly independent, so there are real numbers r_1 and r_2 such that

$$r_1 \begin{pmatrix} s \\ s-1 \end{pmatrix} + r_2 \begin{pmatrix} k \\ k \end{pmatrix} = \begin{pmatrix} 1 \\ 0 \end{pmatrix}.$$

We have that

$$\Psi_\beta(r_1 \begin{bmatrix} 1 \\ 0 \\ w^T \end{bmatrix} + r_2 \begin{bmatrix} 0 \\ 0 \\ v^T \end{bmatrix}) = r_1 \Psi_\beta \begin{bmatrix} 1 \\ 0 \\ w^T \end{bmatrix} + r_2 \Psi_\beta \begin{bmatrix} 0 \\ 0 \\ v^T \end{bmatrix}$$

$$= r_1 \cdot \begin{bmatrix} s \\ s-1 \\ w^T \end{bmatrix} + r_2 \cdot \begin{bmatrix} k \\ k \\ v^T \end{bmatrix}$$

$$= \begin{bmatrix} 1 \\ 0 \\ r_1 w^T + r_2 v^T \end{bmatrix}$$

By the property (ii), we must have $a = 1$ which contradicts with (i). So the only non-zero solution in this case is $(0, 0, v_3, \cdots, v_n)$. Thus the parameters v_1 and v_2 forms a 2-dimensional solution space of the equation $\Psi_\beta v^T = w^T$. In this case, the slope of $p(\overline{V}_n \cap \hat{\Gamma}_\beta)$ at A is ∞. Thus our result follows from the above discussions. $\square$

Proposition 5.7.5. *Let s_β be the slope of $p(\overline{V}_n \cap \hat{\Gamma}_\beta)$ at A. Then $s_\beta \neq \pm 1$ for $K = \overline{\beta}$.*

Proof: By (1) in the proof of Proposition 5.7.4, if $s_\beta = 1$, $v^T = (1, 0, v_3, \cdots, v_n)^T$ will be fixed by Ψ_β in (5.16), then the property (i) in the proof of Proposition 5.7.4 will be contradicted.

Since the Burau matrix of σ_1 with parameter -1 is given by $\begin{pmatrix} 2 & -1 \\ 1 & 0 \end{pmatrix}$ and

$$\begin{pmatrix} 2 & -1 \\ 1 & 0 \end{pmatrix}^2 \cdot \begin{bmatrix} s \\ s-1 \end{bmatrix} = \begin{pmatrix} 2 & -1 \\ 1 & 0 \end{pmatrix}^2 \cdot \begin{bmatrix} -1 \\ (-1)-1 \end{bmatrix} = \begin{bmatrix} 1 \\ 0 \end{bmatrix}.$$

If $s_\beta = -1$, then the slope of $p(\overline{V}_n \cap \hat{\Gamma}_{\sigma_1^2 \beta})$ at A is 1 which contradicts with the above since the closure of $\sigma_1^2 \beta$ is also a knot. Therefore the slope s_β cannot be ± 1 for the knot $K = \overline{\beta}$. $\square$

Proposition 5.7.4 and Proposition 5.7.5 show that the intersection number of the skein crossing is determined by the absolute value of s_β.

Lemma 5.7.6. *We have the following identity*

$$\lambda_{CL}(\sigma_1^2 \beta) - \lambda_{CL}(\beta) = \begin{cases} 1 & \text{if } |s_\beta| > 1 \\ 0 & \text{if } |s_\beta| < 1. \end{cases}$$

Denote the left side by $\lambda_{CL}(K_0)$.

Proof: We show that there exists a neighborhood U_D of D in $\overline{H}_2$ such that $U_D \cap p(\overline{V}_n \cap \hat{\Gamma}_\beta) = \emptyset$. Suppose the contrary. Then the limit of $p(\overline{V}_n \cap \hat{\Gamma}_\beta)$ is another reducible fixed point $(X_1, \cdots, X_n) \in Q_n$ of $\phi_\beta|_{Q_n}$ such that $(X_1, X_2) = (\pi, \theta_2)$ for $\theta_2 = 0, \pm\pi$. This is impossible since the permutation induced by β among X_i's has no nontrivial subcycle. In fact there always exists a neighborhood U_C of C in $\overline{H}_2$ such that $U_C \cap p(\overline{V}_n \cap \hat{\Gamma}_\beta) = \emptyset$.

By a small perturbation relative to a neighborhood of A in $\overline{H}_2$ changing $\hat{\Gamma}_\beta$ to $\tilde{\Gamma}_\beta$ such that $p(\overline{V}_n \cap \tilde{\Gamma}_\beta)$ is a 1-dimensional submanifold of $\overline{H}_2$. Without more notation, we use s_β again to denote the slope of the curve $p(\overline{V}_n \cap \tilde{\Gamma}_\beta)$ approaching A. By Lemma 5.7.1 and Lemma 5.7.2, we have

$$\lambda_{CL}(\sigma_1^2\beta) - \lambda_{CL}(\beta) = \langle \hat{\Gamma}_{\sigma_1^{-2}} - \overline{\Gamma}_2, p(\overline{V}_n \cap \tilde{\Gamma}_\beta) \rangle_{\overline{H}_2}.$$

The difference cycle separates the space $\overline{H}_2$ into two disjoint halves. With the above discussion on the 1-dimensional submanifold $p(\overline{V}_n \cap \tilde{\Gamma}_\beta)$, the result follows. $\qquad\square$

Lemma 5.7.7.

$$\frac{s_\beta - 1}{s_\beta + 1} = \frac{\det(Id - m_{B_n}(\beta))}{\det(Id - m_{B_n}(\sigma_1^2\beta))},$$

where $s_\beta = \infty$ *and the left hand side is defined to be 1.*

Proof: The Burau matrices of β and $\sigma_1^2\beta$ with parameter -1 are given by

$$\Psi_\beta = \begin{pmatrix} a & b \\ c & d \end{pmatrix}, \quad \Psi_{\sigma_1^2\beta} = \begin{pmatrix} G^2 & 0 \\ 0 & Id_{(n-2)\times(n-2)} \end{pmatrix} \Psi_\beta,$$

where $G = \Psi_{\sigma_1} = \begin{pmatrix} 2 & -1 \\ 1 & 0 \end{pmatrix}$. We follow the proof of Proposition 5.7.4 to set the case $\det(Id_{(n-2)\times(n-2)} - d) \neq 0$ and the case $\det(Id_{(n-2)\times(n-2)} - d) = 0$.

If $\det(Id_{(n-2)\times(n-2)} - d) \neq 0$, then we set $v = (Id_{(n-2)\times(n-2)} - d)^{-1}c\begin{bmatrix} 1 \\ 0 \end{bmatrix}$. Thus

$$\begin{pmatrix} s_\beta \\ s_\beta - 1 \end{pmatrix} = \{a + b(Id_{(n-2)\times(n-2)} - d)^{-1}c\}\begin{pmatrix} 1 \\ 0 \end{pmatrix},$$

and $\{a + b(Id_{(n-2)\times(n-2)} - d)^{-1}c\}(0,1)^T = (r_1, r_2)^T$ for some real numbers

r_1 and r_2. Then

$$(t-1)\det(Id - m_{B_n}(\beta))$$

$$= \det\left(\begin{pmatrix} T & 0 \\ 0 & Id_{(n-2)\times(n-2)} \end{pmatrix} - \Psi_\beta\right)$$

$$= \det(T - \{a + b(Id_{(n-2)\times(n-2)} - d)^{-1}c\}) \cdot \det(Id_{(n-2)\times(n-2)} - d)$$

$$= \det\begin{pmatrix} 1 - s_\beta & -r_1 \\ 1 - s_\beta & t - r_2 \end{pmatrix} \cdot \det(Id_{(n-2)\times(n-2)} - d),$$

where $T = \begin{pmatrix} 1 & 0 \\ 0 & t \end{pmatrix}$. From the assumption and for $t = 1$, one gets

$$\det\begin{pmatrix} 1 - s_\beta & -r_1 \\ 1 - s_\beta & 1 - r_2 \end{pmatrix} = 0.$$

Since $s_\beta \neq 1$, then $1 - r_2 + r_1 = 0$. Substituting $r_1 = r_2 - 1$, we have

$$\det\begin{pmatrix} 1 - s_\beta & -r_1 \\ 1 - s_\beta & t - r_2 \end{pmatrix} = (t - r_2 + r_1)(1 - s_\beta) = (t-1)(1 - s_\beta).$$

Thus this gives

$$\det(Id - m_{B_n}(\beta)) = (1 - s_\beta)\det(Id_{(n-2)\times(n-2)} - d). \qquad (5.17)$$

Similarly, using $\Psi_{\sigma_1^2 \beta}$,

$$\det(Id - m_{B_n}(\sigma_1^2 \beta)) = (1 - s_{\sigma_1^2 \beta})\det(Id_{(n-2)\times(n-2)} - d).$$

The calculation of $s_{\sigma_1^2 \beta}$ is given by

$$\begin{pmatrix} s_{\sigma_1^2 \beta} \\ s_{\sigma_1^2 \beta} - 1 \end{pmatrix} = G^2 \cdot \begin{pmatrix} s_\beta \\ s_\beta - 1 \end{pmatrix}$$

$$= \begin{pmatrix} 2 & -1 \\ 1 & 0 \end{pmatrix}^2 \cdot \begin{pmatrix} s_\beta \\ s_\beta - 1 \end{pmatrix}$$

$$= \begin{pmatrix} s_\beta + 2 \\ s_\beta + 1 \end{pmatrix}.$$

Therefore

$$\det(Id - m_{B_n}(\sigma_1^2 \beta)) = -(1 + s_\beta)\det(Id_{(n-2)\times(n-2)} - d), \qquad (5.18)$$

follows from $s_{\sigma_1^2 \beta} = s_\beta + 2$. Hence the result follows from (5.17) and (5.18) for the case $\det(Id_{(n-2)\times(n-2)} - d) \neq 0$.

(2) The other case $\det(Id_{(n-2)\times(n-2)} - d) = 0$ follows from the same argument in Proposition 5.7.4. □

Exercise 5.7.8. Complete the proof of Lemma 5.7.7 for the case $\det(Id_{(n-2)\times(n-2)} - d) = 0$.

Lemma 5.7.9. *The following identity holds:*

$$\lambda_{CL}(K_0) = \begin{cases} 1 \text{ if } \Delta_{K_+}(-1) \cdot \Delta_{K_-}(-1) < 0 \\ 0 \text{ if } \Delta_{K_+}(-1) \cdot \Delta_{K_-}(-1) > 0. \end{cases}$$

Proof: From Lemma 5.7.6, we know that $\lambda_{CL}(K_0)$ depends on the absolute value of s_β. For $n \geq 3$ odd, the difference of the exponent sums of $\sigma_1^2 \beta$ and β is 2, by (4.14),

$$\Delta_{K_+}(-1) = i^{e_+ - n + 1} \det(Id - m_{B_n}(\sigma_1^2 \beta)),$$

$$\Delta_{K_-}(-1) = i^{e_- - n + 1} \det(Id - m_{B_n}(\beta)),$$

where $e_+ - e_- = 2$. Therefore

$$\frac{\Delta_{K_-}(-1)}{\Delta_{K_+}(-1)} = -\frac{\det(Id - m_{B_n}(\beta))}{\det(Id - m_{B_n}(\sigma_1^2 \beta))} = -\frac{s_\beta - 1}{s_\beta + 1}.$$

Note that the sign of $a \cdot b$ is the same as the sign of a/b. Therefore $\Delta_{K_+}(-1) \cdot \Delta_{K_-}(-1) > 0$ if and only if $\frac{s_\beta - 1}{s_\beta + 1} < 0$ which is equivalent to $|s_\beta| < 1$. Hence $\lambda_{CL}(K_0) = 0$ if $\Delta_{K_+}(-1) \cdot \Delta_{K_-}(-1) > 0$. Similarly $\Delta_{K_+}(-1) \cdot \Delta_{K_-}(-1) < 0$ if and only if $\frac{s_\beta - 1}{s_\beta + 1} > 0$, and if and only if $|s_\beta| > 1$, so $\lambda_{CL}(K_0) = 1$ by Lemma 5.7.6. $\qquad\square$

Recall that the signature of a knot is an even integer invariant of (unoriented) knots [Rolfsen, 1976]. Let S be a Seifert surface of K which is a compact connected and oriented surface embedded in S^3 with boundary K. For any two oriented simple loops l_1, l_2 on S, define $q(l_1, l_2) = lk(l_1, l_2^+)$, where $lk(,)$ is the linking number of l_1 and l_2 in S^3, and l_2^+ is the pushoff of l_2 away from the Seifert surface along a positive normal direction. Thus q defines a bilinear form on $H_1(S; \mathbb{Z})$. If V is a matrix representation of this bilinear form q under some basis of $H_1(S; \mathbb{Z})$, then the signature of the symmetric matrix $V + V^T$ is defined to be the signature $\sigma(K)$ of the knot K. The number $\sigma(K)$ is independent of choices made in the definition [Rolfsen, 1976, Chapter 8. E], and is an invariant of the knot K.

Proposition 5.7.10. *(1) If $r : S^3 \to S^3$ is an orientation-reversing homeomorphism, then $\sigma(r(K)) = -\sigma(K)$.*

(2) $\sigma(K_1 \# K_2) = \sigma(K_1) + \sigma(K_2)$ and $\sigma(K) \equiv 0 \pmod{2}$ for all knots.

(3) If K is a slice knot, then $\sigma(K) = 0$.

(4) $|\sigma(K)| \leq 2u(K)$, where $u(K)$ is the unknotting number of the knot K.

See [Rolfsen, 1976; Murasugi, 1965] for the proof of Proposition 5.7.10. By the result of [Conway, 1970], we have the property for $\sigma(K)$:

$$\sigma(K) \equiv 0 \quad (\text{mod } 4) \text{ if } \Delta_K(-1) > 0; \quad \sigma(K) \equiv 2 \quad (\text{mod } 4) \text{ if } \Delta_K(-1) < 0;$$

$$0 \le \sigma(K_+) - \sigma(K_-) \le 2.$$

By Lemma 5.7.9, if $\Delta_{K_+}(-1) > 0$ and $\Delta_{K_-}(-1) > 0$, then $\sigma(K_+) - \sigma(K_-) \equiv 0 \pmod 4$ and $0 \le \sigma(K_+) - \sigma(K_-) \le 2$. We must have $\sigma(K_+) - \sigma(K_-) = 0 = \lambda_{CL}(K_0)$. If $\Delta_{K_+}(-1) < 0$ and $\Delta_{K_-}(-1) < 0$, then $\sigma(K_+) - \sigma(K_-) \equiv 0 \pmod 4$ and $0 \le \sigma(K_+) - \sigma(K_-) \le 2$ again implies $\sigma(K_+) - \sigma(K_-) = 0 = \lambda_{CL}(K_0)$.

If $\Delta_{K_+}(-1) > 0$ and $\Delta_{K_-}(-1) < 0$, then $\sigma(K_+) - \sigma(K_-) \equiv 2 \pmod 4$ and $0 \le \sigma(K_+) - \sigma(K_-) \le 2$. That is $\frac{1}{2}\sigma(K_+) - \frac{1}{2}\sigma(K_-) \equiv 1 \pmod 2$ and $0 \le \frac{1}{2}\sigma(K_+) - \frac{1}{2}\sigma(K_-) \le 1$. This forces $\frac{1}{2}\sigma(K_+) - \frac{1}{2}\sigma(K_-) = 1 = \lambda_{CL}(K_0)$. The other case follows exactly the same argument.

Theorem 5.7.11. *For a knot* $K = \bar{\beta}$, $\lambda_{CL}(K) = \pm\frac{1}{2}\sigma(K)$.

Proof: The signature $\sigma(K)$ is also characterized by the skein relation from the result of [Conway, 1970]. Surgery relation of the Casson-Lin invariant $\lambda_{CL}(K)$ leads to the above identification. The orientations for $\sigma(K)$ and $\lambda_{CL}(K)$ may not be the same. We are going to determine the sign in the next section. $\square$

5.8 Other extensions of the $SU(2)$ Casson-Lin invariants

In this section, we are going to follow [Li, 1997] to give an extension of the Casson-Lin invariant to a symplectic Floer homology. We only present the symplectic point of view here and refer the symplectic Floer homology to [Li, 1997]. Other extensions will be mentioned in the end of this section.

Let (M, ω) be a $2n$-dimensional symplectic manifold and $\phi : M \to M$ be a symplectic diffeomorphism, i.e., ω is a nondegenerate closed 2-form and $\phi^*\omega = \omega$. By choosing an almost complex structure J on (M, ω) such that $\omega(\cdot, J\cdot)$ defines a Riemannian metric, we have an integer valued cohomology class $c_1(M) \in H^2(M, \mathbb{Z})$ (i.e. the first Chern class of (M, ω)). Note that the space of all almost complex structures which are compatible with ω is connected, so $c_1(M)$ is uniquely determined by ω. There are two homomorphisms $I_\omega : \pi_2(M) \to \mathbb{R}$ and $I_{c_1} : \pi_2(M) \to \mathbb{Z}$ defined by $I_\omega(u) = \int_{S^2} u^*(\omega) \in \mathbb{R}$ and $I_{c_1}(u) = \int_{S^2} u^*(c_1) \in \mathbb{Z}$ for $u \in \pi_2(M)$.

Definition 5.8.1. *[Floer, 1989]* The symplectic manifold (M, ω) is called *monotone* if $\pi_2(M) = 0$ or if there exists a non-negative $\alpha \geq 0$ such that $I_\omega = \alpha I_{c_1}$ on $\pi_2(M)$.

Lemma 5.8.2. *Let Q_n be the space $\{(X_1, \cdots, X_n) \in SU(2)^n | \operatorname{tr}(X_i) = 0, i = 1, \cdots, n\}$. Then Q_n is a monotone symplectic manifold of dimension $2n$.*

Proof: An element in $SU(2)$ can be viewed as $\begin{bmatrix} a + bi & c + di \\ -c + di & a - bi \end{bmatrix}$, with $a^2 + b^2 + c^2 + d^2 = 1$. For $(X_1, \cdots, X_n) \in Q_n$, each $X_j = \begin{bmatrix} a_j + b_j i & c_j + d_j i \\ -c_j + d_j i & a_j - b_j i \end{bmatrix}$ is a matrix with $\operatorname{tr}(X_j) = 2a_j = 0$ and $b_j^2 + c_j^2 + d_j^2 = 1$. Hence Q_n is the product $S^2 \times S^2 \cdots \times S^2$ of n copies of the 2-sphere with radius 1. So all the S^2 have the same area 4π. It is known that $S^2 \times S^2 \cdots \times S^2$ is a monotone symplectic manifold if and only if all the S^2 have the same area [Floer, 1989, p. 577]. Thus the result follows. □

Lemma 5.8.3. *The space $\mathcal{R}^*(S^2 \backslash K)^{[i]}$ is a monotone symplectic manifold of dimension $4n - 6$.*

Proof: It suffices to show that $\mathcal{R}^*(S^2 \backslash K)^{[i]}$ is monotone. Note that $\mathcal{R}^*(S^2 \backslash K)^{[i]} = (H_n \backslash S_n)/SU(2)$ in Lin's [1992] notation, where

$$H_n = \{(X_1, \cdots, X_n, Y_1, \cdots, Y_n) \in Q_n \times Q_n | \ X_1 \cdots X_n = Y_1 \cdots Y_n\},$$

S_n is the subspace of H_n consisting of all the reducible points, and $H_n \backslash S_n$ is the total space of a $SU(2)$-fiber bundle over $\mathcal{R}^*(S^2 \backslash K)^{[i]}$. It is remarkable that the points in H_n where the $SU(2)$-action is not locally free are precisely those points in S_n.

Note that the Cartesian product of a monotone symplectic manifold with itself is again monotone. So we have $I_\omega = \alpha I_{c_1}$ for $\alpha \geq 0$ over $(Q_n \times Q_n, \omega)$. The space (H_n, ω) is a pre-symplectic space since the closed 2-form ω is degenerate along the $SU(2)$-fiber. For $n \geq 2$, S_n has codimension bigger than 2. The tangent space of $H_n \backslash S_n$ splits into a direct sum of tangent spaces along vertical and horizontal directions:

$$T(H_n \backslash S_n) = T\mathcal{R}^*(S^2 \backslash K)^{[i]} \oplus TSU(2). \tag{5.19}$$

Here $TSU(2) \cong T^*SU(2)$ carries a canonical symplectic form $\omega_0 = -d\lambda$, where λ is the Liouville 1-form . Thus $\omega - \omega_0$ is a nondegenerate closed 2-form on $\mathcal{R}^*(S^2 \backslash K)^{[i]}$. Let $\pi : H_n \backslash S_n \to \mathcal{R}^*(S^2 \backslash K)^{[i]}$ be the $SU(2)$-orbit projection. So $\omega - \omega_0 = \pi_*(\omega)$. From (5.19) and the fact that $SU(2)$ is parallelizable, $A = a + \theta$ where A is the Levi-Civita connection on $T(H_n \setminus S_n)$, a is the Levi-Civita connection on $T\mathcal{R}^*(S^2 \setminus K)^{[i]}$, and θ is the trivial connection on $TSU(2)$. Since $2\pi i c_1 = F_A$, we have $\pi^* c_1 = c_1|_{H_n \setminus S_n} \in H^2(H_n \setminus S_n; \mathbb{Z})$, i.e. $c_1|_{H_n \setminus S_n}$ is the natural pullback of the first Chern class c_1 in $H^2(\mathcal{R}^*(S^2 \setminus K)^{[i]}, \mathbb{Z})$. By the homotopy exact sequence of the fibration $\pi : H_n \setminus S_n \to \mathcal{R}^*(S^2 \setminus K)^{[i]}$:

$$0 = \pi_2(SU(2)) \to \pi_2(H_n \setminus S_n) \overset{\pi_*}{\to} \pi_2(\mathcal{R}^*(S^2 \setminus K)^{[i]}) \to \pi_1(SU(2)) = 0,$$

we obtain the natural isomorphism $\pi_* : \pi_2(H_n \setminus S_n) \cong \pi_2(\mathcal{R}^*(S^2 \setminus K)^{[i]})$. By Stokes theorem, $I_{\omega_0} = 0$. Hence for any $v \in \pi_2(\mathcal{R}^*(S^2 \setminus K)^{[i]}) = \pi_2(H_n \setminus S_n)$ (after identification by π_*),

$$I_\omega(v) = I_{\pi_*(\omega)}(v) + I_{\omega_0}(v) = I_{\pi_*(\omega)}(v)$$

$$I_{c_1}(v) = I_{\pi^*(c_1)}(v).$$

Thus $I_{\pi_* \omega} = \alpha I_{c_1}$ for $\alpha \geq 0$ on $\mathcal{R}^*(S^2 \setminus K)^{[i]}$. So we get the monotonicity for $\mathcal{R}^*(S^2 \setminus K)^{[i]}$. □

For $\beta \in B_n$ with $\overline{\beta} = K$, there is an induced diffeomorphism (still denoted by β) from Q_n to itself. Such a diffeomorphism also induces a diffeomorphism $\phi_\beta : \mathcal{R}^*(S^2 \setminus K)^{[i]} \to \mathcal{R}^*(S^2 \setminus K)^{[i]}$ in Lemma 5.5.2.

Lemma 5.8.4. *For $\beta \in B_n$ with $\overline{\beta} = K$, the induced diffeomorphism $\phi_\beta : \mathcal{R}^*(S^2 \setminus K)^{[i]} \to \mathcal{R}^*(S^2 \setminus K)^{[i]}$ is symplectic, and the fixed point set of ϕ_β is $\mathcal{R}^*(S^3 \setminus K)^{[i]}$.*

Proof: Note that β maps each X_j to a conjugate of some $X_{j'}$. So it gives rise to an area-preserving diffeomorphism of Q_n, i.e.

$$Area(S_j^2) = Area(S_{j'}^2) = Area(S_{\pi(\beta)(j)}^2),$$

where $S_j^2 = \{X_j \in SU(2) | \mathrm{tr} X_j = 0\}$. Let $\omega = \sum_{j=1}^n \omega_j$ be the symplectic form on Q_n with the symplectic form ω_j on the j-th given $\mathbf{CP}^1$ by Lemma 5.8.2. Since $\beta \in B_n$ is a complete cycle of n-letters, we get

$$\beta^*(\omega) = \sum_{j=1}^n \omega_{\pi(\beta)(j)} = \omega.$$

Note that $\beta^* \omega_0 = \omega_0$ on the $SU(2)$-orbit. By Lemma 5.8.3, the symplectic form on $\mathcal{R}^*(S^3 \setminus K)^{[i]}$ is $\pi_*(\omega) = \omega - \omega_0$. The induced diffeomorphism ϕ_β satisfies:

$$\phi_\beta^*(\pi_* \omega) = \beta^*(\omega - \omega_0) = \beta^*(\omega) - \beta^*(\omega_0) = \omega - \omega_0 = \pi_* \omega.$$

Hence ϕ_β is symplectic. Note that $\overline{\Gamma}_\beta = (\Gamma_\beta \setminus (\Gamma_\beta \cap S_n))/SU(2)$ is the graph of ϕ_β. Similarly, $\overline{\Gamma}_{id}$ can be thought of as a "diagonal". By (5.6), it is clear that

$$\text{Fix}(\phi_\beta|_{\mathcal{R}^*(S^2\setminus K)^{[i]}}) = \overline{\Gamma}_\beta \cap \overline{\Gamma}_{id} = \mathcal{R}^*(S^3 \setminus K)^{[i]}.$$

Thus the result follows. □

Let $\tilde{D}^3_\pm$ be the double branched covering of $D^3_\pm$ along $D^3_\pm \cap K$. There is an induced $\mathbb{Z}_2$-action on the representation space $\mathcal{R}(\tilde{D}^3_\pm)$. The submanifolds $\overline{\Gamma}_\beta = \mathcal{R}^*(D^3_+ \setminus K)^{[i]}$ and $\overline{\Gamma}_{id} = \mathcal{R}^*(D^3_- \setminus K)^{[i]}$ can be identified as components of the fixed point set $\mathcal{R}(\tilde{D}^3_\pm)^{\mathbb{Z}_2}$ under the $\mathbb{Z}_2$-action. It follows that $\mathcal{R}^*(D^3_+ \setminus K)^{[i]}$ and $\mathcal{R}^*(D^3_- \setminus K)^{[i]}$ have orientations inherited from the orientation on $\mathcal{R}(D^3_\pm)^{\mathbb{Z}_2}$. These oriented submanifolds $\mathcal{R}^*(D^3_+ \setminus K)^{[i]}, \mathcal{R}^*(D^3_- \setminus K)^{[i]}$ intersect each other in a compact subspace of $\mathcal{R}^*(S^2 \setminus K)^{[i]}$ by Lemma 5.5.4. Hence we can perturb ϕ_β via a Hamiltonian vector field if necessary. In other words, we can perturb $\mathcal{R}^*(D^3_+ \setminus K)^{[i]}$ to $\hat{\mathcal{R}}^*(D^3_+ \setminus K)^{[i]}$ by a compactly support isotopy so that $\hat{\mathcal{R}}^*(D^3_+ \setminus K)^{[i]}$ intersects $\mathcal{R}^*(D^3_- \setminus K)^{[i]}$ transversally at a finite number of points. Denote the perturbed symplectic diffeomorphism by $\tilde{\phi}_\beta$. So the fixed points of $\tilde{\phi}_\beta$ are all nondegenerate.

Lemma 5.8.5 (Lemma 5.5.3). *For any* $K = \overline{\beta}$, $\mathcal{S}^*(D^3_+ \setminus K)^{[i]}$ *and* $\mathcal{S}^*(D^3_- \setminus K)^{[i]}$ *intersect transversally in* $T_s\mathcal{S}^*(S^2\setminus K)^{[i]}$. *Also* $\mathcal{S}^*(S^3\setminus K)^{[i]} = \{s\}$ *is an isolated point of the intersection between* $\mathcal{S}^*(D^3_+ \setminus K)^{[i]}$ *and* $\mathcal{S}^*(D^3_- \setminus K)^{[i]}$ *in* $\mathcal{S}^*(S^2 \setminus K)^{[i]}$.

Remark: The isolated point $\{s\} = \mathcal{S}^*(S^3 \setminus K)^{[i]}$ plays the same role as the trivial connection in the Casson invariant for integral homology 3-spheres. But defining an isotopy invariant with $\{s\}$ is more difficult. Nevertheless, it can be done by understanding the Walker type correction term in [Walker, 1992].

The vector spaces $\{H^1(S^2 \setminus K, \sigma_p^{\otimes 2}) | p \in \mathcal{S}^*(S^2 \setminus K)^{[i]}\}$ form a symplectic vector bundle ν over $\mathcal{S}^*(S^2 \setminus K)^{[i]}$. There is a Hermitian structure on ν compatible with its symplectic structure. An isotopy $\{h_t\}_{0 \leq t \leq 1}$ of $\mathcal{R}(S^2 \setminus K)^{[i]}$ is called *special* if (i) $h_t|_{\mathcal{S}(S^2\setminus K)^{[i]}} = Id$ for all t and (ii) in a neighborhood of $\mathcal{S}^*(S^2\setminus K)^{[i]}$, h_t is induced by a complex symplectic bundle automorphism of ν. Walker [1992, Proposition 1.20] shows that there exists a special isotopy $\{h_t\}_{0 \leq t \leq 1}$ of $\mathcal{R}(S^2 \setminus K)^{[i]}$ such that $h_1(\mathcal{R}^*(D^3_+ \setminus K)^{[i]})$ is transverse to $\mathcal{R}^*(D^3_- \setminus K)^{[i]}$.

We are going to discuss a perturbation around the reducible point s in order to compute $\lambda_{CL}(\beta)$. For the surgery formula for $\lambda_{CL}(\beta)$, we consider the expression

$$\lambda_{CL}(\alpha^{-1}\beta) - \lambda_{CL}(\beta) = \#\tilde{\Gamma}_{\alpha^{-1}\beta} \cap \overline{\Gamma}_{id} - \#\tilde{\Gamma}_{\beta} \cap \overline{\Gamma}_{id}, \qquad (5.20)$$

where $\alpha^{-1} = \sigma_1^2$ is a full twist on the first two strands. More generally, we can consider $\lambda_{CL}(\alpha^{-1}\beta) - \lambda_{CL}(\beta)$ for $\alpha \neq \sigma_1^{-2}$. However, for $\overline{\alpha^{-1}\beta}$ to be a knot, α must lie in the pure braid group $\ker(\pi : B_n \to S_n)$. Note that $\ker(\pi : B_n \to S_n)$ is generated by σ_1^2 and its conjugate. It is enough to get the surgery formula for $\lambda_{CL}(\sigma_1^2\beta) - \lambda_{CL}(\beta)$, where $\sigma_1^2\beta$ is obtained from β by doing surgery along a loop encircling the first two strands as in [Akbulut and McCarthy, 1990, §6.7].

Let $\alpha_* : \overline{H}_n \to \overline{H}_n$ be the automorphism on $\overline{H}_n$ induced by $\alpha \in B_n$:

$$\alpha_*(X_1 \cdots X_n, Y_1 \cdots Y_n) = (X_1 \cdots X_n, \alpha(Y_1) \cdots \alpha(Y_n)).$$

So $\alpha_*(\overline{\Gamma}_{id} \cap \overline{\Gamma}_{\alpha^{-1}\beta}) = \overline{\Gamma}_\alpha \cap \overline{\Gamma}_\beta$. As we perturb $\overline{\Gamma}_{\alpha^{-1}\beta}$ to a transverse position $\tilde{\Gamma}_{\alpha^{-1}\beta}$, its image $\alpha_*(\tilde{\Gamma}_{\alpha^{-1}\beta})$ is also transverse to $\overline{\Gamma}_\alpha$. Around a neighborhood of the reducible point s, the counterclockwise motion in perturbing $\tilde{\Gamma}_{\alpha^{-1}\beta}$ is also preserved by α_*. Hence

$$\#(\overline{\Gamma}_{id} \cap \tilde{\Gamma}_{\alpha^{-1}\beta}) = \#(\overline{\Gamma}_\alpha \cap \tilde{\Gamma}_\beta). \qquad (5.21)$$

Using (5.21), we can rewrite (5.20) as

$$\lambda_{CL}(\alpha^{-1}\beta) - \lambda_{CL}(\beta) = \#(\overline{\Gamma}_{id} \cap \tilde{\Gamma}_{\alpha^{-1}\beta}) - \#(\overline{\Gamma}_{id} \cap \tilde{\Gamma}_\beta)$$
$$= \#(\overline{\Gamma}_\alpha - \overline{\Gamma}_{id}) \cap \tilde{\Gamma}_\beta.$$

In some sense, $\overline{\Gamma}_\alpha - \overline{\Gamma}_{id}$ and $\tilde{\Gamma}_\beta$ can be thought of as cycles except they contain the reducible points. To remedy this situation, we consider the difference:

$$[\lambda_{CL}(\alpha^{-1}\beta) - \lambda_{CL}(\beta)] - [\lambda_{CL}(\alpha^{-1}\beta') - \lambda_{CL}(\beta')]$$

$$= \#(\overline{\Gamma}_\alpha - \overline{\Gamma}_{id}) \cap \tilde{\Gamma}_\beta - \#(\overline{\Gamma}_\alpha - \overline{\Gamma}_{id}) \cap \tilde{\Gamma}_{\beta'} = \#(\overline{\Gamma}_\alpha - \overline{\Gamma}_{id}) \cap (\tilde{\Gamma}_\beta - \tilde{\Gamma}_{\beta'}).$$

In a neighborhood of the reducible point s, we choose a 1-parameter family of $\tilde{\Gamma}_\beta(t)$ which brings $(\tilde{\Gamma}_{\beta'})_\rho$ to $(\tilde{\Gamma}_\beta)_\rho$. This isotopy fixes $(\tilde{\Gamma}_\beta - \tilde{\Gamma}_{\beta'})$ on the complement of the neighborhood of s. Inside the neighborhood of s, the isotopy $\partial\tilde{\Gamma}_\beta(t)$ brings $\partial\tilde{\Gamma}_\beta$ to $\partial\tilde{\Gamma}_{\beta'}$. So we can construct a cycle $(\tilde{\Gamma}_\beta - \tilde{\Gamma}_{\beta'})$ globally by using this isotopy.

Let $(\tilde{\Gamma}_\beta - \tilde{\Gamma}_{\beta'})_{global}$ be the cycle in the non-singular piece $\overline{H}_n$. The intersection number

$$I(s) = \#(\overline{\Gamma}_\alpha - \overline{\Gamma}_{id}) \cap (\tilde{\Gamma}_\beta - \tilde{\Gamma}_{\beta'})_{global},$$

makes sense as the intersection of a relative cohomology class with an absolute class. The number $I(s)$ counts irreducibles via the global twisting around the reducible s. So the term $I(s)$ is essentially the Walker correction term. Note that Walker's geometric construction of the correction term is not applicable for our situation since the trace-zero representations contain non-typical elements such as the trivial representation for rational homology 3-spheres. However, we can use an analytic definition [Cappell, Lee and Miller, 1993] for the Walker correction term $I(s)$. The difference is given by

$$[\lambda_{CL}(\alpha^{-1}\beta) - \lambda_{CL}(\beta)] - [\lambda_{CL}(\alpha^{-1}\beta') - \lambda_{CL}(\beta')]$$

$$= \#(\overline{\Gamma}_\alpha - \overline{\Gamma}_{id}) \cap (\tilde{\Gamma}_\beta - \tilde{\Gamma}_{\beta'})_0 + I(s). \tag{5.22}$$

(1) The term $\#(\overline{\Gamma}_\alpha - \overline{\Gamma}_{id}) \cap (\tilde{\Gamma}_\beta - \tilde{\Gamma}_{\beta'})_0$ can be expressed in terms of triple Maslov indices

$$\#(\overline{\Gamma}_{id} - \overline{\Gamma}_\alpha) \cap (\tilde{\Gamma}_\beta - \tilde{\Gamma}_{\beta'})_0 = \frac{1}{2}\text{Mas}(\overline{\Gamma}_\alpha, \tilde{\Gamma}_\beta, \overline{\Gamma}_{id})_s - \frac{1}{2}\text{Mas}(\overline{\Gamma}_\alpha, \tilde{\Gamma}_{\beta'}, \overline{\Gamma}_{id})_s. \tag{5.23}$$

Since $\tilde{\Gamma}_\beta, \tilde{\Gamma}_{\beta'}$ are transverse to $\overline{\Gamma}_\alpha$ and $\overline{\Gamma}_{id}$, the dimension correction terms all disappear [Cappell, Lee and Miller, 1993, 1996]

(2) The term $\lambda_{CL}(\alpha^{-1}\beta') - \lambda_{CL}(\beta')$ in (5.22) can be eliminated by choosing β' as follows. There exists a braid β' such that both $\overline{\beta'}$ and $\overline{\sigma_1^2\beta'}$ represent the trivial knot. For $n = 2$, we choose $\overline{\beta'}$ and $\overline{\sigma_1^2\beta'}$ to be $\overline{\sigma_1^{-1}}, \overline{\sigma_1}$ respectively; for $n > 2$, we choose $\beta' = \sigma_1^{-1}\sigma_2^{-1} \cdots \sigma_n^{-1}$, $\overline{\sigma_1^2\beta'} = \sigma_1\sigma_2^{-1} \cdots \sigma_n^{-1}$. Thus $\lambda_{CL}(\alpha^{-1}\beta') - \lambda_{CL}(\beta') = 0$ and $\text{Mas}(\overline{\Gamma}_\alpha, \tilde{\Gamma}_{\beta'}, \overline{\Gamma}_{id})_s = 0$. The expression (5.22) becomes

$$\lambda_{CL}(\alpha^{-1}\beta) - \lambda_{CL}(\beta) = \frac{1}{2}\text{Mas}(\overline{\Gamma}_\alpha, \tilde{\Gamma}_\beta, \overline{\Gamma}_{id})_s + I(s).$$

(3) Using Wall's nonadditivity of signature, Cappell, Lee and Miller [1993] identified the triple Maslov index with the (twisted) signature. So (5.22) becomes the following:

$$\lambda_{CL}(\alpha^{-1}\beta) - \lambda_{CL}(\beta) = \frac{1}{2}\{\sigma(\overline{\sigma_1^2\beta}) - \sigma(\overline{\beta})\} + I(s). \tag{5.24}$$

(4) In the following, we show by two different methods [Cappell, Lee and Miller, 1993] that the Walker-type correction term $I(s)$ vanishes. Note that the signature of σ_1^2 is clearly zero. So

$$\lambda_{CL}(\beta) = \frac{1}{2}\sigma(\overline{\beta}; [i^2]) = \frac{1}{2}\sigma(\overline{\beta}).$$

Lemma 5.8.6. *The Walker-type correction term vanishes,* $I(s) = 0$.

Proof: It suffices to prove that $I(s) = 0$ for $\alpha = \sigma_1^{-2}$. In this case, we have

$$\alpha(X_1) = (X_1 X_2)^{-1} X_1 (X_1 X_2)$$
$$\alpha(X_2) = (X_1 X_2)^{-1} X_2 (X_1 X_2)$$
$$\alpha(X_j) = X_j, \quad j \geq 3.$$

We define a Casson-deformation from $SU(2) \backslash \{-I\}$ to itself as $X \mapsto X^t$ such that X^t commutes with X, $X^0 = Id$, $X^1 = X$. Denote such a deformation with α by $\phi(t)$. We get

$$\phi(t)(X_1) = [(X_1 X_2)^t]^{-1} X_1 [(X_1 X_2)^t]$$
$$\phi(t)(X_2) = [(X_1 X_2)^t]^{-1} X_2 [(X_1 X_2)^t]$$
$$\phi(t)(X_j) = X_j, \quad j \geq 3.$$

Since $\text{tr}(\phi(t)(X_j)) = \text{tr}(X_j)$ for all j, the formula

$$\overline{\Gamma}_{\phi(t)} = \{(X_1, \cdots, X_n, \phi(t)(X_1), \cdots, \phi(t)(X_n)) \in Q_n \times Q_n\}/SU(2),$$

gives us a 1-parameter family of subspaces such that $\overline{\Gamma}_{\phi(0)} = \Gamma_{id}/SU(2)$ and $\overline{\Gamma}_{\phi(1)} = \Gamma_\alpha/SU(2)$. As $\phi(t)(X_1 X_2) = [(X_1 X_2)^t]^{-1} (X_1 X_2)[(X_1 X_2)^t] = X_1 X_2$, $\phi(t)(X_j) = X_j, j \geq 3$ we have

$$\phi(t)(X_1 X_2 \cdots X_n) = (X_1 X_2 \cdots X_n).$$

So $\Gamma_{\phi(t)}$ is a subspace in $\mathcal{R}(S^2 \setminus K)^{[i]}$. By the definition of $\phi(t)$, we have

$$\phi(t)|_{\mathcal{S}^*(S^2 \setminus K)^{[i]}} = id.$$

Thus $\phi(t)$ is a special isotopy. It follows that $\{\Gamma_{\phi(t)} : 0 \leq t \leq 1\}$ gives us a cycle in $\mathcal{R}(S^2 \setminus K)^{[i]}$ whose boundary is $\Gamma_{id}/SU(2) - \Gamma_\alpha/SU(2) = \mathcal{R}(D^3_- \setminus K)^{[i]} - \Gamma_\alpha/SU(2)$. Thus

$$-I(s) = \#(\mathcal{R}(D^3_- \setminus K)^{[i]} - \Gamma_\alpha/SU(2)) \cap (\tilde{\Gamma}_\beta - \tilde{\Gamma}_{\beta'})_{global}$$
$$= \#(\partial\{\Gamma_{\phi(t)} : 0 \leq t \leq 1\} \cap (\tilde{\Gamma}_\beta - \tilde{\Gamma}_{\beta'})_{global}$$
$$= 0.$$

$\square$

Let $\rho \in \mathcal{S}^*(S^2 \setminus K)^{[i]}$ be a $U(1)$ reducible representation of $\pi_1(S^2 \setminus K)$ with trace zero condition. A normal neighborhood of ρ is isomorphic to the cone of $H^1(S^2 \setminus K, \mathbf{h}^\perp_{Ad\rho})/U(1)$. Here $\mathbf{h}$ is the Lie algebra of the fixed oriented maximal torus of $SU(2)$ and $\mathbf{h}^\perp$ is the orthogonal complement of $\mathbf{h}$ with respect to the Killing form of $SU(2)$. The cone bundle $E(\mathcal{S}^*(S^2 \setminus K)^{[i]}) \to \mathcal{S}^*(S^2 \setminus K)^{[i]}$ is isomorphic to a neighborhood of $\mathcal{S}^*(S^2 \setminus K)^{[i]}$

in $\mathcal{R}(S^2 \setminus K)^{[i]}$ via an exponential map $\exp : \mathcal{N}(\mathcal{S}^*(S^2 \setminus K)^{[i]}) \to Q_n \times Q_n$. The exponential map allows us to identify a germ of functions $\phi \in C^\infty(Q_n \times Q_n, \mathbb{R})$ near $\mathcal{S}^*(S^2 \setminus K)^{[i]}$ with a function on $\mathcal{N}(\mathcal{S}^*(S^2 \setminus K)^{[i]})$. For example, choose a partition of unity $\chi : Q_n \times Q_n \to \mathbb{R}$ which is 0 outside $\mathcal{N}(\mathcal{S}^*(S^2 \setminus K)^{[i]})$ and 1 near the 0-section. We consider a function $g : \mathcal{N}(\mathcal{S}^*(S^2 \setminus K)^{[i]}) \to \mathbb{R}$ induced by a Hermitian pairing on each fiber $H^1(S^2 \setminus K, \mathbf{h}^\perp_{Ad\rho})/U(1)$. Locally, $g(\rho, z) = \sum a_{ij}(\rho) z_i \overline{z_j}$, $z = (z_1, \cdots, z_d) \in H^1(S^2 \setminus K, \mathbf{h}^\perp_{Ad\rho})/U(1)$,

$$H_g(\exp(\rho, z)) = \chi(\rho, z) g(\rho, z) \qquad (5.25)$$

gives us a function $H_g \in C^\infty(Q_n \times Q_n, \mathbb{R})$. The vector field $\mathrm{grad} H_g$ has the property that $g(\rho, z)$ is quadratic in the normal z-direction. So $\mathrm{grad} H_g = 0$ when it is restricted to the zero section. The Hessian $\mathrm{Hess}(H_g)$ at each fiber $H^1(S^2 \setminus K, \mathbf{h}^\perp_{Ad\rho})/U(1)$ is given by the Hermitian pairing $\sum a_{ij}(\rho) z_i \overline{z_j}$.

Let $\mathcal{L}ag(W)$ be the space of complex Lagrangians in $W = H^1(S^2 \setminus K, \mathbf{h}^\perp_{Ad\rho})$. The space $\mathcal{L}ag(W)$ is a homogeneous manifold whose tangent space can be identified with the space of Hermitian pairings on $H^1(S^2 \setminus K, \mathbf{h}^\perp_{Ad\rho})$ [Walker, 1992, pp. 13–14]. We consider the special perturbations Π_0 such that for $H_g \in \Pi_0$,

(1) $\mathrm{grad} H_g(\rho) = 0$ for $\rho \in \mathcal{S}^*(S^2 \setminus K)^{[i]}$,
(2) $(\phi_\beta + H_g)(\overline{\Gamma}_{id})$ is mapped injectively into a complex Lagrangian subspace in $H^1(S^2 \setminus K, \mathbf{h}^\perp_{Ad\rho})$.

Let $\mathrm{Lag}(\mathcal{W})$ denote the smooth fiber bundle over $\mathcal{S}^*(S^2 \setminus K)^{[i]}$ whose fiber at a point $\rho \in \mathcal{S}^*(S^2 \setminus K)^{[i]}$ is the homogeneous space $\mathcal{L}ag(W)$. Let $\Gamma(\mathrm{Lag}(\mathcal{W}))$ be the space of smooth sections of this bundle completed with respect to an appropriate Sobolev norm. Then by assigning to π the section of Lagrangian $\rho \mapsto (\phi_\beta + H_g)(\overline{\Gamma}_{id})$ in $H^1(S^2 \setminus K, \mathbf{h}^\perp_{Ad\rho})$ there exists a mapping $p : \Pi_0 \to \Gamma(\mathrm{Lag}(\mathcal{W}))$.

Proposition 5.8.7. *(i) The mapping $p : \Pi_0 \to \Gamma(Lag(\mathcal{W}))$ is a submersion whose image is the entire space $\Gamma(Lag(\mathcal{W}))$.*

(ii) Π_0 is path connected, i.e. any two different perturbations can be homotopically connected to each other through a 1-parameter family in Π_0.

Proof: The argument follows exactly as in the proofs of Theorem B and Proposition 3.3 in [Lee and Li, 1995a]. Since the correction term $I(s)$ vanishes, there is no obstruction to connecting any two different perturbations. $\qquad \square$

Now the vanishing property of the Walker-type correction term implies that the compact support perturbations H_g do not effect the invariant $\lambda_{CL}(\beta)$. Unlike the situation studied by Lee and Li [1994], the symplectic Floer homology defined by Li [1997, §4] is independent of the Lagrangian perturbations around the reducible s (see also [Lee and Li, 1995b]). So we can pick a positive definite quadratic form H_g to orient the perturbation $\mathcal{R}^*(D_+^3 \setminus K)^{[i]}$ (as the graph of ϕ_β) into $\hat{\mathcal{R}}^*(D_+^3 \setminus K)^{[i]}$ (as the graph of $\tilde{\phi}_\beta$) counterclockwisely. In particular, $\hat{\mathcal{R}}^*(D_+^3 \setminus K)^{[i]}$ meets $\mathcal{R}^*(D_-^3 \setminus K)^{[i]}$ transversely. Therefore Casson-Lin's invariant $\lambda_{CL}(\beta)$ is also the Casson-Walker type invariant with trivial correction term $I(s)$.

Now we are going to compute the trefoil knot in order to fix the sign in Theorem 5.7.11. For the right handed trefoil knot $3_1 = \overline{\sigma_1^3}$ (or the $(2, 3)$ torus knot), by the previous calculation $\mathcal{R}^*(S^2 \setminus K)^{[i]} = \overline{H}_2$ is a 2-sphere with four punctured points deleted. Thus

$$\pi_1(\mathcal{R}^*(S^2 \setminus \overline{\sigma_1^3})^{[i]}) = F_3, \quad \pi_2(\mathcal{R}^*(S^2 \setminus \overline{\sigma_1^3})^{[i]}) = 0,$$

where F_3 is a free group of 3-generators. As in the previous section, we can assume that up to conjugation,

$$X_1 = \begin{pmatrix} i\cos\theta_1 & \sin\theta_1 \\ -\sin\theta_1 & -i\cos\theta_1 \end{pmatrix}, \quad X_2 = \begin{pmatrix} i & 0 \\ 0 & -i \end{pmatrix}, \quad 0 \le \theta_1 \le \pi.$$

The conditions $X_1 X_2 = Y_1 Y_2$ and $\mathrm{tr}(Y_i) = 0 (i = 1, 2)$ provide

$$Y_1 = \begin{pmatrix} i\cos\theta_2 & \sin\theta_2 \\ -\sin\theta_2 & -i\cos\theta_2 \end{pmatrix}, \quad -\pi \le \theta_2 \le \pi.$$

So $\mathcal{R}^*(S^2 \setminus \overline{\sigma_1^3})^{[i]}$ is parametrized by (θ_1, θ_2) modulo the involution $(\theta_1, \theta_2) \to (-\theta_1, -\theta_2)$ (see Figure 5.5). The orientation is given by the counterclockwise orientation of the (θ_1, θ_2)-plane.

$$\mathcal{R}^*(D_-^3 \setminus \overline{\sigma_1^3})^{[i]} = \{(\theta_1, \theta_1)| \ 0 < \theta_1 < \pi\},$$

$$\mathcal{R}^*(D_+^3 \setminus \overline{\sigma_1^3})^{[i]} = Graph(\phi_{\overline{\sigma_1^3}}) = \{(X_1, X_2, \sigma_1^3(X_1), \sigma_1^3(X_2))\},$$

where $Y_1 = \sigma_1^3(X_1) = (X_1 X_2)^2 X_1^{-1}(X_1 X_2)^{-1}$.

$$
\begin{aligned}
Y_1 &= \begin{pmatrix} i\cos\theta_2 & \sin\theta_2 \\ -\sin\theta_2 & -i\cos\theta_2 \end{pmatrix} \\
&= (X_1 X_2)^2 X_1^{-1}(X_1 X_2)^{-1} \\
&= \begin{pmatrix} -\cos\theta_1 & -i\sin\theta_1 \\ -i\sin\theta_1 & -\cos\theta_1 \end{pmatrix}^2 \begin{pmatrix} -i\cos\theta_1 & -\sin\theta_1 \\ \sin\theta_1 & i\cos\theta_1 \end{pmatrix} \begin{pmatrix} -\cos\theta_1 & -i\sin\theta_1 \\ -i\sin\theta_1 & -\cos\theta_1 \end{pmatrix}^{-1} \\
&= \begin{pmatrix} i\cos 4\theta_1 & \sin 4\theta_1 \\ -\sin 4\theta_1 & -i\cos 4\theta_1 \end{pmatrix},
\end{aligned}
$$

by a straightforward matrix multiplication with given X_1 and X_2. Thus

$$Graph(\phi_{\overline{\sigma_1^3}}) = \{(\theta_1, 4\theta_1) \mid 0 < \theta_1 < \pi\},$$

and that

$$\text{Fix}(\phi_{\overline{\sigma_1^3}}) = \{\rho = (\frac{2\pi}{3}, \frac{2\pi}{3})\} = \mathcal{R}^*(D_+^3 \setminus \overline{\sigma_1^3})^{[i]} \cap \mathcal{R}^*(D_-^3 \setminus \overline{\sigma_1^3})^{[i]} \quad (5.26)$$

is a single element.

The Maslov index of ρ, $\mu(\rho) = \mu(\rho, s)$, is the same Maslov index of the two Lagrangian submanifolds $L_0 = \mathcal{R}^*(D_-^3 \setminus \overline{\sigma_1^3})^{[i]}$ and $L_1 = Graph(\phi_{\overline{\sigma_1^3}})$ by Cappell, Lee and Miller [1994]. This index can be computed as the winding number of a path of L_1 about the diagonal L_0 as in [Arnold, 1967]. Note that in [Cappell, Lee and Miller, 1994] L_0 is chosen to be $\{(\theta_1, 0)\}$. So L_1 *clock-wisely* winds around L_0 once: $\mu(\rho) = -1$ [Li, 1997, Figure 3]. Hence

$$HF_i^{\text{sym}}(\phi_{\sigma_1^3}) = \begin{cases} \mathbb{Z} & \text{if } i = -1 \\ 0 & \text{if } i \neq -1. \end{cases} \quad (5.27)$$

The Euler number $\chi(HF_*^{\text{sym}}(\phi_{\overline{\sigma_1^3}})) = -1 = -\lambda_{CL}(\overline{\sigma_1^3})$ [Li, 1997]. Note that the left-handed trefoil has signature $+2$ in [Rolfsen, 1976, p. 220]. This example fixes the sign in Theorem 5.7.11. For a knot $K = \overline{\beta}$, we have

$$\lambda_{CL}(\beta) = -\frac{1}{2}\sigma(K).$$

Two natural questions arise from the Casson-Lin invariant $\lambda_{CL}(\beta)$:

(1) Does there exist a Floer homology generalization for the Casson-Lin invariant ?
(2) What kind of invariants can one get for representations of $\pi_1(S^3 \setminus K)$ with the trace of all meridians fixed (not necessarily zero) ?

The first question is addressed by Collin and Steer [1999] from the gauge theory point of view and by Li [1997, 2001a] from the symplectic topology point of view. Cappell, Lee and Miller [1993] studied the second question from the symplectic theory point of view and defined an equivariant Casson invariant for 3-manifolds with cyclic group actions by Heusener and Kroll [1998] for the generalization of Casson-Lin invariant for fixing trace on meridians. Independently, Herald [1997] studied the same problem from the gauge theory point of view. Then Li [2001b] extended the Casson-Lin invariant to links.

5.9   Calculation of the $SU(2)$ Casson-Lin invariants

Example 5.9.1. For a unknotted knot $K_0 = \overline{\sigma_1^{\pm}}$, we have

$$\mathcal{R}^*(S^3 \setminus K_0)^{[i]} = \emptyset \quad \text{(empty set)}, \quad \mathcal{S}^*(S^3 \setminus K_0)^{[i]} = \{s\},$$

by the unknotting theorem. By Markov's result, any unknotted knot can be obtained by a finite sequence of Markov moves from $\sigma_1^{\pm}$. Hence $\lambda_{CL}(\sigma_1^{\pm}) = 0$.

For the $(2, q)$ torus knot $\overline{\sigma_1^q}$ (q must be odd), we use the same method to calculate:

$$Graph\phi_{\sigma_1^q} = \{(\theta_1, (q+1)\theta_1) \mid 0 < \theta_1 < \pi\},$$

$$\text{Fix}(\phi_{\sigma_1^q}) = \{\rho_k = (\frac{2\pi k}{q}, \frac{2\pi k}{q}) \mid k = 1, 2, \cdots, \frac{q-1}{2}\}.$$

The Maslov indices are determined by $\mu(\rho_1) = -1$. Let $\mu(\rho_{k-1}, \rho_k)$ be the Maslov index of two Lagrangian submanifolds intersecting transversally at two smooth points ρ_{k-1} and ρ_k. By [Arnold, 1967, §1.4], $\mu(\rho_{k-1}, \rho_k)$ is the number of rotations of Det^2. Thus $\mu(\rho_{k-1}, \rho_k) = -2$ (orientation). By the additivity of the Maslov index, each fixed point has a negative sign. So

$$\lambda_{CL}(\sigma_1^q) = -\frac{q-1}{2}.$$

Example 5.9.2. The figure eight knot 4_1 has the braid representative $\sigma_1 \sigma_2^{-1} \sigma_1 \sigma_2^{-1}$. The knot 4_1 has signature zero since 4_1 is equivalent (by an orientation preserving homeomorphism) to its mirror image $\overline{4_1}$. So the figure eight knot is amphicheiral. Also it is well-known that the figure eight knot is not a slice knot, and represents an element of order 2 in the knot cobordism group [Rolfsen, 1976].

Let $\mathcal{R}^*(S^2 \setminus 4_1)^{[i]}$ be the subset of $\mathcal{R}(S^2 \setminus 4_1)^{[i]}$ consisting of irreducible representations. Then $\mathcal{R}^*(S^2 \setminus 4_1)^{[i]}$ can also be identified with the quotient space $(H_3 \setminus S_3)/SU(2)$ in Lin's [1992] notation, i.e., the set of 6-tuple $(X_1, X_2, X_3, Y_1, Y_2, Y_3) \in SU(2)^6$ satisfying $\text{tr}(X_j) = \text{tr}(Y_j) = 0$ ($j = 1, 2, 3$) and

$$X_1 X_2 X_3 = Y_1 Y_2 Y_3.$$

By operating the conjugation on X_3 and Y_3, we may assume that

$$X_3 = \begin{pmatrix} i & 0 \\ 0 & -i \end{pmatrix}, \quad Y_3 = \begin{pmatrix} i \cos \theta & \sin \theta \\ -\sin \theta & -i \cos \theta \end{pmatrix}, \quad 0 \le \theta \le \pi.$$

If $\theta = 0$ and π, then we get two copies of $(H_2 \setminus S_2)/SU(2)$ which is the pillow case (a 2-sphere with four cone points deleted [Li, 1997; Lin, 1992]). For $0 < \theta < \pi$, the identification reduces down to the following

$$X_1 X_2 \begin{pmatrix} \cos\theta & -i\sin\theta \\ -i\sin\theta & \cos\theta \end{pmatrix} = Y_1 Y_2.$$

Let R_θ be the representations in $\mathcal{R}^*(S^2 \setminus 4_1)^{[i]}$ satisfying the above equation. So the space R_θ is the non-singular piece in $\mathcal{R}^*(S^2 \setminus K)^{[i]}$. For $0 < \theta, \theta' < \pi$, the space R_θ is diffeomorphic to the space $R_{\theta'}$. In particular, they are all diffeomorphic to $R_{\pi/2}$. So $\mathcal{R}^*(S^2 \setminus 4_1)^{[i]}$ is a generalized pillow case:

$$\mathcal{R}^*(S^2 \setminus 4_1)^{[i]} = \bigcup_{0 \le \theta \le \pi} R_\theta.$$

It is not easy to determine the $\pi_2(\mathcal{R}^*(S^2 \setminus 4_1)^{[i]})$ and $N(4_1)$.

The fixed point set of ϕ_{4_1} is $\mathcal{R}^*(S^3 \setminus 4_1)^{[i]}$ Fix$(\phi_{4_1}) = \{(X_1, X_2, X_3) \in SU(2)^3 | \sigma(X_j) = X_j, j = 1, 2, 3\}$ up to conjugation. Let B_n be the braid group of rank n with the standard generators $\sigma_1, \cdots, \sigma_{n-1}$, and F_n be the free group of rank n generated by $x_1, \cdots, x_n$. Then the automorphism of F_n representing σ_k is given by (still denote it by σ_k)

$$
\begin{aligned}
\sigma_k : \quad & x_k \mapsto x_k x_{k+1} x_k^{-1} \\
& x_{k+1} \mapsto x_k \\
& x_l \mapsto x_l, \quad l \neq k, k+1.
\end{aligned}
\tag{5.28}
$$

For instance, the actions in B_3 can be given by

$$
\sigma_1 : \begin{cases} x_1 \mapsto x_1 x_2 x_1^{-1} \\ x_2 \mapsto x_1 \\ x_3 \mapsto x_3 \end{cases}
\qquad
\sigma_2^{-1} : \begin{cases} x_1 \mapsto x_1 \\ x_2 \mapsto x_3 \\ x_3 \mapsto x_3^{-1} x_2 x_3 \end{cases}
$$

$$\sigma_1\sigma_2^{-1}\sigma_1\sigma_2^{-1}(x_1) = \sigma_1\sigma_2^{-1}\sigma_1(x_1) = \sigma_1\sigma_2^{-1}(x_1x_2x_1^{-1})$$
$$= \sigma_1(x_1x_3^{-1}x_1^{-1})$$
$$= (x_1x_2x_1^{-1})x_3^{-1}(x_1x_2x_1^{-1})^{-1}$$
$$= x_1x_2x_1^{-1}x_3^{-1}x_1x_2^{-1}x_1^{-1}.$$

$$\sigma_1\sigma_2^{-1}\sigma_1\sigma_2^{-1}(x_2) = \sigma_1\sigma_2^{-1}\sigma_1(x_3)$$
$$= \sigma_1\sigma_2^{-1}(x_3)$$
$$= \sigma_1(x_3^{-1}x_2x_3)$$
$$= x_3^{-1}x_1x_3.$$

$$\sigma_1\sigma_2^{-1}\sigma_1\sigma_2^{-1}(x_3) = \sigma_1\sigma_2^{-1}\sigma_1(x_3^{-1}x_2^{-1}x_3)$$
$$= \sigma_1\sigma_2^{-1}(x_3^{-1}x_1x_3)$$
$$= \sigma_1(x_3^{-1}x_2^{-1}x_3x_1x_3^{-1}x_2x_3)$$
$$= x_3^{-1}x_1^{-1}x_3x_1x_2x_1^{-1}x_3^{-1}x_1x_3.$$

Note that Li's [2000, bottom of p. 347] error arose from mis-calculating σ_2^{-1}. Therefore the fixed point set of ϕ_{4_1} is the set of points $(X_1, X_2, X_3) \in SU(2)^3$ such that

$$\text{tr}(X_j) = 0, \quad j = 1, 2, 3,$$
$$X_1X_2X_1^{-1}X_3X_1X_2^{-1}X_1^{-1} = X_1,$$
$$X_3^{-1}X_1X_3 = X_2,$$
$$X_3^{-1}X_1^{-1}X_3X_1X_2X_1^{-1}X_3^{-1}X_1X_3 = X_3,$$

up to conjugation. Up to conjugation, we can assume that

$$X_1 = \begin{pmatrix} i\cos\theta & \sin\theta \\ -\sin\theta & -i\cos\theta \end{pmatrix}, \quad X_3 = \begin{pmatrix} i & 0 \\ 0 & -i \end{pmatrix}, \quad 0 \le \theta \le \pi.$$

From the second equation $X_3^{-1}X_1X_3 = X_2$, we have

$$X_2 = X_3^{-1}X_1X_3$$
$$= \begin{pmatrix} -i & 0 \\ 0 & i \end{pmatrix} \begin{pmatrix} i\cos\theta & \sin\theta \\ -\sin\theta & -i\cos\theta \end{pmatrix} \begin{pmatrix} i & 0 \\ 0 & -i \end{pmatrix}$$
$$= \begin{pmatrix} i\cos\theta & -\sin\theta \\ \sin\theta & -i\cos\theta \end{pmatrix}.$$

Thus we have $X_1X_2 = \begin{pmatrix} -\cos 2\theta & -i\sin 2\theta \\ -i\sin 2\theta & -\cos 2\theta \end{pmatrix}$ and $X_1^{-1} = \begin{pmatrix} -i\cos\theta & -\sin\theta \\ \sin\theta & i\cos\theta \end{pmatrix}$ from a straightforward calculation.

$$X_1X_2X_1^{-1} = \begin{pmatrix} -\cos 2\theta & -i\sin 2\theta \\ -i\sin 2\theta & -\cos 2\theta \end{pmatrix} \begin{pmatrix} -i\cos\theta & -\sin\theta \\ \sin\theta & i\cos\theta \end{pmatrix}$$

$$= \begin{pmatrix} i\cos 3\theta & \sin 3\theta \\ -\sin 3\theta & -i\cos 3\theta \end{pmatrix}.$$

From this, $X_1 X_2^{-1} X_1^{-1} = (X_1 X_2 X_1^{-1})^{-1} = \begin{pmatrix} -i\cos 3\theta & -\sin 3\theta \\ \sin 3\theta & i\cos 3\theta \end{pmatrix}$. From

the first equation $\sigma_1 \sigma_2^{-1} \sigma_1 \sigma_2^{-1}(X_1) = X_1$, we have

$$X_1 = (X_1 X_2 X_1^{-1}) X_3 (X_1 X_2 X_1^{-1})^{-1}$$

$$= \begin{pmatrix} i\cos 3\theta & \sin 3\theta \\ -\sin 3\theta & -i\cos 3\theta \end{pmatrix} \begin{pmatrix} i & 0 \\ 0 & -i \end{pmatrix} \begin{pmatrix} -i\cos 3\theta & -\sin 3\theta \\ \sin 3\theta & i\cos 3\theta \end{pmatrix}$$

$$= \begin{pmatrix} i\cos 6\theta & \sin 6\theta \\ -\sin 6\theta & -i\cos 6\theta \end{pmatrix}.$$

Hence the first equation is equivalent to the relation

$$\cos 6\theta = \cos\theta, \quad \sin 6\theta = \sin\theta. \tag{5.29}$$

Similarly, from the last equation $\sigma_1 \sigma_2^{-1} \sigma_1 \sigma_2^{-1}(X_3) = X_3$, we need to compute the following:

$$X_3^{-1} X_1^{-1} X_3 = \begin{pmatrix} -i\cos\theta & \sin\theta \\ -\sin\theta & i\cos\theta \end{pmatrix}$$

$$X_3^{-1} X_1^{-1} X_3 X_1 = \begin{pmatrix} \cos 2\theta & -i\sin 2\theta \\ -i\sin 2\theta & \cos 2\theta \end{pmatrix}$$

$$X_1^{-1} X_3^{-1} X_1 X_3 = (X_3^{-1} X_1^{-1} X_3 X_1)^{-1}$$

$$= \begin{pmatrix} \cos 2\theta & i\sin 2\theta \\ i\sin 2\theta & \cos 2\theta \end{pmatrix}.$$

Now the last equation is

$$X_3 = (X_3^{-1} X_1^{-1} X_3 X_1) X_2 (X_1^{-1} X_3^{-1} X_1 X_3)$$

$$= \begin{pmatrix} \cos 2\theta & -i\sin 2\theta \\ -i\sin 2\theta & \cos 2\theta \end{pmatrix} \begin{pmatrix} i\cos\theta & -\sin\theta \\ \sin\theta & -i\cos\theta \end{pmatrix} \begin{pmatrix} \cos 2\theta & i\sin 2\theta \\ i\sin 2\theta & \cos 2\theta \end{pmatrix}$$

$$= \begin{pmatrix} i\cos 5\theta & -\sin 5\theta \\ \sin 5\theta & -i\cos 5\theta \end{pmatrix}.$$

This leads to an equivalent relation

$$\cos 5\theta = 1, \quad \sin 5\theta = 0. \tag{5.30}$$

Thus the fixed point of ϕ_{4_1} can be identified with

$$X_1 = \begin{pmatrix} i\cos\theta & \sin\theta \\ -\sin\theta & -i\cos\theta \end{pmatrix}, \quad X_2 = \begin{pmatrix} i\cos\theta & -\sin\theta \\ \sin\theta & -i\cos\theta \end{pmatrix},$$

$$X_3 = \begin{pmatrix} i & 0 \\ 0 & -i \end{pmatrix}, \quad 0 \le \theta \le \pi,$$

subject to equations (5.29) and (5.30).

There are two solutions corresponding to $\theta = \frac{2\pi}{5}$ and $\frac{4\pi}{5}$. Hence there are two irreducible representations of the figure eight knot with trace zero along meridians. From the known fact of $\sigma(4_1) = 0$, we know that these two irreducible representations has opposite signs, i.e., one with odd Maslov index and the other with even Maslov index. Theorem 2.2 of [Li, 2000] may still be corrected by understanding the symplectic Floer boundary maps. It is hard to determine the nontrivial symplectic Floer boundary maps. As far as this author knows, there is only one example in this direction [Li, 1999a] in which the calculation involves the connected sum in [Li, 1999b].

Exercise 5.9.3. (1) Find out $\lambda_{CL}(\sigma_1^5)$;
    (2) Compute the $\mathcal{R}^*(S^3 \setminus 6_2)$.

Bibliography

S. Akbulut and J. McCarthy, (1990). Casson's invariant for oriented homology 3-spheres. An Exposition. Math. Notes, **36**., Princeton Univ. Press, Princeton, NJ.

V. Arnold, (1967). *On a characteristic class entering into conditions of quantization*, Funt. Anal. Appl. **1**, 1-8.

V. Arnold, (1970). *On some topological invariants of algebraic functions*, Trans. Moscow Math. Soc. **21**, 30–52.

V. Arnold, (1973). *The asymptotic Hopf invariant and its applications*, Proc. Summer School in Diff. Equations at Dilizhan, 1973; English transl: Sel Math. Sov **5** (1986), 327-345.

E. Artin, (1925). *Theorie der Zöpfe*, Abh. Math. Sem, Univ. Hamburg, **4**, 47-72.

E. Artin, (1947). *Theory of braids*, Ann. of Math., **48**, 101-126.

E. Artin and R. Fox, (1948). *Some wild cells and spheres in three-dimensional space*, Ann. of Math. (2) **49**, 979-990.

J. Baez, (1992). *Link invariants of finite type and perturbation theory*, Lett. Math. Phys., **26**, No. 1, 43-51.

S. Bigelow, (1999). *The Burau representation is not faithful for n = 5*, Geometry and Topology, Vol. **3**, 397-404.

S. Bigelow, (2001a). *Braid groups are linear*, J. Amer. Math. Soc. **14**, 471–486.

S. Bigelow, (2001b). *A homological definition of the Jones polynomial*, Geometry and Topology Monographs, Volume 4: Invariants of knots and 3-manifolds (Kyoto 2001), 29–41.

J. S. Birman, (1974). *Braids, links and mapping class groups*, Ann. Math. Studies, No. **82**, Princeton University Press.

J. S. Birman, (1976). *On the stable equivalence of plat representations of knots and links*, Canad. J. Math. 28, 264–290.

J. S. Birman, (1993). *New points of view in knot theory*, Bulletin of AMS, Vol **28**, No. 2, 253-287.

J. S. Birman and W. W. Menasco, (1991). *Studying Links Via Closed Braids II: On A Theorem Of Bennequin*, Topology and its applications.

S. A. Bleiler, (1984). *A note on unknotting number*. Math. Proc. Camb. Phil. Soc., **96**(3), 469–471.

223

M. Borodzik and S. Friedl, (2014). *On the algebraic unknotting number*, Trans. London. Math. Soc., 1(1), 57–84.

W. Burau, (1936). *Über Zopfgruppen und gleichsinnig verdrillte Verkettungen*, Abh. Math. Sem. Hamburg, 11, 179–186.

G. Burde and H. Zieschang, (1985). Knots, de Gruyter Studies in Math., **5**, Walter de Gruyter.

S. Cappell, R. Lee and E. Miller, (1993). *Equivariant Casson invariant*, preprint.

S. Cappell, R. Lee and E. Miller, (1994). *On the Maslov index*, Comm. Pure. Appl. Math, Vol XLVII, 121-186.

S. Cappell, R. Lee and E. Miller, (1996). *Self-adjoint elliptic operators and manifold decompositions: I, II, III*, Comm. Pure. Appl. Math., Vol. 49 (8), 825–866; Vol. 49 (9), 869–909,

J. Cerf, (1974).*Sur les Difféomorphisms de la Sphére de Dimensional Trois (Γ₄ = 0)*, Lecture Notes in Math, **Vol 53**, Springer-Verlag, New York.

W.-L. Chow, (1948). *On the algebraic braid group*, Ann of Math., (2) **49**, 654-658.

O. Collin and B. Steer, (1999). *Instanton Floer homology for knots via 3-orbifolds*, J. Diff. Geometry, **51** (1), 149–202.

J. Conway, (1970). *An enumeration of knots and links and some of their algebraic properties*, Computational Problems in Abstract Algebra, Pergamon Press, New York, 329-358.

R. Crowell and R. Fox, (1977). Introduction to knot theory, Ginn and Co.; Reissue, Grad. Texts Math., **57**, Springer Verlag.

M. Dehn, (1987). *Papers on group theory and topology*. Translated from German and with introductions and an appendix by John Stillwell, New York. viii, pp396.

Y. Diao, (2004). *The additivity of crossing numbers*, Journal of Knot Theorem and Its Ramifications, Vol. 13, No. 7, 857–866.

E. Fadell and L. Neuwirth, (1962). *Configuration spaces*, Math. Scand. **10**, 119-126.

R. Fenn, E. Keyman and C. Rourke, (1998). *The singular braid monoid embeds in a group*, Journal of Knot Theory and Its Ramifications, Vol. 7, 881–892.

A. Floer, (1989). *Symplectic fixed points and holomorphic spheres*, Comm. Math. Phys. **120**, 575-611.

M. E. Fogel, (1994). *Knots with algebraic unknotting number one*, Pacific J. Math., **163**, 277–295.

R. Fox, (1952). *On the complementary domains of a certain pair of inequivalent knots*, Kon. Nederl. Akad. van Wetenschappin. Proc. Series A, Vol **55**, 37-40.

M. Freedman and Z-X. He, (1991). *Divergence-free fields: energy and asymptotic crossinf number*, Ann. of Math. (2) **134**, No. 1, 189-229.

R. Fricke and F. Klein, (1897). *Vorlesungen über die Theorie der automorphen Funktionen, Bd.I. Gruppentheoretischen Grundlagen*, Teubner, Leipzig 1897 (Johnson, New York 1965).

D. B. Fuks, (1970). *Cohomology of the braid group mod 2*, Funct. Anal. Apple. 4, 143–151.

N. Gilbert and T. Porter, (1994). Knots and Surfaces, Oxford Science Publications.

W. Goldman, (1984). *The symplectic nature of fundamental groups of surfaces,* Advances in Math., **54**, 200-225.

C. Gordon and J. Luecke, (1989). *Knots are determined by their complements,* J. Amer. Math. Soc. **2**, 371-415.

P. D. Harpe, M. Kervaire and C. Weber, (1986). *On the Jones polynomial,* L'Enseignement Math., t. **32**, 271-335.

J. Hass, J. C. Lagarias and N. Pippenger, (1999). *The computational complexity of knot and link problems,* Journal of the ACM, **46**(2), 185–211.

C. Herald, (1997). *Flat connections, the Alexander invariant and Casson's invariant,* Comm. Anal and Geom., **5**, 93-120.

M. Heusener and J. Kroll, (1998). *Deforming abelian SU(2)-representations of knot groups,* Comment. Math. Helv. **73**, 480-498.

J. Hoste, M. Thistlethwaite and J. Weeks, (1998). *The First 1,701,936 Knots,* The Math. Intelligencer, Vol **20**, No. 4, 33-48.

A. Hurwitz, (1891). *Über Riemannsche Flächen mit gegebenen Verweigungspunkten,* Math. Ann., **39**, 1-61.

V. Jones, (1987). *Hecke algebra representations of braid groups and link polynomials,* Annals of Math., **126**, 335-388.

R. Kashaev, (1997). *The hyperbolic volume of knots from the quantum dilogarithm,* Lett. Math. Phys., **39**(3), 269–275.

L. Kauffman, (1987). *State Models and the Jones Polynomial,* Topology **26**, 395-407.

L. Kauffman, (1987). *On Knots,* Annals of Math. Studies, **115**, Princeton Uni. Press.

M. Kontsevich, (1993). *Vassiliev's knot invariants,* Adv. in. Sov. Math., 16 (2), 137–150.

D. Krammer, (2000). *The braid group B_4 is linear,* Invent. Math. **142**, 451–486.

D. Krammer, (2002). *Braid groups are linear,* Ann. of Math. **155**, 131–156.

P. B. Kronheimer and T.S. Mrowka, (1993). *Gauge theory for embedded surfaces I,* Topology **32**, No. 4, 773–826.

S. Lang, (1996). Topics in Cohomology of groups, Lecture Notes in Mathematics, 1625, Springer-Verlag, Berlin.

R. J. Lawrence, (1990). *Homological representations of the Hecke algebra,* Comm. Math. Phy. 135, 141–191.

R. J. Lawrence, (1993). *A functorial approach to the one-variable Jones polynomial,* J. Differential Geom. 37, 689–710.

R. Lee and W. Li, (1994). *Floer homology for rational homology 3-spheres,* MSRI preprint #04394.

R. Lee and W. Li, (1995a). *Symplectic Floer homology of SU(2)-representation spaces (I),* OSU preprint.

R. Lee and W. Li, (1995b). *Floer homologies for Lagrangian intersections and instantons (II),* OSU preprint.

W. Li, (1997). *Casson-Lin's invariant and Floer homology,* J. Knot and its ramifications, Vol. 6, No. 6, 851-877.

W. Li, (1999a). *The symplectic Floer homology of the square knot and granny knots*, Acta Math. Sinica (Series B), 15:1, 1-10.

W. Li, (1999b). *The symplectic Floer homology of composite knots*, Forum Mathematicum, **11**, 617-646.

W. Li, (2000). *The symplectic Floer homology of the figure eight knot*, Asian J. Math., Vol 4, 345-350.

W. Li, (2001a). *Equivariant knot signatures and Floer homologies*, J. Knot and its ramifications, Vol. 10, No. 5, 687-701

W. Li, (2001b). *Knot and link invariants and moduli space of parabolic bundles*, Comm. Contemporary Math., Vol **3**, 501-531.

W. Li and L. Xu, (2003). *Counting $SL_2(F_{2^s})$ representations of torus knot groups*, Acta Mathematica Sinica, English version, **Vol. 19 (2)**, 233-244.

W. Lickorish, (1985). *The unknotting number of a classical knot*, Contemp. Math. **44**, 117-121.

W. Lickorish and K. Millet, (1986). *Some evaluations of link polynomials*, Comment. Math. Helv. **61**, 349-359.

X-S. Lin and Z. Wang, (1996). *Integral geometry of plane curves and knot invariants*, J. Diff. Geom. **44**, No. 1, 74-95.

X-S. Lin, (1992). *A knot invariant via representation spaces*, J. Diff. Geometry, **35**, 337-357.

X-S. Lin, (2001). *Representations of Knot groups and twisted Alexander polynomials*, Acta Mathematica Sinica, English version, Vol **17**, No 3, 361-380.

D. Long and M. Patton, (1993). *The Burau representation is not faithful for $n \geq 6$*, Topology, **32**, 439-447.

R. Lyndon and P. Schupp, (1977). Combinatorial Group Theory, A series of Modern Surveys in Math, **89**, Springer-Verlag, Berlin, Heidelberg, New York.

W. Magnus and A. Peluso, (1969). *On a theorem of V. I. Arnol'd*, Comm. Pure Appl. Math. **22**, 683-692.

A. Markov, (1945). Foundations of the algebraic theory of tresses, Trudy Mat. Inst. Steklov. **16**, 3-53.

W. Massey, (1977). Algebraic Topology: An Introduction, Grad. Texts in Math, **56**, Springer-Verlag, New York.

W. Menasco and M. Thistlethwaite, (1993). *The classification of alternating links*, Ann. of Math., **138**, 113-171.

J. Milnor, (1952). *On the total curvature of knots*, Ann of Math., Vol **52**, 248-257.

J. Milnor, (1956). *On manifolds homeomorphic to the 7-sphere*, Annals of Math., **64**, 399-405.

J. Milnor, (1968). *Infinite cyclic coverings*, Conference on the Topology of Manifolds, Prindle Weber and Schnitdt, Boston, 115-133.

J. Moody, (1993). *The faithfulness question for the Burau representation*, Proceedings of the American Mathematical Society, **119**(2), 671-679.

J. Munkres, (1960). *Differentiable isotopies on the 2-sphere*, Mich. Math. J., **7**, 193-197.

J. Murakami and H. Murakami, (2001). *The colored Jones polynomials and the simplicial volume of a knot*, Acta Math., **186**(1), 85-104.

K. Murasugi, (1965). *On a certain numerical invariant of link types*, Trans, AMS,

117, 387-422.

K. Murasugi, (1987). *The Jones polynomial and classical conjectures in knot theory*, Topology **26**, 187-194; *Jones polynomials and classical conjectures in knot theory II*, Math. Proc. Camb. Phil. Soc. **102**, 317-318.

Y. Nakanishi, (1983). *Unknotting numbers and knot diagrams with the minimum crossings.* Math. Sem. Notes Kobe Univ., **11**(2), 257-258.

G. Perelman, (2002), *The entropy formula for the Ricci flow and its geometric applications,* arXiv.math.DG/0211159.

M. Polyak and O. Viro, (1994). *Gauss Diagram Formulas for Vassiliev invariants,* Int. Math. Research Notices, **11**, 445-453.

N. Reshetikhin and V. G. Turaev, (1991). *Invariants of 3-manifolds via link polynomials and quantum groups,* Invent. Math. 103, 547-597.

D. Rolfsen, (1976). Knots and Links, Publish or Perish, Inc, (1976, 1990).

Y. W. Rong, (1993). *Some knots not determined by their complements,* Quantum Topology, Series Knots Everything **3**, 339-353, World Sci. Publishing, River Edge, NJ.

M. Scharlemann, (1985). *Unknotting number one knots are prime,* Invent. Math., **82**, 37-55.

H. Schubert, (1949). *Die eindeutige Zerlegbarkeit eines Knoten in Primknoten,* Sitzungsber. Akad. Wiss. Heidenberg, Math.-Nat. Kl., **3**, 57-104.

H. Schubert, (1956). *Knoten mit zwei Brücken,* Math. Z., **65**, 133-170.

S. Smale, (1959). *Diffeomorphisms of the 2-sphere,* Proc. Amer. Math. Soc. **10**, 621-626.

A. B. Sossinsky, (1992). Preparation Theorems for isotopy invariants of links in 3-manifolds, Lecture Notes in Math. **1510**, 354-362.

M. Thistlethwaite, (1987). *A panning tree expansion of the Jones polynomial,* Topology **26**, 297-309.

M. Thistlethwaite, (1988). *Kauffman's polynomial and alternating links,* Topology **27**, 311-318.

W. Thomson, (1867). *On Vortex Atoms,* Proceedings of the Royal Society of Edinburgh, **6**, 94-105.

M. Wada, (1994). *Twisted Alexander polynomial for finitely presentable groups,* Topology, Vol **33**, No 2, 241-256.

F. Waldhausen, (1968). *On Irreducible 3-manifolds which are sufficiently large,* Annals of Math., **87**, 56-88.

F. Waldhausen, (1970). *On mappings of handlebodies and of Heegaard splittins,* 1970 Topology of Manifolds (Proc. Inst., Univ. of Georgia, Athens, Ga., pp. 205-211 Markham, Chicago, Ill.

K. Walker, (1992). An extension of Casson's invariant. Annals of Mathematics Studies, 126. Princeton University Press, Princeton, NJ.

H. Wenzl, (1985). *Representations of hecke algebras and subfactors,* Thesis, Univ. of Pennsylvania.

E. Witten, (1989). *Quantum field theory and the Jones polynomial,* Comm. Math. Phys. 121, 351-399.

S. Yamada, (1987). *The minimal number of Seifert circles equals the braid index of a link,* Invent. Math. 89, no. 2, 347-356.

Index

Printed in the United States
By Bookmasters